Samuel Jorge Marques Cartaxo
Tubular Heat Exchangers

Also of interest

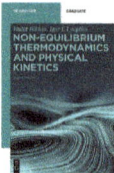

Non-equilibrium Thermodynamics and Physical Kinetics
Bikkin, Lyapilin, 2021
ISBN 978-3-11-072706-7, e-ISBN 978-3-11-072719-7

Physics of Energy Conversion
Krischer, Schönleber, 2015
ISBN 978-1-5015-0763-2, e-ISBN 978-1-5015-1063-2

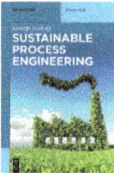

Sustainable Process Engineering
Szekely, 2022
ISBN 978-3-11-071712-9, e-ISBN 978-3-11-071713-6

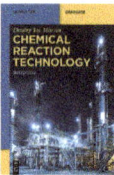

Chemical Reaction Technology
Murzin, 2022
ISBN 978-3-11-071252-0, e-ISBN 978-3-11-071255-1

Process Technology.
An Introduction
De Haan, Padding, 2022
ISBN 978-3-11-071243-8, e-ISBN 978-3-11-071244-5

Samuel Jorge Marques Cartaxo

Tubular Heat Exchangers

―

for Chemical Engineers

DE GRUYTER

Author
Prof. Dr. Samuel Jorge Marques Cartaxo
Department of Chemical Engineering
Center of Technology
Campus do Pici Bloco 709
Federal University of Ceará
60440-900 Fortaleza-Ce
Brazil
samuel@ufc.br

ISBN 978-3-11-058573-5
e-ISBN (PDF) 978-3-11-058587-2
e-ISBN (EPUB) 978-3-11-058591-9

Library of Congress Control Number: 2021950113

Bibliographic information published by the Deutsche Nationalbibliothek
The Deutsche Nationalbibliothek lists this publication in the Deutsche Nationalbibliografie;
detailed bibliographic data are available on the Internet at http://dnb.dnb.de.

Cover image: Dimpleflo heat exchanger, with friendly permission from Teralba Industries PTY LTD
Typesetting: Integra Software Services Pvt. Ltd.
Printing and binding: CPI books GmbH, Leck

www.degruyter.com

To my beloved parents, for the birth,
and all the immense love
that came afterwards:
João Vianey Assis Cartaxo
Judith Maria Marques Cartaxo

Acknowledgments

"No man is an island". I believe poet John Donne grasped much wisdom in his celebrated phrase. This endeavor could never be undertaken by alone. Firstly, I thank to every one of the many hundreds of students that passed the Unit Operations II (chemical process heat transfer) course. The material within this book was developed and refined over several years for them, as part of that course.

I am grateful to all my friends and colleagues from the Department of Chemical Engineering at the Federal University of Ceará, for making such faculty environment that allows the deep and longtime dedication required to finish this book. For each word of motivation over the various years of writing, gratitude to professors André Casimiro de Macedo and Maria Valderez Ponte Rocha, and the students (at the time) Andréa da Silva Pereira and Madson Linhares Magalhães.

Finding the perfect cover (among thousands of options) for the book was a challenge. Thanks to Andrew Baker (Managing Director, Teralba Industries Pty Ltd) for the kind permission to use one of their beautiful heat exchangers as the cover.

Pictures and design drafts of actually constructed heat exchangers are invaluable resource for the learner. I thank very much Hesham Derazi (President and CEO, Heat Exchanger Design, Inc.) for the kind consent to use their website content as illustrative material in the book.

I want also to express my gratitude for the many great people at De Gruyter Publishing House, namely, dear Mareen Pagel (Editor), Ria Sengbusch (Editor), Lena Stoll (Project Editor), Dorothea Wunderling (Content Editor), Anne Stroka (Full Service Project Manager) and David Jüngst (Full Service Project Manager).

Lastly, for the countless hours that were stolen from them during the preparation of this book, I thank my wife Silvia Rebeca Guimarães Cartaxo for all the care, patience, and support, along with my beloved godsend kids Mateus Guimarães Cartaxo and my sweet darling Melissa Guimarães Cartaxo.

Samuel Jorge Marques Cartaxo

https://doi.org/10.1515/9783110585872-202

Disclaimer

All information provided in this document is provided "as is" without warranty of any kind, either expressed or implied. Every effort has been made to ensure accuracy and conformance to standards accepted at the time of publication. The reader is advised to research other sources of information on these topics.

The material in this book is a result of many hours of work and was prepared in good faith and carefully revised and edited. However, like with any book, the author and any collaborators cannot be held responsible for errors of any sort in this content. Furthermore, since it is not possible for the author to verify the correctness and reliability of information referenced from public literature, though can only examine them for suitability for the intended purpose herein, this information cannot be warranted. Also, because the author cannot assure the experience or technical capability of the reader/user of the information and the suitability of the information for the intended purpose, the use of the contents must be based on the best judgment of the user. Therefore, the final user is responsible for any damage or loss caused by the use or misuse of the information provided herein.

https://doi.org/10.1515/9783110585872-203

Preface

The making of a book like this may easily become a true journey, for many reasons. Process heat transfer is among the three bedrocks of the chemical engineering (and related ones), and heat transfer equipment (i.e., heat exchangers), per si, is a large branch in this river. The variety of types is immense, and some compromise of the content coverage becomes mandatory. Another important decision regards the presentation depth. Hence, this volume lays down an in-depth discussion of the thermal design of selected types of tubular heat exchangers. The detailed mechanical design is not targeted herein, however mechanical features and specifications that interferes with the thermal performance are tackled as part of the thermal design. For sure, there exist a myriad of amazing books on this arena; the first one I could mention is the seminal masterpiece from Donald Q. Kern, which deserves a place in any engineer's library. While no pretension of replacement exists here, some novelty contributions to the field are made. The reader will find methods and equations not available in any other book. The precise analysis and new formulas developed for wall temperatures in tubular exchangers are a good example. The served approach for performance evaluation of finned heat exchangers and the pressure drop calculations, derived from the Hooper's 2-K method, are another one. Not to mention new formulas to predict the exit temperatures from a heat exchanger.

Concerning the audience, this text was purposely prepared with the student in mind. Every effort was applied to clarify the presented subjects, both by balancing the depth of the description and the material format yet preserving the technical accuracy. Accordingly, it is targeted to facilitate the learning process and understanding of the student challenged with the subject of heat exchangers in chemical engineering, mechanical engineering, and other related engineering fields. For turning the text volume manageable, introductory chapters on the very fundamentals of heat transfer mechanisms were not included, therefore a background from a basic heat transfer (or transport phenomena) course is recommended as prerequisite. The content is presented in chapters arranged with the purpose of promoting the understanding of the subject matter by the undergraduate student that is touching it for the first time, what makes the book not primarily intended for use as a reference by experts in the field of heat exchangers. However, absolutely no compromise is made regarding the precision and reliability of the techniques and methods described. Despite its pedagogical format, the knowledge presented herein envisions to be applied in professional practice, being aligned with proven information found in the scientific-technical literature and up-to-date industrial practice.

Regarding the heat exchangers design, the Kern's alike methods (i.e., based on the logarithmic mean of the temperature differences) suffices for most applications. I have seen many exchangers designed with the Kern's method, and they performed quite satisfactorily with respect to the design specifications, although it tends to overdesign the equipment to some extent. Of course, the contemporary industrial

https://doi.org/10.1515/9783110585872-204

practice involves the use of closed commercial software packages that implement more sophisticated methods such as Tinker, Bell-Delaware, or derived variations. In fact, those mentioned more sophisticated methods are built upon the physical rationale of the Kern's method, which contains the fundamental "physics" to approach the problem of designing or analyzing a heat exchanger. The knowledge of the physics supporting the heat exchanger design method is a key factor when the student learns the operation of specialized heat transfer software. Many times, the student is not aware of the underlying fundamentals behind the design method inside the software, and simply operates it in "black box" mode. In the view of the author, this move is not beneficial for someone learning engineering and involves risks when the student goes professional, since heat exchanger design software should be used uncritically in no circumstance. Usually, there are so many considerations, hypotheses and simplifications involved, which sometimes are violated, generating unpredictable errors. The engineer must have enough technical comprehension to identify such anomalies minimally.

One distinct feature of this book is the abundant didactical use of many worked examples. This resulted from my attempt to address a much recurrent complaint from the students in my courses: lacking of step-by-step solved problems in existing textbooks. Almost every calculation intensive section has been equipped with detailed stepwise solutions exemplifying the presented methods, and all the rational involved. With the aim to improve the knowledge retention, another notable aspect is the availability of several comprehensive proposed problems for student practice at the end of each chapter. As an added benefit, each problem is accompanied by its answers, which is a dramatic aid for promoting much desired self-learning skills.

Also, special care was taken with the units. Although, many systems of measurement are available, students and field engineers worldwide have mostly to deal with SI or some derivative of the British Imperial System (e.g., US customary). For that reason, both systems are equally used in the many example calculations and proposed problems. To avoid in some extent the tedious (and error prone) work of unit conversions, which is another frequent student complaint, equations are presented in dimensionless form whenever possible. If not, SI and British versions are provided. The same approach is adopted for pipe/tube specifications (Appendix A), Properties of Materials (Appendix B) and Physical Properties of Chemicals (Appendix C), where both units sets are explicitly offered for easing the calculations.

Samuel Jorge Marques Cartaxo

Contents

Conversion factors

Physical quantity	Value
Acceleration	$1.0000 \text{ m/s}^2 = (4.2520e+7) \text{ ft/h}^2 = 3.2808 \text{ ft/s}^2$
Area	$1.0000 \text{ m}^2 = 10.764 \text{ ft}^2 = 1,550.0 \text{ in}^2 = 0.00024711 \text{ acre} = 0.00010000 \text{ ha}$
Density	$1.0000 \text{ kg/m}^3 = 0.062428 \text{ lb/ft}^3 = 0.0010000 \text{ g/cm}^3$ $= 0.0010000 \text{ kg/L} = 1.0000 \text{ g/L}$
Dynamic viscosity	$1.0000 \text{ Pa} \cdot \text{s} = 1,000.0 \text{ cP} = 10.000 \text{ P} = 1.0000 \text{ kg}/(\text{m} \cdot \text{s})$
Energy	$1.0000 \text{ J} = 0.23901 \text{ cal} = 0.00094782 \text{ BTU} = (1.0000e+7) \text{ erg} = (6.2415e+18) \text{ eV}$ $= 0.00027778 \text{ Wh} = (2.7778e-7) \text{ kWh} = (2.3901e-10) \text{ tTNT}$ $= (9.4782e-9) \text{ thm} = 1.0000 \text{ (kg} \cdot \text{m}^2)/\text{s}^2$
Force	$1.0000 \text{ N} = 0.22481 \text{ lbf} = (1.0000e+5) \text{ dyn} = 0.10197 \text{ kgf} = 3.5969 \text{ ozf}$ $= 0.00022481 \text{ kip} = 1.0000 \text{ (kg} \cdot \text{m)}/\text{s}^2$
Fouling factor	$1.0000 \text{ (m}^2 \cdot \text{K)}/\text{W} = 1.0000 \text{ (m}^2 \cdot °\text{C)}/\text{W} = 5.6783 \text{ (ft}^2 \cdot \text{h} \cdot °\text{F)}/\text{BTU}$ $= 5.6783 \text{ (ft}^2 \cdot \text{h} \cdot °\text{R)}/\text{BTU} = 1.0000 \text{ (s}^3 \cdot °\text{C)}/\text{kg}$
Heat flux	$1.0000 \text{ W/m}^2 = 0.31700 \text{ BTU}/(\text{ft}^2 \cdot \text{h}) = 1.0000 \text{ kg/s}^3$
Heat transfer coefficient	$1.0000 \text{ W}/(\text{m}^2 \cdot \text{K}) = 1.0000 \text{ W}/(\text{m}^2 \cdot °\text{C}) = 0.17611 \text{ BTU}/(\text{ft}^2 \cdot \text{h} \cdot °\text{F})$ $= 0.17611 \text{ BTU}/(\text{ft}^2 \cdot \text{h} \cdot °\text{R}) = 1.0000 \text{ kg}/(\text{s}^3 \cdot °\text{C})$
Heat volumetric rate	$1.0000 \text{ W/m}^3 = 0.096621 \text{ BTU}/(\text{ft}^3 \cdot \text{h}) = (2.3901e-7) \text{ cal}/(\text{cm}^3 \cdot \text{s})$ $= 1.0000 \text{ kg}/(\text{m} \cdot \text{s}^3)$
Kinematic viscosity	$1.0000 \text{ m}^2/\text{s} = 38,750 \text{ ft}^2/\text{h} = 10.764 \text{ ft}^2/\text{s}$
Latent heat	$1.0000 \text{ kJ/kg} = 0.23901 \text{ cal/g} = 0.42992 \text{ BTU/lb} = 1000.0 \text{ m}^2/\text{s}^2$
Length	$1.0000 \text{ m} = 3.2808 \text{ ft} = 39.370 \text{ in} = (1.0000e+9) \text{ nm} = (1.0000e+12) \text{ pm}$ $= 0.00062137 \text{ mi} = 1.0936 \text{ yd} = 0.00053996 \text{ nmi} = 236.22 \text{ pica}$
Mass	$1.0000 \text{ kg} = 2.2046 \text{ lb} = 35.274 \text{ oz} = 0.15747 \text{ st} = 0.023453 \text{ bag}$ $= 0.068522 \text{ slug}$
Mass flux	$1.0000 \text{ kg}/(\text{m}^2 \cdot \text{s}) = 737.34 \text{ lb}/(\text{ft}^2 \cdot \text{h}) = 0.20482 \text{ lb}/(\text{ft}^2 \cdot \text{s})$
Power	$1.0000 \text{ W} = 3.4121 \text{ BTU/h} = 0.0013410 \text{ hp} = 1.0000 \text{ (kg} \cdot \text{m}^2)/\text{s}^3$
Pressure	$1.0000 \text{ kPa} = 1000.0 \text{ N/m}^2 = 10,000 \text{ dyn/cm}^2 = 20.885 \text{ lbf/ft}^2$ $= 0.010197 \text{ kgf/cm}^2 = 1.0000 \text{ kPa} = 0.010000 \text{ bar} = 0.0098692 \text{ atm}$ $= 0.14504 \text{ psi} = 7.5006 \text{ torr} = 7.5006 \text{ mmHg} = 10.197 \text{ cmH}_2\text{O}$ $= 1,000.0 \text{ kg}/(\text{m} \cdot \text{s}^2)$

https://doi.org/10.1515/9783110585872-206

(continued)

Physical quantity	Value
Surface tension	$1.0000 \ \text{N/m} = 1,000.0 \ \text{dyn/cm} = 0.10197 \ \text{kgf/m} = 0.068522 \ \text{lbf/ft}$ $= 1.0000 \ \text{kg/s}^2$
Temperature	$K = {}^\circ C + 273.15 = \dfrac{5}{9}\,{}^\circ R = \dfrac{5}{9}\,({}^\circ F + 459.67)$
Temperature difference	$1.0000 \ K = 1.0000 \ {}^\circ C = 1.8000 \ {}^\circ F = 1.8000 \ {}^\circ R$
Thermal conductivity	$1.0000 \ \text{W/(m·K)} = 0.57779 \ \text{BTU/(ft·h·}{}^\circ\text{F)} = 1.0000 \ (\text{kg·m})/(\text{s}^3\text{·}{}^\circ\text{C})$
Thermal resistance	$1.0000 \ \text{K/W} = 0.52753 \ (\text{h·}{}^\circ\text{F})/\text{BTU} = 1.0000 \ (\text{s}^3\text{·}{}^\circ\text{C})/(\text{kg·m}^2)$
Volume	$1.0000 \ \text{m}^3 = 35.315 \ \text{ft}^3 = 61,024 \ \text{in}^3 = 1,000.0 \ \text{L} = (1.0000e+6) \ \text{mL}$ $= (1.0000e+6) \ \text{cc} = 264.17 \ \text{gallon} = 1,056.7 \ \text{quart} = 4,226.8 \ \text{cup}$ $= 2,113.4 \ \text{pt} = 67,628 \ \text{tbsp} = (2.0288e+5) \ \text{tsp}$

Physical constants

Physical constant	Value
Avogadro constant	$(6.0221e+23)\ \text{mol}^{-1}$
Boltzmann constant (k)	$(1.3807e-23)\ \text{J/K} = (7.2700e-27)\ \text{BTU/}^\circ\text{R} = (1.3807e-16)\ \text{erg/}^\circ\text{C}$
Molar gas constant (R)	$8.3145\ \text{J/(mol}\cdot\text{K)} = 0.0043781\ \text{BTU/(mol}\cdot{}^\circ\text{F)}$ $= (8.2057e-5)\ (\text{m}^3\cdot\text{atm})/(\text{mol}\cdot\text{K})$ $= (8.3145e+7)\ \text{erg/(mol}\cdot{}^\circ\text{C)}$ $= 8.3145\ (\text{kg}\cdot\text{m}^2)/(\text{s}^2\cdot\text{mol}\cdot{}^\circ\text{C})$
Molar mass constant	$0.0010000\ \text{kg/mol} = 1.0000\ \text{g/mol} = 0.0022046\ \text{lb/mol}$
Molar Planck's constant	$(3.9903e-10)\ (\text{J}\cdot\text{s})/\text{mol} = (3.9903e-10)\ (\text{kg}\cdot\text{m}^2)/(\text{s}\cdot\text{mol})$
Molar volume of ideal gas (273.15 K, 101.325 kPa)	$0.022414\ \text{m}^3/\text{mol} = 0.79154\ \text{ft}^3/\text{mol} = 22{,}414\ \text{cm}^3/\text{mol}$
Newtonian constant of gravitation	$(6.6743e-11)\ \text{m}^3/(\text{kg}\cdot\text{s}^2) = 0.013856\ \text{ft}^3/(\text{lb}\cdot\text{h}^2)$ $= (6.6743e-8)\ \text{cm}^3/(\text{g}\cdot\text{s}^2)$
Standard acceleration of gravity	$9.8067\ \text{m/s}^2 = 32.174\ \text{ft/s}^2 = (4.1698e+8)\ \text{ft/h}^2 = 980.67\ \text{cm/s}^2$
Standard atmosphere	$101.32\ \text{kPa} = 1.0000\ \text{atm} = 14.696\ \text{psi} = 760.00\ \text{mmHg}$ $= 1033.2\ \text{cmH}_2\text{O} = 1.0332\ \text{kgf/cm}^2 = 14.696\ \text{lbf/in}^2$ $= (1.0133e+5)\ \text{kg/(m}\cdot\text{s}^2)$
Stefan–Boltzmann constant	$(5.6704e-8)\ \text{W/(m}^2\cdot\text{K}^4) = (1.7140e-9)\ \text{BTU/(ft}^2\cdot\text{h}\cdot{}^\circ\text{R}^4)$ $= (5.6704e-5)\ \text{erg/(cm}^2\cdot\text{s}\cdot\text{K}^4)$ $= (5.6704e-8)\ \text{kg/(s}^3\cdot\text{K}^4)$

https://doi.org/10.1515/9783110585872-207

Nomenclature

Subscripts

0	Constant fluid properties
1	Inlet (entrance)
2	Outlet (exit)
eq	Equivalent
i	Inner (internal) fluid, channel or surface
o	Outer (external) fluid, channel or surface
w	Evaluated at the wall

Greek

μ	Viscosity at bulk temperature (Pa s or lb/ft h)
μ_w	Viscosity at wall temperature (Pa s or lb/ft h)
ε	Absolute roughness (m or ft)
ρ	Fluid specific mass (kg/m^3 or lb/ft^3)
τ_w	Shear stress at the wall (Pa or lb/ft h^2)
ϕ	Correction factor for variable physical properties

Symbols

ΔP	Pressure difference (Pa or psi)
ΔP_{dist}	Distributed pressure drop (Pa or psi)
$\Delta P_{i,dist}$	Distributed pressure drop in the tube (Pa or psi)
$\Delta P_{i,loc}$	Localized pressure drop in the tube (Pa or psi)
ΔP_{loc}	Localized pressure drop (Pa or psi)
$\Delta P_{o,dist}$	Distributed pressure drop in the annulus (shell) (Pa or psi)
$\Delta P_{o,loc}$	Localized pressure drop in the annulus (shell) (Pa or psi)
ΔT_m	Mean temperature difference (°C or °F)
ΔT_1	Temperature difference in the terminal (side) 1 (°C or °F)
ΔT_2	Temperature difference in the terminal (side) 2 (°C or °F)
$\Delta T_{lm,c}$	Logarithmic mean temperature difference for countercurrent flow (°C or °F)
$\Delta T_{lm,p}$	Logarithmic mean temperature difference for parallel flow (°C or °F)
ΔT_{lm}	Logarithmic mean temperature difference (°C or °F)$T_{c,in}$
$T_{c,in}$	Inlet temperature of the cold fluid (K or °F)
$T_{c,out}$	Outlet temperature of the cold fluid (K or °F)
$T_{c,sat}$	Boiling temperature of the cold fluid (K or °F)
$T_{h,in}$	Inlet temperature of the hot fluid (K or °F)
$T_{h,out}$	Outlet temperature of the hot fluid (K or °F)
$T_{h,sat}$	Boiling temperature of the hot fluid (K or °F)
A	Accumulative heat transfer surface (m^2 or ft^2)
A	Actual (final) heat transfer area (m^2 or ft^2)

https://doi.org/10.1515/9783110585872-208

A	Heat transfer area of the heat exchanger (m² or ft²)
A	Reference heat transfer area of the heat exchanger (m² or ft²)
A_c	Clean heat transfer area (m² or ft²)
A_d	Design (fouled) heat transfer area (m² or ft²)
A_i	Area of the inner surface of the internal pipe (m² or ft²)
A_o	Area of the outer surface of the internal pipe (m² or ft²)
Cp_c	Specific heat of cold fluid in liquid or vapor phase (J/kg K or BTU/lb °F)
$Cp_{c,l}$	Specific heat of cold fluid in liquid phase (J/kg K or BTU/lb °F)
$Cp_{c,l}$	Specific heat of hot fluid in liquid phase (J/kg K or BTU/lb °F)
$Cp_{c,v}$	Specific heat of cold fluid in vapor phase (J/kg K or BTU/lb °F)
Cp_h	Specific heat of hot fluid in liquid or vapor phase (J/kg K or BTU/lb °F)
$Cp_{h,v}$	Specific heat of hot fluid in vapor phase (J/kg K or BTU/lb °F)
c	Specific heat of the cold fluid (J/kg K or BTU/lb °F)
C	Specific heat of the hot fluid (J/kg K or BTU/lb °F)
Cp_1	Specific heat of the fluid 1 (J/kg K or BTU/lb °F)
Cp_2	Specific heat of the fluid 2 (J/kg K or BTU/lb °F)
D_h	Hydraulic diameter (m or ft)
D_i	Inner (internal) diameter of the internal pipe (m or ft)
D_i	Inner diameter of the internal pipe (m or ft)
D_o	External diameter of the internal pipe (m or ft)
D_o	Outer (external) diameter of the internal pipe (m or ft)
D_o	Outer diameter of the internal pipe (m or ft)
D_s	Inner (internal) diameter of the shell (m or ft)
D_s	Internal diameter of the external pipe (shell) (m or ft)
f_D	Darcy friction factor
f_F	Fanning friction factor
f_{FO}	Fanning friction factor for constant fluid properties
f_{Fi}	Fanning friction factor of the tube
f_{Fo}	Fanning friction factor of the annulus (shell)
F_{sp}	Temperature difference factor for series-parallel arrangement
G	Mass flux (kg/s m² or lb/h ft²)
G_i	Mass flux in the tube (kg/s m² or lb/h ft²)
G_o	Mass flux in the annulus (shell) (kg/s m² or lb/h ft²)
Gr	Grashof number
h_i	Convective heat transfer coefficient of the inner surface of the internal pipe (W/m² K or BTU/h ft² °F)
h_{io}	Internal convective heat transfer coefficient corrected for the external (reference) heat transfer surface (W/m² K or BTU/h ft² °F)
h_o	Convective heat transfer coefficient of the outer surface of the internal pipe (W/m² K or BTU/h ft² °F)
ID	Inner (internal) diameter (m or ft)
K	Friction loss coefficient
k	Thermal conductivity of the material of the internal pipe (W/m K or BTU/h ft°F)
K_i	Friction loss coefficient of the tube
K_o	Friction loss coefficient of the annulus (shell)
L	Effective length of the heat exchanger (m or ft)
L	Pipe length (m or ft)
$L_{fd,T}$	Fully developed thermal entrance length (m or ft)
L_{fd}	Fully developed entrance length (m or ft)

L_{hp}	Length of the "hairpin" heat exchanger (m or ft)
LMTD	Logarithmic mean temperature difference (°C or °F)
n	Number of parallel streams in a series-parallel arrangement
n_c	Number of parallel streams of the cold fluid in a series-parallel arrangement
n_h	Number of parallel streams of the hot fluid in a series-parallel arrangement
N_{hp}	Number of "hairpin" heat exchangers
Nu	Nusselt number based on pipe diameter
Nu_0	Nusselt number of constant fluid properties
Nu_a	Nusselt number for the annulus, based on the hydraulic diameter
Nu_b	Nusselt number with fluid properties at bulk temperature
Nu_{fd}	Nusselt number for fully developed flow
Nu_t	Nusselt number for a circular tube, based on the hydraulic diameter
Pe	Péclet number
Pr	Prandtl number
Pr_w	Prandtl number evaluated at the wall temperature
q	Heat transfer rate (W or BTU/h)
R_{di}	Internal fouling coefficient (m^2 °C/W or h ft^2 °F/BTU)
R_{do}	External fouling coefficient (m^2 °C/W or h ft^2 °F/BTU)
Re	Reynolds number
Re_{Dh}	Reynolds number based on the hydraulic diameter
Re_f	Reynolds number at fluid film temperature T_f
Re_i	Reynolds number of the tube
R_{eq}	Equivalent resistance of the whole thermal circuit (°C/W or h °F/ BTU)
Re_w	Reynolds number evaluated with fluid properties at wall temperature
R_{fi}	Internal fouling resistance (°C/W or h °F/ BTU)
R_{fo}	External fouling resistance (°C/W or h °F/ BTU)
R_i	Convective resistance of the fluid flowing inside the inner pipe (°C/W or h °F/ BTU)
R_k	Conductive resistance of the wall between fluids (°C/W or h °F/ BTU)
R_o	Convective resistance for the fluid in the annular channel (°C/W or h °F/ BTU)
S	Cross flow area (m^2 or ft^2)
S_i	Cross flow area of the tube (m^2 or ft^2)
S_o	Cross flow area of the annulus (shell) (m^2 or ft^2)
T	Fluid bulk temperature (°C or °F)
T	Temperature (°C or °F)
t	Temperature of the cold fluid (°C or °F).
T	Temperature of the hot fluid (°C or °F).
$T_{1,in}$	Entrance temperature of the fluid 1 (°C or °F)
$T_{1,out}$	Exit temperature of the fluid 1 (°C or °F)
t_1	Entrance temperature of the cold fluid (°C or °F).
T_1	Entrance temperature of the hot fluid (°C or °F).
$T_{2,in}$	Entrance temperature of the fluid 2 (°C or °F)
$T_{2,out}$	Exit temperature of the fluid 2 (°C or °F)
t_2	Exit temperature of the cold fluid (°C or °F).
T_2	Exit temperature of the hot fluid (°C or °F).
$T_{b,in}$	Fluid bulk temperature at the inlet (°C or °F)
$T_{b,out}$	Fluid bulk temperature at the outlet (°C or °F)
T_b	Fluid bulk temperature (°C or °F)
t_i	Temperature of the inner (internal) fluid (°C or °F).

t_o	Temperature of the outer (external) fluid (°C or °F).
t_w	Wall temperature (°C or °F)
T_w	Wall temperature (°C or °F)
t_{wi}	Temperature of the inner surface of heat transfer (°C or °F).
t_{wo}	Temperature of the outer surface of heat transfer (°C or °F).
U	Actual (final) overall heat transfer coefficient (W/m^2 °C or BTU/h ft^2 °F)
$\bar{U}$	Mean overall heat transfer coefficient (W/m^2 K or BTU/h ft^2 °F)
U	Overall heat transfer coefficient (W/m2 °C or BTU/h ft^2 °F)
U_c	Clean overall heat transfer coefficient (W/m^2 °C or BTU/h ft^2 °F)
U_d	Design (fouled) overall heat transfer coefficient (W/m^2 °C or BTU/h ft^2 °F)
U_i	Overall heat transfer coefficient referenced to the inner surface of the internal pipe (W/m^2 K or BTU/h ft^2 °F)
U_m	Mean overall heat transfer coefficient (W/m^2 °C or BTU/h ft^2 °F)
U_o	Overall heat transfer coefficient referenced to the outer surface of the internal pipe (W/m^2 K or BTU/h ft^2 °F)
v	Fluid velocity (m/s or ft/s)
W	Mass flow rate (kg/s or lb/h)
w	Mass flow rate of the cold fluid (kg/s or lb/h)
W	Mass flow rate of the hot fluid (kg/s or lb/h).
W_1	Mass flow rate of the fluid 1 (kg/s or lb/h)
W_2	Mass flow rate of the fluid 2 (kg/s or lb/h)
W_c	Mass flow rate of the cold fluid (kg/s or lb/h)
W_h	Mass flow rate of the hot fluid (kg/s or lb/h)
W_i	Mass flow rate in the tube (kg/s or lb/h).
W_o	Mass flow rate in the annulus (shell) (kg/s or lb/h).
$\Delta h_{vap,c}$	Enthalpy change of vaporization of the cold fluid (J/kg or BTU/lb)
$\Delta h_{vap,h}$	Enthalpy change of vaporization of the hot fluid (J/kg or BTU/lb)

1 An overview of the equipment

Double-pipe exchangers are the simplest type of heat transfer equipment that you might encounter in a chemical process plant. Of course, a hot fluid flowing inside a straight pipe standing in the ambient air may work as a heat exchanger device; however, this arrangement is not the most effective way to transfer a heat load.

In essential terms, double-pipe heat exchangers are an assembly of two pipes mounted concentrically. A cross section of a double pipe is shown in Fig. 1.1. This configuration forms two physically separated channels, where the fluids involved in the heat transfer service are allocated. The central channel, simply called "pipe" or "tube," is delimited by the inner diameter (ID) of the internal (smaller) pipe. The peripheral channel is usually referred to as "annulus" and is defined by the outer diameter (OD) of the internal pipe and ID of the external (larger) pipe. Differently from a circle, which is specified by the single measure of a diameter, the annulus of the double-pipe exchanger requires two diameters, D_o and D_s, to be completely defined. Generally, the OD of the external pipe is not relevant for the performance of the heat exchanger, except when the heat transfer with the surroundings cannot be neglected. In such case, the impairment of the equipment performance should be minimized with a layer of thermal insulation.

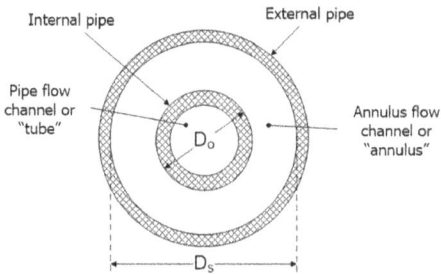

Fig. 1.1: Assemble of two concentric pipes in a double-pipe heat exchanger.

Pipe sections of fixed lengths varying from 4 to 16 m can be found from manufacturers [1–4]; however, shorter nonstandard lengths can be arranged upon special ordering. The length of the exchanger is recommended to not exceed 6 m in order to avoid bending of the inner pipe and consequent bad flow distribution inside the annulus [5]. With relatively short pipes, the unit price of the exchanger is likely to increase significantly, since the flange connections are an important part of the cost [6].

A typical device has four ports or nozzles for feeding and withdrawing the process fluids. The manner the fluid inlets are connected may have a significant impact

https://doi.org/10.1515/9783110585872-001

on the heat exchanger performance. As depicted in Fig. 1.2, when both fluids enter the equipment on the same side, the obtained arrangement is called parallel flow or concurrent. On the other hand, the insertion of the fluids by nonadjacent nozzles creates a counterflow or countercurrent flow pattern.

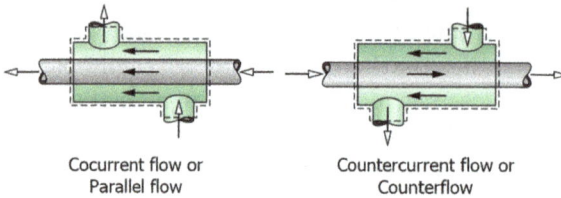

Cocurrent flow or
Parallel flow

Countercurrent flow or
Counterflow

Fig. 1.2: Schematic representation of a double-pipe heat exchanger (adapted from Ref. [13]).

The flow arrangement is an engineer's decision. Usually, the choice should be in favor of the countercurrent flow because there is an advantage of the mean temperature difference produced between the streams, leading to a more efficient and less expensive design. However, there are exceptions favoring the concurrent flow, as will be discussed in Section 2.5.9.

Although "double-pipe heat exchanger" is the most common designation, this equipment is referred by several other names in the field jargon and specialized literature, including variations such as double-tube heat exchanger, concentric heat exchanger, concentric pipes heat exchanger, and tube-in-tube heat exchanger [7–11]. The engineer should be aware of various synonyms, to communicate well in practical situations.

1.1 Process applicability

Double-pipe heat exchangers are very robust and reliable. They can operate for long periods with high fouling streams, at severe process conditions, involving extreme temperature crossings, high fluid temperatures, high pressures, and large temperature ranges between the inlet and outlet of the fluids. Pressure capabilities are full vacuum to over 96,000 kPa (about 14,000 psi), limited by size, material, and design conditions [12]. These characteristics allow its use for very specialized services. I can recall of a gas-phase polymerization tubular reactor that I saw in a polyethylene production plant, which was designed as a concentric heat exchanger to handle the high temperature and pressure required for performing the chemical reaction. Figure 1.3 shows a pipe section of this reactor. This is an example of process requirements that other type of exchanger could hardly match.

Fig. 1.3: A section cropped from the inner pipe of a high-pressure double-pipe heat exchanger (source: the author).

Double-pipe exchangers are very suited for corrosive, extremely fouling services, such as high viscosity fluids or slurries. There are construction adaptations available on the market with scrapping mechanisms capable of removing even crusts deposited on the heat transfer surface. The usual design is to place the most corrosive, fouling, higher pressure, and higher temperature fluid in the inner pipe [5].

A heat exchanger may transfer sensible and/or latent heat between fluids. While in more integrated plants, heat exchangers are found to contact two process streams, the most common service is transferring energy between a process and a utility fluid, such as cooling water, chilled water, commercial thermal fluids, saturated vapor, or air. The flow rate of the process fluid is determined by the production requirements of the plant, whereas the flow rate of the utility stream is usually manipulated to control the outlet temperature of the process stream. The flow rate of the utility stream may be regulated by a valve at the inlet or outlet of the heat exchanger, however, in the case of cooling water, it is recommended to restrict its flow at the entrance, because the higher temperature and vapor pressure may produce cavitation on the outlet valve [5].

1.1.1 Types of hairpins

Generally, two types of regular double-pipe or multi-tube heat exchangers are readily found in the market. The main distinction is the way the annulus fluid flows between the heat exchanger legs.

Figure 1.4 shows a hairpin exchanger in which the annulus fluid passes through the internal chamber built with a flanged cover or bonnet attached to the heat exchanger outer pipes. In this case, the inner fluid usually flows in a single or multiple "U"-shaped tube, which is loaded inside the outer tube, or shell.

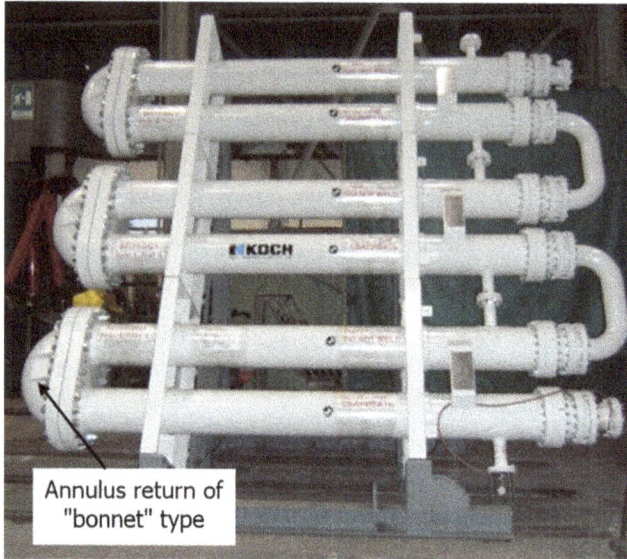

Fig. 1.4: Hairpin heat exchangers with integral bonnet (left end) covering the inner tubes and providing the return chamber for the fluid in the annulus (adapted from Ref. [14]).

For more fouling and severe services, manufacturers offer a so-called ROD-THRU hairpin exchanger type in which the inner fluid passes through a simple 180° return bend [15]. With this design, the internal tubes can be accessed from both heat exchanger ends for cleaning operations, which is not possible for "U" tubes. The annulus fluid flows from one leg to another through a plain straight tube; no bonnet is used in this construction, as shown in Fig. 1.5.

Fig. 1.5: Return bends for pipe and annulus of "ROD-THRU" hairpin heat exchangers used for food processing. These bends do not have a bonnet-type design, being just plain pipes (adapted from Ref. [9]).

1.1.2 Primary concerns

When selecting and designing a heat exchanger, there are issues beyond costs. Certainly, costs do matter a lot, however, considering that, historically, major breakthroughs in chemical and petrochemical revenues are due to the development of new products and processes, rather than from more inexpensive designs, I endorse the point of view of Ref. [6] for asserting that safety and reliability come in first place. In this regard, some parameters should be carefully considered while selecting a heat exchanger type for a given service, namely:

- Process safety
- Environment protection
- Operation reliability
- Operating pressure and temperature
- Corrosivity of the fluid streams
- Fouling

Only after considering these aspects, the engineer should give a chance to a possible design configuration. Among those feasible designs, there will be one option that is more cost-effective for performing the service.

1.1.3 Where to use a double-pipe exchanger?

The answer for this question is not straightforward as it might appear. Generally, the design decision for using a double-pipe heat exchanger might be driven by one of the following approaches:

1. Rigorous optimization
2. Rules of thumb (guidelines based on previous experiences)

1.1.3.1 A few guidelines for fast selection of a double-pipe exchanger

Double-pipe exchangers are especially appropriate for small-to-moderate duties, typically 10–20 m^2 (about 100–200 ft^2) [16], in extreme conditions. The services in which double-pipe exchangers outperform other heat exchanger types are those requiring severe operating conditions, such as high temperatures and pressures above 3 MPa (400 psi) in the tube side [17]. Although double-pipe heat exchangers are not indicated for very large duties, because the number of necessary hairpins may grow unmanageably, they are highly versatile, being able to work from high vacuum to fairly high pressures, typically not greater than 30 MPa (4,350 psi) inside the annulus and 140 MPa (20,000 psi) on the tube side [18].

The structural simplicity enables its use in highly corrosive and erosive services. Because of the robust design with a relatively small number of parts, hairpin exchangers are well suited for greater corrosion and erosion allowances, once the inner pipe may be specified with increased thickness in several millimeters, which could make other heat exchanger types impracticable to build. The ASME B31.3 code provides a formula for calculating the pipe wall thickness as the sum of the pressure design thickness with an added thickness provisioned for compensation of mass loss due to corrosion or erosion [19].

The existence of royalty-free, open, and dependable design/analysis methods counts in favor of double-pipe exchangers. The availability of proven and reliable design methods is an important factor for selecting a given type of heat exchanger. Accordingly, the double-pipe heat exchanger is among the types with very strong publicly available design procedures, which can be used by anyone without incurring in royalties. Of course, there are satisfactory design methods for most heat exchanger types performing eventually any service; however, the best methods for narrowly used models are often proprietary to manufacturers or research organizations [20].

1.1.3.2 Possible applications
In the absence of a more rigorous selection process, such as using optimization methods, a hairpin double-pipe or multi-tube heat exchanger should be more economical by the following guidelines [21, 22]:
- Heating or cooling of a process stream, when a relatively small heat duty is involved (normally not >50 m^2) [20]
- The process results in a temperature cross
- High-pressure tube-side application
- High terminal temperature differences (300 °F/149 °C or greater)
- A low-allowable pressure drop is required on one side
- When the exchanger is subject to thermal shock
- When flow-induced vibration may be a problem
- When solids are present in the process stream
- Cyclic service
- High flow rate ratios between shell-side and tube-side fluids
- When an augmentation device will enhance the heat transfer coefficient
- When heating or cooling vapors
- When complete vaporization is required
- When the mechanical advantages of a hairpin are preferred
- Suitable for partial boiling or condensations

1.1.3.3 Advantages and disadvantages
Some advantages favoring the use of double-pipe heat exchangers may be listed as follows:
-   Low capital cost
-   Low installation cost
-   Flexibility and versatility
-   Simple piping arrangements [20]
-   Ease of cleaning on both sides, provided that the end fittings are removable [20]
-   The heat transfer area can be increased with relative simplicity in existing equipment [20]
-   Adaptable flow distribution controlled by insertion of pumps between banks in series, if necessary [20]
-   Very cost-effective for services requiring 10–20 m^2 (about 100–200 ft^2) of heat transfer surface
-   Do not have temperature crosses between streams, since they are inherently true countercurrent devices
-   For shell-and-tube units with shell-side heat transfer coefficient lower than half the tube side, the use of double pipes may equate the heat transfer coefficients, avoiding an undesirable controlling fluid design [5]
-   Straightforward and very reliable design and rating methods are widely available

Another heat exchanger type might be considered if the following disadvantages are important issues in each service:
-   A large heat duty may require an unacceptable number of double-pipe exchangers
-   A bank with many exchangers is more susceptible to leakages, due to the large number of connections
-   Not practical for services involving low heat transfer coefficients
-   The plant space occupied by the installation may be considerable, which is especially undesirable in offshore applications
-   The time required for disassembling and cleaning a large exchanger bank may be unacceptable for certain plant operations

1.1.4 About the optimal design

When specifying a heat exchanger, the designer may achieve a satisfactory configuration through the recursive application of the design method, guided using rules of thumb and recommended best practices. Such situation is illustrated in Fig. 1.6, in which case each possible configuration is selected and tested manually by the designer. Though this manual approach can lead to a reasonable design, in practice, it has limited range in search for an optimal solution, which is more feasible using an automated optimization method.

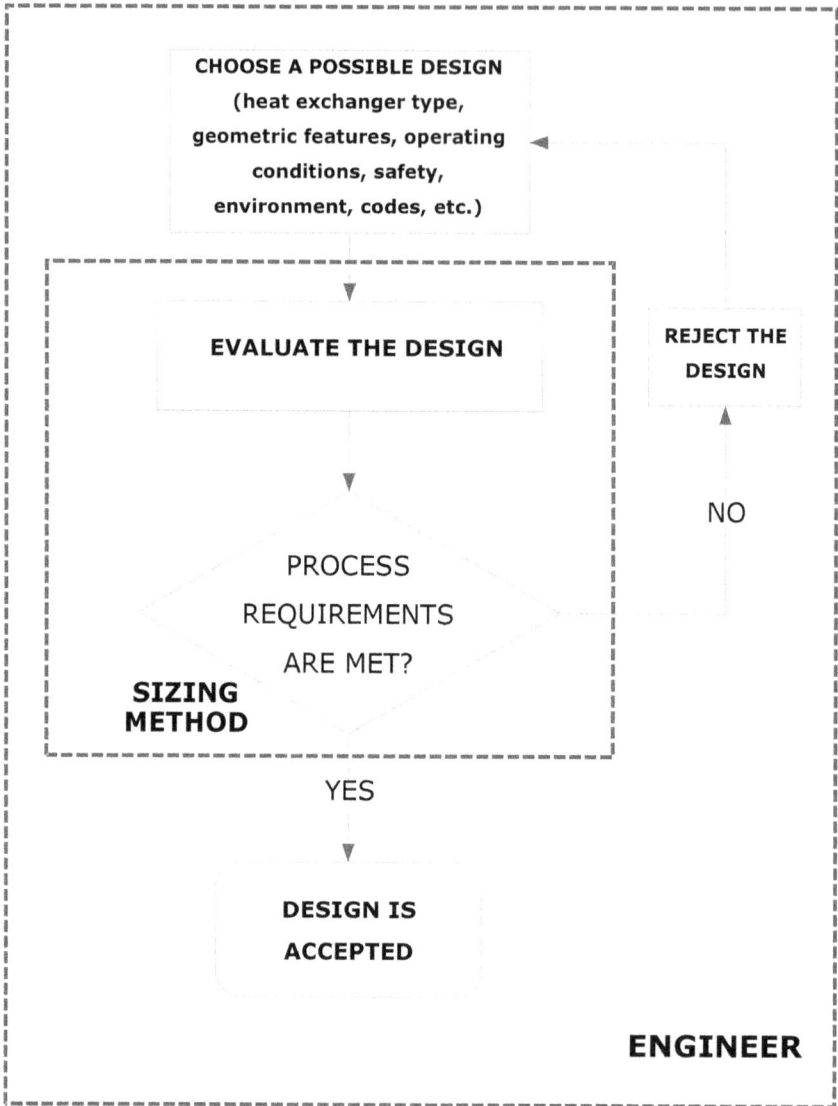

Fig. 1.6: Procedure for designing a heat exchanger without optimization. The process engineer is responsible for every decision in the design routine.

CHOOSE A POSSIBLE
DESIGN (heat exchanger
type, geometric features,
operating conditions, safety,
environment, codes, etc.)

EVALUATE THE DESIGN

PROCESS
REQUIREMENTS   NO   REJECT
ARE MET?   THE
                              DESIGN

SIZING
METHOD

YES

EVALUATE
OBJECTIVE   NO
FUNCTION

OBJECTIVE
FUNCTION IS
MINIMIZED (OR
MAXIMIZED)?

YES

DESIGN IS
ACCEPTED   OPTIMIZATION
                    METHOD

ENGINEER

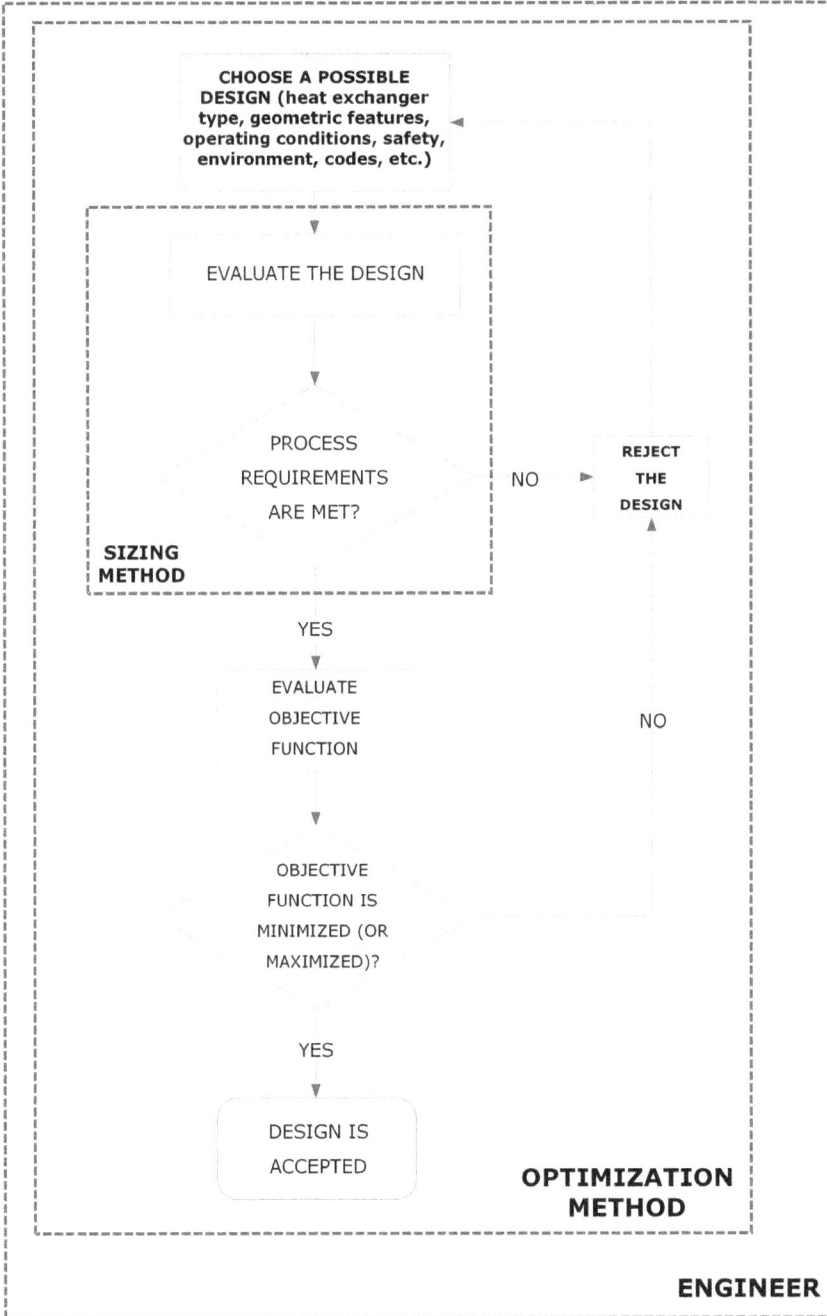

Fig. 1.7: Procedure for designing a heat exchanger with the external optimization loop in place. The process engineer is responsible for setting and fine-tuning of optimizer parameters.

Although not being the first choice for "old school" process engineers, if you are given enough time and resources (knowledge and tools), optimization is the way to go. The precise assessment for using or not a particular type of heat exchanger should be based on a comparative analysis among all possible types, considering an "overall" quantitative criterion, for example:

- Capital cost
- Total cost (capital and operational cost)
- Available ground area for installation
- Spatial constraints like height and width of the equipment
- Total weight of the equipment, which impact on transportation and cargo costs, along with support structure costs
- Reduction of fluid holdup inside the exchanger because the process fluid may be hazardous for people or for the environment, and its retained volume should be kept small

To achieve the finest design founded on a quantitative decision, the use of optimization algorithms is required. Essentially, what an optimization method does is an "intelligent" search and evaluation of alternatives in a set of possible designs. Considering the possibilities of equipment types, geometric features for each type and operating conditions, the number of design options capable of accomplishing a given heat transfer service may be enormous, and not manageable through engineer's direct insight or judgment. Of course, previous experience, some proven heuristics, and rules of thumb should lead to a feasible design of the heat exchanger; however, no guarantees are warranted that they will be the most effective for the proposed service.

In summary, within an automatic optimization scheme, the quantitative criterion is expressed as a mathematical relationship called objective function, which is evaluated for each attempted design configuration. The optimization method is responsible for the rational selection of successive configurations that should converge to an optimum heat exchanger design. As represented schematically in Fig. 1.7, the optimization method assumes the role played by the engineer engaged in a "handmade" design. The subject of optimization techniques is vast, and out of the scope of this text. If you are interested or feel the need for going deeper in this field, there are very good sources of information in the literature, and some of them are Refs. [23–27].

1.2 The double-pipe exchanger is actually a hairpin!

The most used type of double-pipe exchangers is not simply a pipe inside another, but a "U-shaped" device mounted with two straight legs connected by a 180° return bend. Each leg is a set of concentric pipes (Fig. 1.8). Due to its characteristic shape, this kind of double-pipe exchanger is normally entitled "Hairpin".

Large batteries of bare-tube double pipes are typically needed to meet heat transfer areas exceeding 100 m² (about 1,000 ft²). It is possible to attenuate this limitation by using finned tubes. In fact, there is the commercial availability of multi-tube hairpins with area as large as 1,100 m² (about 12,000 ft²) [28].

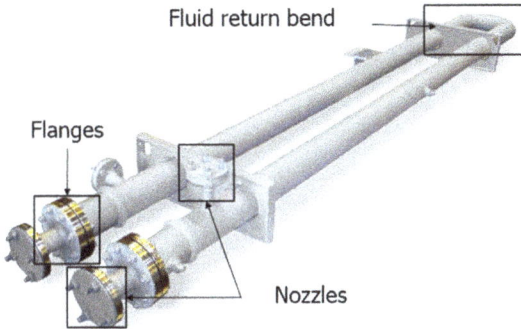

Fig. 1.8: Hairpin (double-pipe) heat exchanger (adapted from Ref. [29]).

The hairpin double-pipe exchanger has some practical advantages. Frequently, a single exchanger is not enough to perform a given heat duty. Therefore, several hairpins are usually associated, forming a bank or battery of exchangers, in which the U-shape is convenient because the outlet nozzles of one exchanger are spatially close to the inlet nozzles of the next exchanger in the bank.

The proximity of nozzles implies that minimum additional pipeline is necessary to connect the units. Also, there is the marginal consequence that not every hairpin in the bank needs to be of the same length, which may be a key advantage for maintenance purposes, since individual exchangers in the bank may be temporarily replaced with units of different lengths, possibly available in warehouse. In case the exchanger bank is a critical equipment in the chemical process, a procedure such as that may reduce significantly the downtime of the plant, meaning sometimes a fair amount of saved revenue [30].

The structure with few connections accommodates differential thermal expansion that creates mechanical tension between the internal and external pipes. The double-pipe exchanger maintenance usually involves less time and reduced costs, because in comparison to other models, it can be easily dismounted for inspection and cleaning. If a process is discontinued, the flexibility of the double-pipe exchanger favors its use in another process of the same plant. Hairpins may be readily encountered commercially with external pipe diameter varying from 50 to 200 mm, while the internal pipe is normally in the range of 20–150 mm nominal diameter ([6, 28]).

Nowadays, some manufacturers offer a bank of double-pipe exchangers composed of conjugated straight concentric pipes. The bank shown in Fig. 1.9 is not strictly an association of hairpins, being just a sequence of several straight double

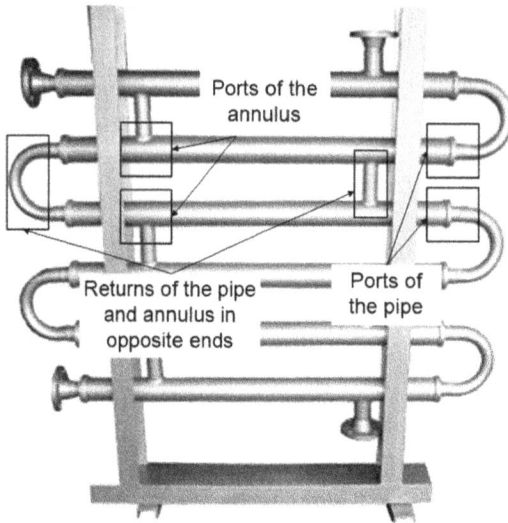

Fig. 1.9: Conjugated straight multi-tube heat exchanger associated in series (adapted from Ref. [31]).

pipes connected in series by welded connections. The feature that distinguishes this exchanger from a hairpin is the installation of the pipes and annuli ports at opposite ends of any couple of consecutive pipes. In contrast, Figs. 1.10 and 1.11 show batteries of hairpin heat exchangers from another manufacturer, with the particularity that the connections between the annuli of neighboring hairpins are

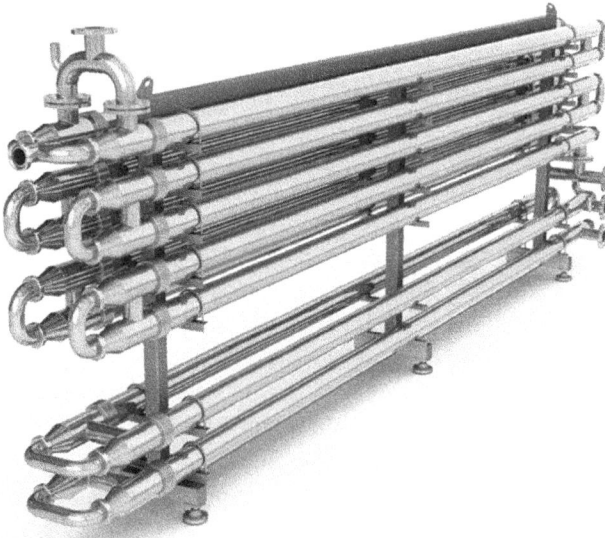

Fig. 1.10: Bank of multi-tube hairpin heat exchangers (adapted from Ref. [32]).

welded, not flanged like in Fig. 1.8. Roughly speaking, the capital cost of a designed double-pipe exchanger may range from 80% to 140% of the cost of a shell-and-tube fixed tube sheet unit with the same capacity [17].

1.3 Pipes and tubes

1.3.1 Is there a difference?

Currently, the difference between pipes and tubes is just a matter of classification and the way in which they are specified and ordered from the supplier. In plain terms, a tube is simply a hollow cylinder. Decades ago, "pipes" and "tubes" were not made by the same process. "Tubes" had more careful finish and, accordingly, their surfaces were smoother. At that time, "pipes" were less finished, and their internal surface had a greater relative roughness in comparison with tubes.

Any hollow cylinder produced by modern techniques may be regarded as a tube [33]. However, for classification purposes, both terms "tubes" and "pipes" are still used in industry. "Tubes" are used for mechanical and structural applications (e.g., welded together in a tubular structure) and typically specified by an outside diameter given in inches or millimeters. The wall thickness of tubes is determined by standard "scales" such as Birmingham Wire Gauge (BWG) and Standard Wire Gauge (SWG). A representation of the BWG scale is shown in Tab. 2 of Appendix A.2, and a full dataset for specification of heat transfer tubes is found in Tab. 3 of Appendix A.2.

Fig. 1.11: Bank of AISI 304 stainless steel double-pipe hairpin heat exchangers (Adapted from Ref. [34]).

"Pipes" are used for fluid transportation, usually in the form of pipelines. The key dimension of pipes is the ID, since it defines the cross-flow area. Accordingly, pipe sizes are designated by a "nominal pipe size" (NPS) or "diameter nominal" (DN), normally measured in inches and millimeters, respectively. NPS and DN tables can be found in Appendix A.1. For nominal sizes below 12 in, the designated size is approximately equal to the internal pipe diameter. For sizes from 14 to 36 in, the nominal size designates exactly the OD. Distinctly from "tubes," whose thickness comes directly from the BWG scale, the pipe thickness is defined by "schedules" (e.g., 40, 60, 80, and 120), which for certain values are equivalent to mass classes such as STD (standard weight), XS (extra strong), and XXS (double extra strong).

1.3.2 Practical implications of pipe and tube distinction

Besides the fact that "tubes" and "pipes" are essentially the same thing today, the exchanger designer should be aware of the former distinction when using equations for estimating heat transfer coefficients and friction factors. Some traditional and reliable correlations are based on experimental data measured from rough pipes and smooth tubes separately. Good examples of this situation are the Sieder–Tate equations. As a result, the equations derived from rough pipe data give overestimated heat transfer coefficients and friction factors when applied to modern tubes. The engineer must be aware of these errors because they penalize the heat transfer area, increasing the risk of a failed heat exchanger design. However, in ordinary practical cases, such deviations are balanced by the conservative approximations and overdesign performed on the stages of the heat exchanger sizing procedure.

1.3.3 Commercial diameters and lengths

1.3.3.1 Heat transfer tubes

Tubes specially manufactured for heat transfer services can be acquired in a variety of metals and alloys from major suppliers in several standardized sizes, e.g., from 16 to 51 mm according to DIN 28180, and ranging from ½ to 3 in by the ASTM A179 standard [33]. Usually, it is possible to arrange the specification in any size standard upon previous agreement with the supplier. The offered lengths of heat transfer tubes vary according to the manufacturer; however, the buyer is normally allowed to request the tubes cut at any length smaller than the full length, defined by the manufacturing process. Heat transfer tubes in about 18 m is easily found in the market [35].

1.3.3.2 Ordinary pipes

The inner and outer tubes of double-pipe heat exchangers, like the outer tube (shell) of shell-and-tube heat exchangers, are commonly built from general-purpose "regular" pipes, specified under NPS or DN standards; however, ordering in accordance with other size standards is possible upon agreement with the manufacturer. Typical commercial lengths are pipe continuous sections of 5–10 m (16–32 ft) [36].

2 Heat exchanger design equations

2.1 Heat load

The heat load of a heat exchanger is the heat transfer rate required to match a specified thermal service. In general, both fluids could undergo a phase change, which means that the cold fluid may enter as liquid and leave the exchanger as a vapor, and that the hot fluid may enter the exchanger as a vapor and leave it as a liquid. Hence, the heat gained by the cold fluid (q_c) and heat lost by the hot fluid (q_h) are expressed in eqs. (2.1) and (2.2). Note that $\Delta h_{vap,c}$ and $\Delta h_{vap,h}$ are the enthalpy change (or latent heat) of vaporization for cold and hot fluids, respectively; both were taken as positive quantities:

$$q_c = W_c \left(Cp_{c,l}(T_{c,sat} - T_{c,in}) + Cp_{c,v}(T_{c,out} - T_{c,sat}) + \Delta h_{vap,c} \right) \tag{2.1}$$

$$q_h = W_h \left(Cp_{h,l}(T_{h,out} - T_{h,sat}) + Cp_{h,v}(T_{h,sat} - T_{h,in}) - \Delta h_{vap,h} \right) \tag{2.2}$$

where W_c is the mass flow rate of the cold fluid (kg/s or lb/h), W_h is the mass flow rate of the hot fluid (kg/s or lb/h), $Cp_{c,v}$ is the specific heat of cold fluid in vapor phase (J/kg K or BTU/lb °F), $Cp_{c,l}$ is the specific heat of cold fluid in liquid phase (J/kg K or BTU/lb °F), $Cp_{h,v}$ is the specific heat of hot fluid in vapor phase (J/kg K or BTU/lb °F), $Cp_{h,l}$ is the specific heat of hot fluid in liquid phase (J/kg K or BTU/lb °F), $T_{c,sat}$ is the boiling temperature of the cold fluid (K or °F), $T_{h,sat}$ is the boiling temperature of the hot fluid (K or °F), $T_{c,in}$ is the inlet temperature of the cold fluid (K or °F), $T_{c,out}$ is the outlet temperature of the cold fluid (K or °F), $T_{h,in}$ is the inlet temperature of the hot fluid (K or °F), $T_{h,out}$ is the outlet temperature of the hot fluid (K or °F), $\Delta h_{vap,c}$ is the enthalpy change of vaporization of the cold fluid (J/kg or BTU/lb), and $\Delta h_{vap,h}$ is the enthalpy change of vaporization of the hot fluid (J/kg or BTU/lb).

Admitting that all the energy leaving one fluid goes to the other one, no heat is transmitted to the surroundings; therefore, the enthalpy balance accounting for both sensible and latent heat may be expressed as eq. (2.3), and using eqs. (2.1) and (2.2), we write eq. (2.4):

$$q_c + q_h = 0 \tag{2.3}$$

$$W_c \left(Cp_{c,l}(T_{c,sat} - T_{c,in}) + Cp_{c,v}(T_{c,out} - T_{c,sat}) + \Delta h_{vap,c} \right) + W_h \left(Cp_{h,l}(T_{h,out} - T_{h,sat}) \right.$$
$$\left. + Cp_{h,v}(T_{h,sat} - T_{h,in}) - \Delta h_{vap,h} \right) = 0 \tag{2.4}$$

For the special case where the hot fluid does not change phase, we may set $Cp_{h,l} = Cp_{h,v} = Cp_h$, then eq. (2.4) reduces to the following equation:

https://doi.org/10.1515/9783110585872-002

$$Cp_h W_h\left(T_{h,out} - T_{h,in}\right) + W_c\left(Cp_{c,l}\left(T_{c,sat} - T_{c,in}\right) + Cp_{c,v}\left(T_{c,out} - T_{c,sat}\right) + \Delta h_{vap,c}\right) = 0 \quad (2.5)$$

where Cp_h is the specific heat of hot fluid in liquid or vapor phase (J/kg K or BTU/ lb °F).

Correspondingly, if the cold fluid does not undergo a phase change, then $Cp_{c,l} = Cp_{c,v} = Cp_c$, and eq. (2.4) reduces to the following equation:

$$Cp_c W_c\left(T_{c,out} - T_{c,in}\right) + W_h\left(Cp_{h,l}\left(T_{h,out} - T_{h,sat}\right) + Cp_{h,v}\left(T_{h,sat} - T_{h,in}\right) - \Delta h_{vap,h}\right) = 0 \quad (2.6)$$

where Cp_c is the specific heat of cold fluid in liquid or vapor phase (J/kg K or BTU/ lb °F).

Finally, the simplest situation is a process with none of the fluids changing phase, for which the enthalpy balance takes the following form:

$$Cp_c W_c\left(T_{c,out} - T_{c,in}\right) + Cp_h W_h\left(T_{h,out} - T_{h,in}\right) = 0 \quad\quad\quad (2.7)$$

Designing a heat exchanger means essentially a two-step procedure: (1) select a type or model of the equipment and (2) determine its size. The selection of the heat exchanger type is usually a qualitative step, where the process engineer should consider several elements such as safety, environment, operating conditions, corrosion, fouling, and risks. For the specific case of a double-pipe heat exchanger selection, this issue is discussed in more detail in Section 1.1.3. The second step, namely, the sizing of the exchanger heat transfer surface, is strictly quantitative, and achieved via any design method, which should be appropriate to the model type chosen in the first step.

Worked Example 2.1

A stream of vaporized phenol and flow rate of 0.38 kg/s should be cooled from 215 to 93 °C using cold water at 32 °C. The maximum exit temperature of the water should not exceed 60 °C to avoid fouling formation. The phenol boiling temperature is 181 °C, and its enthalpy change of vaporization is $\Delta h_{vap} = (6.55e + 5)$ J/kg. Determine the heat load required from the heat exchanger and the water flowrate to be used in this service. Additionally, estimate the percentage of the heat load that is required just to condensate the phenol vapor.

Solution

The hot fluid phenol changes phase from vapor to liquid in this process, but the cold fluid water only exchanges sensible heat; therefore, the general heat balance equation (2.4) takes the simplified form of eq. (2.6).

We are going to evaluate the physical properties of the phenol vapor using the average of the temperature range from the inlet to the saturation temperature:

$$T_{h,v} = \frac{T_{h,in}}{2} + \frac{T_{h,sat}}{2} = \frac{488.15}{2} + \frac{454.15}{2} = 471.15 \text{ K}$$

The mean temperature for the liquid phase phenol is

$$T_{h,l} = \frac{T_{h,out}}{2} + \frac{T_{h,sat}}{2} = \frac{366.15}{2} + \frac{454.15}{2} = 410.15 \text{ K}$$

And the mean temperature for the cold water is

$$T_c = \frac{T_{c,in}}{2} + \frac{T_{c,out}}{2} = \frac{305.15}{2} + \frac{333.15}{2} = 319.15 \text{ K}$$

From Appendix C.13, the specific heats for phenol and water are

Phenol (vapor): 471.15 K	$Cp_{h,v} = 1,650.0$ J/kg K
Phenol (liquid): 410.15 K	$Cp_{h,l} = 1,980.0$ J/kg K
Water (liquid): 319.15 K	$Cp_c = 4,090.0$ J/kg K

Solving eq. (2.6) for the flowrate of water, and substituting the known parameters, we obtain

$$W_c = \frac{W_h \left(Cp_{h,l} T_{h,out} - Cp_{h,l} T_{h,sat} - Cp_{h,v} T_{h,in} + Cp_{h,v} T_{h,sat} - \Delta h_{vap,h} \right)}{Cp_c (T_{c,in} - T_{c,out})}$$

$$= \frac{0.38000(-(6.5500e+5) + 1,650.0 \cdot 454.15 - 1,650.0 \cdot 488.15 + 1,980.0 \cdot 366.15 - 1,980.0 \cdot 454.15)}{4,090.0(305.15 - 333.15)}$$

$$= 2.9377 \text{ kg/s}$$

Since the water only transfers sensible heat, we can take the shortcut of using the calculated flowrate to obtain the total heat received as follows:

$$q_c = Cp_c W_c (T_{c,out} - T_{c,in}) = 2.9377 \cdot 4,090.0(333.15 - 305.15) = (3.3643e+5) \text{ W}$$

At this point, we could use q_c to finish the resolution; however, as a pedagogical exercise, let us evaluate the total load from the perspective of the hot fluid, which undergoes the phase change. In this procedure, we must evaluate the latent heat of phenol and the sensible heat exchanged before and after its condensation.

From the heat balance for vapor phase of phenol, the sensible heat lost is

$$q_{h,v} = Cp_{h,v} W_h (T_{h,sat} - T_{h,in}) = 0.38000 \cdot 1,650.0(454.15 - 488.15) = -21,318 \text{ W}$$

The latent heat lost by the phenol during condensation is calculated from

$$q_{h,lat} = -W_h \Delta h_{vap,h} = -0.38000(6.5500e+5) = (-2.4890e+5) \text{ W}$$

After condensation, the phenol stream is cooled further at a rate of

$$q_{h,l} = Cp_{h,l} W_h (T_{h,out} - T_{h,sat}) = 0.38000 \cdot 1,980.0(366.15 - 454.15) = -66,211 \text{ W}$$

Then, the total heat lost by the phenol stream is the sum of the previous heat transfer rates as follows:

$$q_h = q_{h,lat} + q_{h,l} + q_{h,v} = (-2.4890e+5) + (-21,318) + (-66,211) = (-3.3643e+5) \text{ W}$$

As expected, it is noticeable that the calculated value of q_h numerically equals q_c, except for the conventional negative sign, indicating that the phenol is losing heat.

Finally, the phenol latent heat corresponds to a fraction of the total heat load of

$$100 \frac{q_{h,lat}}{q} = 100 \frac{(-2.4890e+5)}{(-3.3643e+5)} = 73.983\%$$

This result shows the typical situation in which the latent heat load is significantly greater than the sensible heat exchanged.

2.2 The design equation

The most used design method for sizing a heat exchanger is based on a simple integrated equation. The foundations for this method were put in place by a few researchers, pioneered by Nagle [37, 38], Colburn [39], and Bowman [37, 40]. The referred design equation may be written as

$$q = UA\Delta T_m \tag{2.8}$$

where q is the heat transfer rate (W or BTU/h), i.e., the heat load that the equipment must match; U is the overall heat transfer coefficient (W/m^2 °C or BTU/h ft^2 °F), which is a combination of all the thermal resistances between the energy source (the hot fluid) and the energy target (the cold fluid). Typically, there are three thermal resistances inside a heat exchanger, being two of them convective resistances and the third is the conductive thermal resistance of the wall separating the fluid streams. ΔT_m is a kind of mean temperature difference taken over all the local temperature differences between the two fluids exchanging energy (°C or °F), and A is the reference heat transfer area of the heat exchanger (m^2 or ft^2).

Given that the calculation of the overall heat transfer coefficient (U) and mean temperature difference (ΔT_m) is performed appropriately, eq. (2.8) may be used for the design of several types of heat exchangers. In fact, these last two parameters are the only information that restricts eq. (2.8) for a specific type of heat exchanger.

2.3 The overall heat transfer coefficient

The overall heat transfer coefficient is a very useful concept because, in a practical heat exchanger, the heat transfer process involves multiple thermal resistances. Although, in some situations, one or more resistances may be disregarded with respect to another, generally there are five thermal resistances to deal with, namely:
1. Convective resistance on the hot fluid side
2. Fouling resistance on the hot fluid side
3. Conductive resistance on the separation wall
4. Fouling resistance on the cold fluid side
5. Convective resistance on the cold fluid side

In a double-pipe heat exchanger, the geometry is tubular, i.e., cylindrical; therefore, the separation wall is the wall of the internal pipe. This is shown in Fig. 2.1, which shows a zoomed cut of a counterflow double-pipe heat exchanger. The thermal resistances are identified as forming a thermal circuit of five resistances in series.

Flow regimes in industrial process situations are commonly turbulent. Accordingly, due to the mixing effect of turbulence, the temperature gradient in the turbulent core regions of the flow field is greatly reduced. Consequently, most of the temperature variation, starting from the wall and going into the hot or cold fluids, is in a quite thin region referred to as the "thermal boundary layer". This is the place where almost all convective thermal resistance resides. In fact, the very definition of a thermal boundary layer states that it ends when 99% of the temperature variation has undergone [41–43], departing from the solid wall.

The concept of overall heat transfer coefficient comes from the definition of the equivalent thermal resistance for a given heat exchange circuit. Taking advantage from a direct analogy with the equivalent resistance in electrical circuits taught in fundamental physics courses, we could rewrite the design equation (2.8) as follows:

$$q = \frac{\Delta T_m}{\frac{1}{UA}} = \frac{\Delta T_m}{R_{eq}} \tag{2.9}$$

At this point, you should be a little confused. For a defined heat load q in eq. (2.9), the numeric value of the equivalent resistance must be unique; consequently, the value of $(1/UA)$ must be unique also, but for a cylindrical wall, i.e., a pipe, there are two heat transfer surfaces crossed by the radial heat flux. Therefore, to obtain the same heat load q, the conclusion is that the overall heat transfer coefficient is necessarily tied to a given reference heat transfer area A; if the reference area changes, the overall heat transfer coefficient changes to compensate. If the inside surface A_i

and the outside surface A_o of the internal tube in a heat exchanger are possible choices for the reference heat transfer area, then the equivalent thermal resistance may be expressed as

$$R_{eq} = \frac{1}{UA} = \frac{1}{U_i A_i} = \frac{1}{U_o A_o} \tag{2.10}$$

where U is the mean overall heat transfer coefficient (W/m² K or BTU/h ft² °F); U_i, U_o are overall heat transfer coefficients of the inner and outer surfaces of the internal pipe, respectively (W/m² K or BTU/h ft² °F); A is the area of heat transfer surface (m² or ft²); R_{eq} is the mean equivalent thermal resistance for the whole heat transfer surface A; A_i and A_o are the internal and external heat transfer surfaces of the inner tube with effective length L, respectively:

$$A_i = \pi D_i L \tag{2.11}$$

$$A_o = \pi D_o L \tag{2.12}$$

Thermal resistances are, in general terms, local quantities, since they depend on physical properties that have different values from point to point of the heat transfer surface. Therefore, we should notice that the equivalent thermal resistance in eq. (2.10) is a mean value for the total area of heat transfer. This is a direct consequence from the fact that eq. (2.9) correlates extensive physical quantities, which change with the size of the equipment, such as A, ΔT_m, and U.

Considering the thermal circuit shown in Fig. 2.1, there are five resistances associated in series. The equivalent resistance for this circuit may be expressed as follows:

$$R_{eq} = R_i + R_{fi} + R_k + R_{fo} + R_o \tag{2.13}$$

where R_i is the convective resistance of the fluid flowing inside the inner pipe, R_k is the conductive resistance of the wall between fluids, R_o is the convective resistance of the fluid in the annular channel, R_{fi} is internal fouling resistance (°C/W or h °F/BTU), and R_{fo} is external fouling resistance (°C/W or h °F/BTU).

Conventionally, fouling thermal resistances R_{fi} and R_{fo} are expressed using required values of "fouling coefficients" or "fouling factors" R_{di} and R_{do}, which do not depend on the size of the heat transfer surface and correspond conceptually to the inverse of convective heat transfer coefficients. Therefore, the fouling resistances are given by

$$R_{fi} = \frac{R_{di}}{A_i} \tag{2.14}$$

$$R_{fo} = \frac{R_{do}}{A_o} \tag{2.15}$$

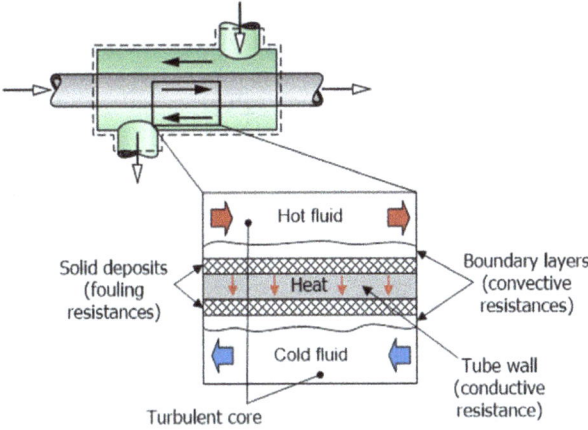

Fig. 2.1: Thermal resistances involved in the heat transfer between a hot and a cold stream inside a heat exchanger.

The wall conductive resistance for a cylindrical duct with thermal conductivity k, inner diameter D_i, and outer diameter D_o can be expressed as follows:

$$R_k = \frac{D_i \ln\left(\frac{D_o}{D_i}\right)}{2A_i k} \tag{2.16}$$

With the heat transfer coefficients h_i and h_o for the tube and annulus fluids, respectively, the convective thermal resistances are

$$R_i = \frac{1}{h_i A_i} \tag{2.17}$$

$$R_o = \frac{1}{h_o A_o} \tag{2.18}$$

With eq. (2.10) and substituting eqs. (2.14)–(2.18) into (2.13), the equivalent thermal resistance becomes

$$R_{eq} = \frac{1}{U_i A_i} = \frac{1}{U_o A_o} = \frac{R_{di}}{A_i} + \frac{1}{h_i A_i} + \frac{D_i \ln\left(\frac{D_o}{D_i}\right)}{2A_i k} + \frac{1}{h_o A_o} + \frac{R_{do}}{A_o} \tag{2.19}$$

A widely accepted (although arbitrary) convention in the technical literature and industry is to define the external surface of the tube as reference for the design and analysis of tubular heat exchangers. Following such convention, the design equation (2.8) is rewritten as follows:

$$q = U_o A_o \Delta T_m \tag{2.20}$$

And the thermal conductance based on the outside area is $U_o A_o = 1/R_{eq}$, where

$$U_o A_o = \left(\frac{R_{di}}{A_i} + \frac{1}{h_i A_i} + \frac{D_i \ln\left(\frac{D_o}{D_i}\right)}{2 A_i k} + \frac{1}{h_o A_o} + \frac{R_{do}}{A_o} \right)^{-1} \tag{2.21}$$

Substituting eqs. (2.11) and (2.12) into eq. (2.21), solving for U_o and canceling terms, we have the overall heat transfer coefficient referenced to the outer heat transfer area as follows:

$$U_o = \left(\frac{1}{h_i}\left(\frac{D_o}{D_i}\right) + \frac{D_o}{2k}\ln\left(\frac{D_o}{D_i}\right) + \frac{1}{h_o} + R_{di}\left(\frac{D_o}{D_i}\right) + R_{do} \right)^{-1} \tag{2.22}$$

Keep in mind that, hereafter, the overall heat transfer coefficient *refers to the outer heat transfer surface* (the outer surface convention), and then let us drop the subscript "o" from eqs. (2.20) and (2.22) to obtain the final equations:

$$q = UA\Delta T_m \tag{2.23}$$

$$U = \left(\frac{1}{h_i}\left(\frac{D_o}{D_i}\right) + \frac{D_o}{2k}\ln\left(\frac{D_o}{D_i}\right) + \frac{1}{h_o} + R_{di}\left(\frac{D_o}{D_i}\right) + R_{do} \right)^{-1} \tag{2.24}$$

Notice that although each term on the right-hand side of eq. (2.24) is related to a type of thermal resistance, they are not thermal resistances at all, at least by their strict definition. Thermal resistances have dimensions of K/W or °F h/BTU, as we may see from eq. (2.10); therefore, it is not the case here, since all summed terms are in m^2K/W or h ft^2 °F/BTU.

2.3.1 "Design" and "clean" overall heat transfer coefficients

Heat exchangers are generally sized with the provision of some additional transfer area to compensate the performance loss due to fouling along their operation cycle. The fouling coefficients R_{di} and R_{do} are responsible for this additional heat transfer area; hence, their values must be judiciously specified.

The actual U value used in the evaluation of the heat transfer area by eq. (2.23) is called the "design" overall heat transfer coefficient (U_d), which considers the equipment fouled condition in the form:

$$U_d = \left(\frac{1}{h_i}\left(\frac{D_o}{D_i}\right) + \frac{D_o}{2k}\ln\left(\frac{D_o}{D_i}\right) + \frac{1}{h_o} + R_{di}\left(\frac{D_o}{D_i}\right) + R_{do} \right)^{-1} \tag{2.25}$$

The "clean" overall heat transfer coefficient (U_c) ignores the fouling effect, being obtained simply by setting $R_{di} = R_{do} = 0$ in eq. (2.25):

$$U_c = \left(\frac{1}{h_i} \left(\frac{D_o}{D_i} \right) + \frac{D_o}{2k} \ln \left(\frac{D_o}{D_i} \right) + \frac{1}{h_o} \right)^{-1} \tag{2.26}$$

With U_c from eq. (2.26), U_d can also be evaluated more conveniently as

$$U_d = \left(\frac{1}{U_c} + R_{di} \left(\frac{D_o}{D_i} \right) + R_{do} \right)^{-1} \tag{2.27}$$

2.3.2 Relatively small wall thermal resistance

Equation (2.26) is a general form for U_c, and it can be simplified in many ways. According to the physical conditions appropriate for a given heat exchanger and chemical process, some reduced versions of the equivalent thermal resistance may be used. For example, in such a case where the pipe material is highly conducive – which is usually the case, because metals are the most common materials for constructing heat exchangers – the thermal resistance of the wall is small in comparison to the sum of the convective resistances, i.e.:

$$\frac{D_o \ln \left(\frac{D_o}{D_i} \right)}{2k} \ll \frac{1}{h_i} \left(\frac{D_o}{D_i} \right) + \frac{1}{h_o} \tag{2.28}$$

This approximation is widely used by Kern [44] in his classic book, in which almost all designs disregard the wall thermal resistance, and the resulting overall heat transfer coefficient becomes

$$U_c = \frac{h_{io} h_o}{h_{io} + h_o} \tag{2.29}$$

where h_{io} is the internal (tube) convective heat transfer coefficient corrected for the external (reference) heat transfer surface:

$$h_{io} = h_i \left(\frac{D_i}{D_o} \right) \tag{2.30}$$

Under this assumption, the design overall coefficient can be calculated from

$$U_d = \left(\frac{1}{U_c} + R_{di} \left(\frac{D_o}{D_i} \right) + R_{do} \right)^{-1} \tag{2.31}$$

2.3.3 Controlling fluid

Sometimes one of the fluids rules the heat transfer rate inside the heat exchanger. When there is one fluid in the heat exchanger for which the convective heat transfer coefficient is remarkably small in comparison with the other fluid, the former is usually said to be the "controlling fluid" or "controlling stream" of the heat transfer operation. The reason is that a change in the convective heat transfer coefficient of the controlling fluid has a direct impact in the overall heat transfer coefficient of the whole exchanger, as we can see from eq. (2.29) by taking for example $h_{io} \gg h_o$ (the annulus fluid is assumed controlling), giving

$$U \approx h_o \qquad\qquad (2.32)$$

Worked Example 2.2

A 1.8 lb/h process stream of acetone is to be cooled from 170.0 to 60.0 °F. Determine the necessary heat load to perform this service.

Solution

$$T_m = \frac{T_{in}}{2} + \frac{T_{out}}{2} = \frac{170.00}{2} + \frac{60.000}{2} = 115.00 \text{ °F}$$

The specific heat the acetone can be taken as $Cp = 0.493$ BTU/lb · °F.

From an enthalpy balance for the given temperature variation, we obtain the heat flow rate as follows:

$$q = CpW(T_{out} - T_{in}) = 0.49300 \cdot 1.8000(60.000 - 170.00) = -97.614 \text{ BTU/h}$$

Worked Example 2.3

About 2 kg/s of methanol needs to be cooled from 57.0 to 30.0 °C using available cold water at 25.0 °C. What is the required water flowrate with respect to the fact that maximum water outlet temperature must not exceed 40.0 °C, to prevent fouling?

Solution

Let us set the water outlet temperature to the maximum allowed value; therefore, the average temperature for the streams is

$$T_{1m} = \frac{T_{1,in}}{2} + \frac{T_{1,out}}{2} = \frac{30.000}{2} + \frac{57.000}{2} = 43.500 \text{ °C}$$

$$T_{2m} = \frac{T_{2,in}}{2} + \frac{T_{2,out}}{2} = \frac{25.000}{2} + \frac{40.000}{2} = 32.500 \text{ °C}$$

From Appendixes C.11 and C.13, the specific heats for methanol and water are:

Methanol: 43.5 °C	$Cp = 3{,}160.0$ J/kg K
Water: 32.5 °C	$Cp = 4{,}070.0$ J/kg K

Assuming that all the energy lost by the methanol find its way to the water stream, then we can write the energy balance in the form:

$$Cp_1 W_1 (T_{1,\text{out}} - T_{1,\text{in}}) + Cp_2 W_2 (T_{2,\text{out}} - T_{2,\text{in}}) = 0$$

Solving for the water flow rate and evaluating with the specified process data give

$$W_2 = \frac{Cp_1 W_1 (T_{1,\text{out}} - T_{1,\text{in}})}{Cp_2 (T_{2,\text{in}} - T_{2,\text{out}})} = \frac{2.0000 \cdot 3,160.0(30.000 - 57.000)}{4,070.0(25.000 - 40.000)} = 2.7951 \text{ kg/s}$$

Worked Example 2.4
A styrene stream of 2 kg/s is used to heat 5.7 kg/s of toluene with initial temperature of 70.0 °C. The styrene ranges from 125.0 °C to 30.0 °C. What is the outlet temperature of toluene? Comment on the effect of variation of the heat capacity with temperature on the accuracy of the calculation.

Solution
The average temperature of styrene is

$$T_{1m} = \frac{T_{1,\text{in}}}{2} + \frac{T_{1,\text{out}}}{2} = \frac{125.00}{2} + \frac{30.000}{2} = 77.500 \text{ °C}$$

The specific heat of styrene at this temperature is $Cp = 1,750.0$ J/kg·K.
 Since the exit toluene temperature is not known, we can start the calculation using the inlet temperature for estimating the specific heat, then $Cp(70.0\,°C) = 1,720.0$ J/kg·K.
 From the enthalpy balance:

$$Cp_1 W_1 (T_{1,\text{out}} - T_{1,\text{in}}) + Cp_2 W_2 (T_{2,\text{out}} - T_{2,\text{in}}) = 0$$

And solving for outlet temperature of toluene:

$$T_{2,\text{out}} = \frac{(Cp_1 T_{1,\text{in}} W_1 - Cp_1 T_{1,\text{out}} W_1 + Cp_2 T_{2,\text{in}} W_2)}{Cp_2 W_2}$$

$$= \frac{(125.00 \cdot 1,750.0 \cdot 2.0000 + 1,720.0 \cdot 5.7000 \cdot 70.000 - 1,750.0 \cdot 2.0000 \cdot 30.000)}{1,720.0 \cdot 5.7000}$$

$$= 103.91 \text{ °C}$$

A more accurate value of the specific heat of toluene is estimated by using this outlet temperature to calculate the average stream temperature:

$$T_{2m} = \frac{T_{2,\text{in}}}{2} + \frac{T_{2,\text{out}}}{2} = \frac{103.91}{2} + \frac{70.000}{2} = 86.955 \text{ °C}$$

With $C_p(87.0\ °C) = 1{,}790.0\ \text{J/kg·K}$, the new outlet temperature becomes

$$T_{2,\text{out}} = \frac{(Cp_1 T_{1,\text{in}} W_1 - Cp_1 T_{1,\text{out}} W_1 + Cp_2 T_{2,\text{in}} W_2)}{Cp_2 W_2}$$

$$= \frac{(125.00 \cdot 1{,}750.0 \cdot 2.0000 - 1{,}750.0 \cdot 2.0000 \cdot 30.000 + 1{,}790.0 \cdot 5.7000 \cdot 70.000)}{1{,}790.0 \cdot 5.7000}$$

$$= 102.59\ °C$$

The deviation between the later outlet temperature and the initial estimate can be evaluated as follows:

$$\text{Error} = \frac{100 \cdot (Cp_{2(a)} - Cp_{2(b)})}{Cp_{2(a)}} = \frac{100 \cdot (1{,}720.0 - 1{,}790.0)}{1{,}720.0} = -4.0698\%$$

2.4 Basic heat exchanger differential equations

When two fluids, with distinct temperatures, exchange energy on a heat exchanger, their temperatures are modified along the fluid path, and a spatial temperature distribution develops, as the heat moves from the hot to cold fluid. A few physical laws, namely, the first and second laws of thermodynamics associated with a heat transfer rate relation, govern this heat exchange.

Considering a differential control volume, with the form of a cross section of infinitesimal thickness, in each stream of a double-pipe heat exchanger, we may apply an energy balance for the hot and cold fluids to obtain the differential equations governing the heat transfer process. A relatively simple, yet representative, mathematical description of this heat exchanger can be obtained by imposing the following assumptions:

1. One-dimensional axial flow path[1]
2. Steady-state operation
3. Only the heat transfer between the fluids changes their enthalpy, i.e., there is no heat loss to the surroundings, and changes of kinetic and potential energies are neglected [45]
4. Constant specific heats and flow rates
5. Heat diffusion in the direction of the flow is ignored [45]

With previously stated simplifications, a one-dimensional representation of a double-pipe heat exchanger is defined using the "accumulative" heat transfer area (A), starting

---

1 Some authors (e.g., [123]) add an explicit assumption for the uniform thermal history of the fluid particles in each stream. Herein, such condition is automatically achieved by the combined assumptions of steady-state and one-dimensional axial flow path.

from the left end and increasing in the direction of the right end of the heat exchanger (Fig. 2.2). Since the inflow and outflow of energy accounted for in the energy balance are dependent on the fluid flow direction for each stream, we need to deal separately with the exchanger operating in parallel flow or counterflow arrangements.

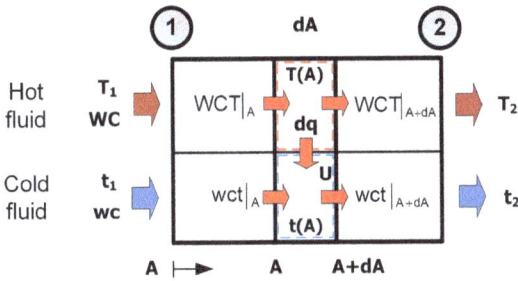

Fig. 2.2: One-dimensional representation of a double-pipe parallel flow heat exchanger.

2.4.1 Parallel flow

The differential equations for the parallel flow heat exchanger are more straightforward to develop, since both fluids flow in the same positive direction of the accumulated heat transfer surface (A). This helps to avoid possible pitfalls involving the signs of the convective terms of the counterflow differential equations, as we will see in Section 2.4.2.

2.4.1.1 Energy balance for hot fluid
From the energy balance on the heat transfer area element identified in Fig. 2.2, evaluating WCT and wct at positions A and $A + dA$ with a Taylor series expansion truncated at the first-order term, we have

$$\overbrace{WCT}^{\substack{\text{flow rate of} \\ \text{enthalpy} \\ \text{INTO the} \\ \text{control volume} \\ \text{at face } A \\ \text{(W or BTU/h)}}} - \overbrace{\left(WCT + \frac{d(WCT)}{dA} dA \right)}^{\substack{\text{flow rate of} \\ \text{enthalpy} \\ \text{OUT of the} \\ \text{control volume} \\ \text{at face } A + dA \\ \text{(W or BTU/h)}}} - \overbrace{UdA(T-t)}^{\substack{\text{heat REMOVED} \\ \text{from the} \\ \text{hot fluid} \\ \text{(W or BTU/h)}}} = 0 \qquad (2.33)$$

where W is the mass flow rate of the hot fluid (kg/s or lb/h), C is the specific heat of the hot fluid (J/kg K or BTU/lb °F), T is the temperature of the hot fluid (°C or °F), and t is the temperature of the cold fluid (°C or °F).

Notice that the term of heat removal must be carefully defined with the proper sign, which is forced by the second law of thermodynamics. Since the energy is subtracted from the hot fluid, it is preceded by the negative sign; therefore, the positivity of the expression $dq = U\,dA\,(T - t)$ must be assured by taking the temperature difference from the hot to cold fluid.

Considering that WC does not change with accumulated area A along the exchanger, canceling terms, and rearranging, we may take out from the derivative to get the differential equation for $T(A)$ as follows:

$$\frac{dT}{dA} + \frac{U}{WC}(T - t) = 0 \tag{2.34}$$

2.4.1.2 Energy balance for cold fluid
As for the hot fluid, doing the same energy balance, to the cold fluid, we obtain

$$\underbrace{wct}_{\substack{\text{flow rate of}\\ \text{enthalpy}\\ \text{INTO the}\\ \text{control volume}\\ \text{at face } A\\ \text{(W or BTU/h)}}} - \underbrace{\left(wct + \frac{d(wct)}{dA}dA\right)}_{\substack{\text{flow rate of}\\ \text{enthalpy}\\ \text{OUT of the}\\ \text{control volume}\\ \text{at face } A + dA\\ \text{(W or BTU/h)}}} + \underbrace{UdA(T - t)}_{\substack{\text{heat ADDED}\\ \text{to the}\\ \text{cold fluid}\\ \text{(W or BTU/h)}}} = 0 \tag{2.35}$$

where w is the mass flow rate of the cold fluid (kg/s or lb/h), c is the specific heat of the cold fluid (J/kg K or BTU/lb °F), and t is the temperature of the cold fluid (°C or °F).

The term for the heat added to the cold fluid control volume is defined accordingly to the second law of thermodynamics. It should be led by a positive sign, provided the expression $dq = U\,dA\,(T - t)$ is guaranteed to be positive also, since $T > t$.

Subsequently, under the consideration that specific heat and mass flow rate (wc) may be assumed constant, then:

$$\frac{dt}{dA} - \frac{U}{wc}(T - t) = 0 \tag{2.36}$$

Equations (2.34) and (2.36) can be solved to allow the prediction of the fluid temperature in each position of the heat transfer surface.

2.4.2 Counterflow

To develop the differential equations for a counterflow exchanger, let us start with the schematic representation in Fig. 2.3. The independent variable accumulated heat transfer area (A) has its axis oriented from left- to right-hand side as indicated. For the one-dimensional description done here, this simple "A-axis" defines a reference frame to which the developed differential equations are tied.

In this counterflow arrangement, the flow of the hot fluid is oriented in the positive A-axis direction, but the cold fluid has the opposite A-axis direction. Such fluid orientations, with respect to which fluid has the same axis orientation, are in fact totally arbitrary! Provided the energy balances are performed correctly, i.e., in strict accordance with the chosen orientations, the final forms of the generated differential equations will be the same for the defined reference frame.

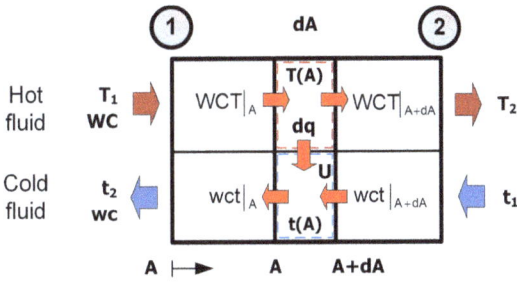

Fig. 2.3: One-dimensional representation of a double-pipe counterflow heat exchanger.

2.4.2.1 Energy balance for hot fluid
With a truncated Taylor series expansion for the enthalpy function WCT, we can draw the following balance:

$$\underbrace{WCT}_{\substack{\text{flow rate of}\\ \text{enthalpy}\\ \text{INTO the}\\ \text{control volume}\\ \text{at face } A\\ \text{(W or BTU/h)}}} - \underbrace{\left(WCT + \frac{d(WCT)}{dA}dA\right)}_{\substack{\text{flow rate of}\\ \text{enthalpy}\\ \text{OUT of the}\\ \text{control volume}\\ \text{at face } A + dA\\ \text{(W or BTU/h)}}} - \underbrace{UdA(T-t)}_{\substack{\text{heat REMOVED}\\ \text{from the}\\ \text{hot fluid}\\ \text{(W or BTU/h)}}} = 0 \qquad (2.37)$$

Assuming the hypothesis that WC does not change with A significantly, we may write

$$\frac{dT}{dA} + \frac{U}{WC}(T-t) = 0 \tag{2.38}$$

2.4.2.2 Energy balance for cold fluid

$$\underbrace{\left(wct + \frac{d(wct)}{dA}dA\right)}_{\substack{\text{flow rate of}\\\text{enthalpy}\\\text{INTO the}\\\text{control volume}\\\text{at face } A + dA\\(\text{W or BTU/h})}} - \underbrace{wct}_{\substack{\text{flow rate of}\\\text{enthalpy}\\\text{OUT of the}\\\text{control volume}\\\text{at face } A\\(\text{W or BTU/h})}} + \underbrace{UdA(T-t)}_{\substack{\text{heat ADDED}\\\text{to the}\\\text{cold fluid}\\(\text{W or BTU/h})}} = 0 \tag{2.39}$$

Taking specific heat and flow rate of the cold fluid as constants, the resulting differential equation is

$$\frac{dt}{dA} + \frac{U}{wc}(T-t) = 0 \tag{2.40}$$

2.5 Log mean temperature difference

The design equation (2.8) requires the determination of a kind of mean temperature difference that is representative for the heat transfer service, specifically for the temperature distributions inside the heat exchanger. A complication arises from the fact that such mean temperature difference is a result of the temperature profiles of both fluids, and, in turn, these profiles depend upon the equipment design and performance. In addition, the geometry and fluid dynamics inside the heat exchanger have direct impact on the value of the overall heat transfer coefficient U, which determine the equipment size (i.e., its geometry). Therefore, the sizing process is inherently iterative.

A detailed derivation of the logarithmic mean temperature difference (LMTD) in a double-pipe heat exchanger may be found in Kern's text [44] and will not be repeated here. Instead, let us take another way to this important definition. Let us start from the descriptive equations for steady-state double-pipe heat exchanger and undertake a mathematical procedure to develop the formula for the mean temperature difference corresponding to the whole area of heat.

2.5.1 Parallel flow double-pipe heat exchanger

In Section 2.4.1, we obtained the differential equations for a double-pipe heat exchanger operating with parallel flow streams as eqs. (2.34) and (2.36), which are rewritten here for convenience:

$$\frac{dT}{dA} + \frac{U}{WC}(T - t) = 0 \tag{2.41}$$

$$\frac{dt}{dA} - \frac{U}{wc}(T - t) = 0 \tag{2.42}$$

A new differential equation may be obtained by subtracting eqs. (2.41) and (2.42), consequently:

$$\frac{d(T - t)}{dA} + \left(\frac{U}{WC} + \frac{U}{wc}\right)(T - t) = 0 \tag{2.43}$$

Let us define the temperature difference between both fluids as a new variable:

$$\Delta T = T - t \tag{2.44}$$

And, substituting in eq. (2.43), we get

$$\frac{d\Delta T}{dA} + \left(\frac{U}{WC} + \frac{U}{wc}\right)\Delta T = 0 \tag{2.45}$$

Equation (2.45) may be readily integrated by separation of variables from terminal (1) to terminal (2) of the heat exchanger, and using the hypothesis of constant specific heat and flow rates, as follows:

$$\int_{\Delta T_1}^{\Delta T_2} \frac{d\Delta T}{\Delta T} = -\left(\frac{1}{WC} + \frac{1}{wc}\right)\int_0^A U dA \tag{2.46}$$

Notice that the assumptions of constant specific heat and flow rates for both fluids are already embedded in the descriptive differential equations (2.41) and (2.42). Thus, we are not placing an additional restriction in this derivation of LMTD.

One should notice that, in the general case, the physical properties affecting the overall heat transfer coefficient U are dependent on the fluid temperature; therefore, the remaining integral on the right-hand side of eq. (2.46) may be rigorously evaluated only if the distribution of $U(A)$ over the heat transfer area is known. At this point, let us use the definition of a mean overall heat transfer coefficient ($\breve{U}$) for the whole heat transfer surface:

$$\bar{U} = \frac{1}{A}\int_0^A U dA \tag{2.47}$$

From eq. (2.47), eq. (2.46) may be rewritten as follows:

$$\ln\left(\frac{\Delta T_2}{\Delta T_1}\right) = -\left(\frac{1}{WC} + \frac{1}{wc}\right)\bar{U}A \tag{2.48}$$

Assuming that all energy lost by the hot fluid is gained by the cold fluid, an energy balance for both streams yields:

$$q = W\int_{T_2}^{T_1} CdT = w\int_{t_1}^{t_2} cdt \tag{2.49}$$

And, using mean values for the specific heats, one may write

$$q = WC(T_1 - T_2) = wc(t_2 - t_1) \tag{2.50}$$

Thus:

$$\frac{1}{WC} = \frac{(T_1 - T_2)}{q} \tag{2.51}$$

$$\frac{1}{wc} = \frac{(t_2 - t_1)}{q} \tag{2.52}$$

$$q\ln\left(\frac{\Delta T_2}{\Delta T_1}\right) = -[(T_1 - t_1) - (T_2 - t_2)]\bar{U}A \tag{2.53}$$

For a parallel flow heat exchanger (see Fig. 2.2), the temperature differences at the terminals are

$$\Delta T_1 = T_1 - t_1 \tag{2.54}$$

$$\Delta T_2 = T_2 - t_2 \tag{2.55}$$

Using eqs. (2.54) and (2.55) in eq. (2.53), and rearranging, results in

$$q = \bar{U}A\frac{(\Delta T_1 - \Delta T_2)}{\ln\left(\frac{\Delta T_1}{\Delta T_2}\right)} \tag{2.56}$$

Comparing eqs. (2.56) and (2.8), we may conclude that the mean temperature difference (ΔT_m) takes the following form:

$$\Delta T_m = \frac{(\Delta T_1 - \Delta T_2)}{\ln\left(\frac{\Delta T_1}{\Delta T_2}\right)} \tag{2.57}$$

This result is an important type of mean temperature difference designated as the Logarithmic (or simply "Log") Mean Temperature Difference (LMTD), hence let us restate its definition:

$$\text{LMTD} = \frac{(\Delta T_1 - \Delta T_2)}{\ln\left(\dfrac{\Delta T_1}{\Delta T_2}\right)} \tag{2.58}$$

2.5.2 Counterflow double-pipe heat exchanger

The descriptive equations for the stream temperature profiles of a counterflow double-pipe heat exchanger were developed in Section 2.4.2, and are reprinted here for convenience:

$$\frac{dT}{dA} + \frac{U}{WC}(T - t) = 0 \tag{2.59}$$

$$\frac{dt}{dA} + \frac{U}{wc}(T - t) = 0 \tag{2.60}$$

A new differential equation may be obtained by subtracting eqs. (2.59) and (2.60), consequently:

$$\frac{d(T - t)}{dA} + \left(\frac{U}{WC} - \frac{U}{wc}\right)(T - t) = 0 \tag{2.61}$$

Using the local temperature difference (eq. (2.44)) as a new variable results in

$$\frac{d\Delta T}{dA} + \left(\frac{U}{WC} - \frac{U}{wc}\right)\Delta T = 0 \tag{2.62}$$

Equation (2.62) may be readily integrated by separation of variables from terminal (1) to terminal (2) of the heat exchanger, and reusing the hypothesis of constant specific heats and flow rates, as follows:

$$\int_{\Delta T_1}^{\Delta T_2} \frac{d\Delta T}{\Delta T} = -\left(\frac{1}{WC} - \frac{1}{wc}\right)\int_0^A U dA \tag{2.63}$$

With the definition of the average overall heat transfer coefficient ($\bar{U}$), eq. (2.47), we can rewrite eq. (2.63) in the form

$$\ln\left(\frac{\Delta T_2}{\Delta T_1}\right) = -\left(\frac{1}{WC} - \frac{1}{wc}\right)\bar{U}A \tag{2.64}$$

Considering that the hot fluid does not lose energy to the surroundings, and using mean values for the specific heats, the integral energy balance for both streams is stated as follows:

$$q = WC(T_1 - T_2) = wc(t_2 - t_1) \tag{2.65}$$

Thus,

$$\frac{1}{WC} = \frac{(T_1 - T_2)}{q} \tag{2.66}$$

$$\frac{1}{wc} = \frac{(t_2 - t_1)}{q} \tag{2.67}$$

Using eqs. (2.66) and (2.67) in eq. (2.64), and rearranging the temperature differences in the right-hand side, yields:

$$q \ln\left(\frac{\Delta T_2}{\Delta T_1}\right) = -[(T_1 - t_2) - (T_2 - t_1)]\bar{U}A \tag{2.68}$$

Notice that, for a counterflow heat exchanger, the temperature differences at the terminals are

$$\Delta T_1 = T_1 - t_2 \tag{2.69}$$

$$\Delta T_2 = T_2 - t_1 \tag{2.70}$$

Therefore, with eqs. (2.69) and (2.70), eq. (2.68) may be rearranged as follows:

$$q = \bar{U}A \frac{(\Delta T_1 - \Delta T_2)}{\ln\left(\dfrac{\Delta T_1}{\Delta T_2}\right)} \tag{2.71}$$

The comparison of this result with eq. (2.8) allows us to conclude that the expression for the mean temperature difference along the heat exchanger has the form

$$\text{LMTD} = \frac{(\Delta T_1 - \Delta T_2)}{\ln\left(\dfrac{\Delta T_1}{\Delta T_2}\right)} \tag{2.72}$$

2.5.3 The restrictions on the overall heat transfer coefficient

The derivation of the LMTD does not require a constant overall heat transfer coefficient. It is a common misconception that the use of LMTD for heat exchanger design imposes the restriction of constant overall heat transfer coefficient (U) for the whole heat transfer surface, which is not true! As shown in Sections 2.5.1 and 2.5.2, the derivation of LMTD for parallel flow and counterflow arrangements could be performed directly from the descriptive differential equations of the one-dimensional heat exchanger, developed under assumptions (1)–(5) in Section 2.4, which do not require a constant U over the area. Therefore, eqs. (2.34) and (2.36) (for parallel flow) and eqs. (2.38) and (2.40) (for counterflow) do not imply a constant U. Additionally, the use of an integrated definition of the averaged U, given by eq. (2.47), may account for any functional variation of the overall heat transfer coefficient over the heat transfer surface $U(A)$; as a result, a uniform heat transfer coefficient is just the special case of a constant function $U(A) = U_{\text{const}}$.

2.5.4 LMTD and fluid flow arrangement

The mean temperature for a heat exchanger is the driving force for the heat transfer when evaluated using the integrated design equation. In general, the greater the mean temperature difference, the more efficient is the heat exchanger, and a smaller heat transfer surface will suffice to perform a specified service. The only exceptions are found when the chosen flow arrangement penalizes the mean temperature difference, but this effect is compensated by a greater increase in the overall heat transfer coefficient.

In the case of a double-pipe heat exchanger, the mean temperature difference, under certain assumptions, was proven to be the LMTD. Surprisingly, the expressions for LMTD have the same form, regardless of the flow orientation of the exchanger, be it in parallel flow or counterflow, as seen in eqs. (2.58) and (2.72). Such verification may lead mistakenly to the conclusion that two heat exchangers operating in parallel flow and counterflow have the same thermal efficiency, which is not true! This equality is only apparent when the LTMD is written using the two temperature differences at the heat exchanger terminals ΔT_1 and ΔT_2; however, we should notice that the definition of the temperature differences at the exchanger terminals is tied to the flow arrangement, and in terms of the four process temperatures T_1, T_2, t_1, and t_2, the LMTD for both situations become distinct.

With eqs. (2.54) and (2.55) substituted into eq. (2.58), we get the LMTD for a parallel heat exchanger as follows:

$$\text{LMTD}_{(\text{parallel flow})} = \frac{(T_1 - t_1) - (T_2 - t_2)}{\ln\left(\dfrac{T_1 - t_1}{T_2 - t_2}\right)} \tag{2.73}$$

The temperature differences for a counterflow heat exchanger are eqs. (2.69) and (2.70); therefore, substituting in the LMTD definition, we have

$$\text{LMTD}_{(\text{counterflow})} = \frac{(T_1 - t_2) - (T_2 - t_1)}{\ln\left(\dfrac{T_1 - t_2}{T_2 - t_1}\right)} \tag{2.74}$$

Considering defined process temperatures, a counterflow heat exchanger will be normally more efficient than a device with the same streams contacted in parallel flow.

The reason for this outcome is that the LMTD value is larger for counterflow than for the parallel flow arrangement. Although we are not going to demonstrate here, it can be proved [46] that eq. (2.74) always produces a greater mean temperature difference than eq. (2.73), except when one of the streams undergoes an isothermal condensation or vaporization. In such a case, there is no distinction between parallel flow and counterflow because both LMTDs are identical.

Worked Example 2.5
In a heat exchanger, the hot fluid is cooled from 70.0 to 50.0 °C, and the cold fluid varies from 20.0 to 45.0 °C. Determine the log mean temperature difference for a parallel flow and counterflow operation. Compare the results.

Solution
– Counterflow
For the counterflow arrangement, the temperature differences at the exchanger ends are

$$\Delta T_1 = T_{1,in} - T_{2,out} = 70.000 - 45.000 = 25.000 \ °C$$

$$\Delta T_2 = T_{1,out} - T_{2,in} = 50.000 - 20.000 = 30.000 \ °C$$

Substituting in definition (2.58), we have

$$\Delta T_{lm,c} = \frac{\Delta T_1 - \Delta T_2}{\ln\left(\dfrac{\Delta T_1}{\Delta T_2}\right)} = \frac{25.000 - 30.000}{\ln\left(\dfrac{25.000}{30.000}\right)} = 27.424 \ °C$$

– Parallel flow
The parallel flow operation gives the following terminal temperature differences:

$$\Delta T_1 = T_{1,in} - T_{2,in} = 70.000 - 20.000 = 50.000 \ °C$$

$$\Delta T_2 = T_{1,out} - T_{2,out} = 50.000 - 45.000 = 5.0000 \ °C$$

The log mean temperature difference becomes

$$\Delta T_{lm,p} = \frac{\Delta T_1 - \Delta T_2}{\ln\left(\dfrac{\Delta T_1}{\Delta T_2}\right)} = \frac{50.000 - 5.0000}{\ln\left(\dfrac{50.000}{5.0000}\right)} = 19.543 \ °C$$

For comparison, let us evaluate the relative change between both values as follows:

$$\text{Change} = \frac{100 \cdot (\Delta T_{lm,c} - \Delta T_{lm,p})}{\Delta T_{lm,p}} = \frac{100 \cdot (27.424 - 19.543)}{19.543} = 40.326\%$$

Therefore, the operation in countercurrent mode increases the temperature driving force in about 40.3%. As a direct consequence, an exchanger designed in this manner would require significant smaller area to accomplish this service.

2.5.5 Special cases of LMTD

LMTD as given by eq. (2.58) can be undetermined in some special cases. For example, consider a heat exchanger where the temperature differences are equal at both terminals. Equation (2.50) shows that such situation is found when the heat capacities of both streams equal each other, i.e., $WC = wc$. Consequently, eq. (2.58) cannot be directly evaluated; however, applying the L'Hopital's rule to eq. (2.58), we find that

$$\text{LMTD} = \Delta T_1 = \Delta T_2 = \frac{\Delta T_1 + \Delta T_2}{2} \tag{2.75}$$

2.5.6 The physical meaning of LMTD

At this point, you might be questioning, why LMTD is distinct from other definitions of mathematical means such as arithmetic or geometric? What advantages it has over other means?

The answer for this question arises from the very derivation of the LMTD. Notice that our starting point for the development of the LMTD formula was the one-dimensional differential equations (2.34) and (2.36), describing the fluid temperature profiles along the heat transfer area of a double-pipe heat exchanger. In their turn, those descriptive equations resulted from the application of physical principles such as the first and second laws of thermodynamics. For that reason, we may say that when such physical principles are applied to a double-pipe heat exchanger, aiming the derivation of a mean temperature difference expression, the single one expression that arises naturally is the log mean temperature difference, the LMTD, and no other! Consequently, the LMTD contains physics of heat transfer phenomena (e.g., conservation of energy) that other mathematical means definitions simply do not.

2.5.7 Do we need LMTD nowadays?

No, we do not! This is the short answer. However, for a longer, more elaborated answer; it can become a sound yes! The choice is all about the engineering task you are doing, what kind of design information is needed, and the required level of detail. Additionally, the available computational tools play an important role.

The development of LMTD provides a very simple, yet effective design method, however, at the cost of accuracy loss in some situations. As we could see from eq. (2.56), the expression of LMTD arises naturally inside a form of the heat exchanger design equation (2.8), suggesting a direct method for sizing or rating a heat exchanger device. The simplifying assumptions embedded in the LMTD are a key element to produce this straightforward design method, comprising only algebraic expressions, which are especially suited for manual calculations.

Besides the clean and elegant design method founded on the LMTD concept, we should remark that the simplifications used to derive LMTD are not imperative anymore. The LMTD concept was developed at ages where nowadays computers were out of reach of the most forward-looking dreams, when numerical calculations were mostly performed by hand, and integrals and differential equations were solved by graphical methods. Today, we have mature numerical methods and personal computers able to do thousands of mathematical operations in less than a fraction of a second, which allow the direct solution of the descriptive differential equations for a given exchanger type and flow arrangement.

2.5.8 Area sizing using the descriptive differential equations

The estimation of the heat transfer area required for a given heat transfer service can be done using the differential equations (2.34) and (2.36), for a parallel flow or eqs. (2.38) and (2.40) for a counterflow heat exchanger. A suitable procedure for sizing the heat transfer area is

1. Choose a plausible initial estimate of the required heat transfer area.
2. Solve the differential equations over the estimated area to obtain the temperature profile of the fluid streams.
3. Integrate the total heat transfer load from the local temperature differences between each pair of streams.
4. Compare the calculated heat transfer load with the heat load required by the process. If the calculated heat load is:
    a. smaller than the process requirement, then increase heat transfer area and solve the differential equations again;
    b. greater than the process requirement, then decrease the heat transfer area and solve the differential equations again;
    c. approximately equal to the process requirement, a feasible heat exchanger size was achieved.

2.5.9 Parallel flow or counterflow: what is the difference?

Given the task of designing a single or a battery of hairpins to fulfill a specified service, it is up to the designer to decide how the fluids are to be thermally contacted: in parallel or counterflow. Depending on the process fluids and operating conditions, one arrangement or another will yield the most effective heat exchanger.

In almost all cases, the counterflow operation should be the right choice because it provides the highest mean temperature difference driving the heat transfer process. Consequently, for a specified heat load, the designed heat exchanger will be smaller and less expensive. However, a parallel flow arrangement may be favored if the cold fluid is highly viscous. In such situation, the feed of the cold fluid at the same hot fluid terminal allows an early heating, which lowers its viscosity more quickly along the flow path. As a result, an increase in the overall heat transfer coefficient may be significant to the point of compensating for the lower mean temperature difference inherent to the parallel flow.

Worked Example 2.6

Consider the same service from Worked Example 2.5, but this time the cold fluid needs to be heated to the exit temperature of 60.0 °C. Determine the log mean temperature difference for counterflow and parallel flow and draw conclusions from the results.

Solution

– Counterflow

The temperature differences at the exchanger terminals are

$$\Delta T_1 = T_{1,in} - T_{2,out} = 70.000 - 60.000 = 10.000 \ °C$$

$$\Delta T_2 = T_{1,out} - T_{2,in} = 50.000 - 20.000 = 30.000 \ °C$$

Then, the log mean for these values is calculated as

$$\Delta T_{lm,c} = \frac{\Delta T_1 - \Delta T_2}{\ln\left(\dfrac{\Delta T_1}{\Delta T_2}\right)} = \frac{10.000 - 30.000}{\ln\left(\dfrac{10.000}{30.000}\right)} = 18.205 \ °C$$

– Parallel flow

The operation in parallel flow leads to the following temperature differences:

$$\Delta T_1 = T_{1,in} - T_{2,in} = 70.000 - 20.000 = 50.000 \ °C$$

$$\Delta T_2 = T_{1,out} - T_{2,out} = 50.000 - 60.000 = -10.000 \ °C$$

Substituting in the log mean temperature difference definition gives an improper operation as follows.

$$\Delta T_{\text{lm,p}} = \frac{\Delta T_1 - \Delta T_2}{\ln\left(\dfrac{\Delta T_1}{\Delta T_2}\right)} = \frac{50.000 - (-10.000)}{\ln\left(\dfrac{50.000}{(-10.000)}\right)} = [\text{undefined}]$$

The practical meaning of such a result is that a counter flow heat exchanger is able to perform this service. Conversely, for an exchanger designed to operate in parallel flow, this service is an impossible task.

Worked Example 2.7
A double-pipe exchanger is used to heat ethanol using saturated steam at 100.0 °C, which only exchanges latent heat. Determine the log mean temperature difference for an exchanger operating in counterflow and parallel flow and discuss the results.

Solution
Since the steam only exchanges latent heat, it is going to perform isothermal condensation at 100 °C. Under this consideration, the log mean temperature differences are evaluated as follows:

– Counterflow
From the temperature differences in each terminal, we have

$$\Delta T_1 = T_{1,\text{in}} - T_{2,\text{out}}$$

$$\Delta T_2 = T_{1,\text{out}} - T_{2,\text{in}}$$

$$\Delta T_1 = T_{1,\text{in}} - T_{2,\text{out}} = 100.00 - 70.000 = 30.000\ °\text{C}$$

$$\Delta T_2 = T_{1,\text{out}} - T_{2,\text{in}} = 100.00 - 50.000 = 50.000\ °\text{C}$$

Therefore, eq. (2.58) or (2.72) gives

$$\Delta T_{\text{lm,c}} = \frac{\Delta T_1 - \Delta T_2}{\ln\left(\dfrac{\Delta T_1}{\Delta T_2}\right)} = \frac{30.000 - 50.000}{\ln\left(\dfrac{30.000}{50.000}\right)} = 39.152\ °\text{C}$$

– Parallel flow
Terminal temperature differences:

$$\Delta T_1 = T_{1,\text{in}} - T_{2,\text{in}} = 100.00 - 50.000 = 50.000\ °\text{C}$$

$$\Delta T_2 = T_{1,\text{out}} - T_{2,\text{out}} = 100.00 - 70.000 = 30.000\ °\text{C}$$

Then, from

$$\Delta T_{\text{lm,p}} = \frac{\Delta T_1 - \Delta T_2}{\ln\left(\dfrac{\Delta T_1}{\Delta T_2}\right)} = \frac{-30.000 + 50.000}{\ln\left(\dfrac{50.000}{30.000}\right)} = 39.152\ °\text{C}$$

The condition that the hot fluid undergoes isothermal heat transfer makes the direction of the flow arrangement inconsequential for the temperature driving force of the heat exchanger.

2.6 Exit temperatures of fluids

There are two common occasions in which the heat exchanger specifications (type, size, material, etc.) are known and the exit temperatures of both fluids must be determined.

The first case is for a freshly designed heat exchanger with a large built-in design margin. Typically, a new heat exchanger placed into service delivers a heat load greater than the nominal value used in its sizing calculations. This is a consequence of the various conservative approximations done during the calculations, of the fouling factors included in the overall heat transfer coefficient and of a final safety factor that might be incorporated in the design. The excess in the heat load makes the exit temperatures of the hot and cold fluids become lower and higher than the design levels, respectively. An exchanger with a large design margin may shift expressively the operating exit temperatures from the reference values applied in its design. If the actual exit temperatures yielded by the exchanger are considerably distinct from the design temperatures, it is advisable to evaluate the impact on the operation of the connected equipment downstream.

The second situation is when a spare heat exchanger is reused to perform a given service, which has been carried out by a primary heat exchanger. Sometimes, using a standby heat exchanger for replacing another one that is shut down for cleaning or repairing is a valuable way of decreasing the process downtime. Besides other process requirements that the fallback exchanger must match, such as operating pressure, corrosion resistance, and space restrictions, it must be able to achieve the exit temperatures demanded for the cold and hot streams, i.e., being able to supply at least the design heat load.

The estimation of the exit temperatures is inherently an iterative procedure because their calculation requires the knowledge of the mean overall heat transfer coefficient for the exchanger, which is itself a function of the entrance and exit temperatures of the fluids.

2.6.1 Counterflow

To develop more general equations, let us label the fluids exchanging heat simply as fluids 1 and 2. With this approach, the resulting equations remain valid independently of what stream is the hot or cold fluid. Using the definition of LMTD in a counterflow heat exchanger (eq. (2.73)) of area A and average overall heat transfer coefficient U, the heat transfer rate can be written as:

$$q = \frac{AU(T_{1,in} - T_{1,out} + T_{2,in} - T_{2,out})}{\ln\left(\dfrac{T_{1,in} - T_{2,out}}{T_{1,out} - T_{2,in}}\right)} \tag{2.76}$$

Under the same assumptions made in the development of LMTD, the enthalpy variation of each fluid is numerically equal to the respective heat transfer rates q_1 and q_2, respectively:

$$q_1 = Cp_1 W_1 (T_{1,out} - T_{1,in}) \tag{2.77}$$

$$q_2 = Cp_2 W_2 (T_{2,out} - T_{2,in}) \tag{2.78}$$

Therefore, the summation of the enthalpy changes must be zero:

$$Cp_1 W_1 (T_{1,out} - T_{1,in}) + Cp_2 W_2 (T_{2,out} - T_{2,in}) = 0 \tag{2.79}$$

It can be observed that the LMTD term in eq. (2.76) and the right-hand side of eq. (2.78) both assume a positive sign when fluid 1 is the hot fluid, or a negative sign if fluid 1 is the cold fluid. Hence, the following equality is valid:

$$Cp_2 W_2 (T_{2,out} - T_{2,in}) = \frac{AU(T_{1,in} - T_{1,out} + T_{2,in} - T_{2,out})}{\ln\left(\dfrac{T_{1,in} - T_{2,out}}{T_{1,out} - T_{2,in}}\right)} \tag{2.80}$$

The system composed by eqs. (2.77)–(2.80) can be solved for the exit temperature of the fluids ($T_{1,out}$ and $T_{2,out}$) and the heat transfer rates (q_1 and q_2). After some algebraic manipulation, the exit temperatures are found to be weighted averages of the entrance temperatures in the form:

$$T_{1,out} = \frac{R_2(R-1)T_{1,in} + (R_2 - R_1)T_{2,in}}{RR_2 - R_1} \tag{2.81}$$

$$T_{2,out} = \frac{R(R_2 - R_1)T_{1,in} + R_1(R-1)T_{2,in}}{RR_2 - R_1} \tag{2.82}$$

$$q_1 = \frac{Cp_1 W_1 (R_1 - R_2)(T_{1,in} - T_{2,in})}{RR_2 - R_1} \tag{2.83}$$

$$q_2 = \frac{Cp_1 W_1 (R_2 - R_1)(T_{1,in} - T_{2,in})}{RR_2 - R_1} \tag{2.84}$$

where

$$R = \frac{Cp_1 W_1}{Cp_2 W_2} \tag{2.85}$$

$$R_1 = e^{-\frac{AU}{Cp_1 W_1}} \tag{2.86}$$

$$R_2 = e^{-\frac{AU}{Cp_2 W_2}} \tag{2.87}$$

Notice that, alternatively to eqs. (2.83) and (2.84), the heat transfer rate can be calculated from eqs. (2.77) and (2.78). As expected, if the hot stream is fluid 1, eqs. (2.77) and (2.83) yield a negative heat transfer rate.

2.6.2 Parallel flow

Following the same procedure from the previous section, using the LMTD for a parallel flow exchanger, its heat load is given by

$$q = \frac{AU(T_{1,in} - T_{1,out} - T_{2,in} + T_{2,out})}{\ln\left(\dfrac{T_{1,in} - T_{2,in}}{T_{1,out} - T_{2,out}}\right)} \tag{2.88}$$

Again, the right-hand side of eq. (2.88) follows the same sign of eq. (2.78); therefore, it is valid to write

$$Cp_2 W_2(T_{2,out} - T_{2,in}) = \frac{AU(T_{1,in} - T_{1,out} - T_{2,in} + T_{2,out})}{\ln\left(\dfrac{T_{1,in} - T_{2,in}}{T_{1,out} - T_{2,out}}\right)} \tag{2.89}$$

The combination of eqs. (2.77), (2.78), (2.79), and (2.89) results in a set, whose solutions for $T_{1,out}$, $T_{2,out}$, q_1, and q_2 can be reduced to the following form:

$$T_{1,out} = \frac{(RR_1 R_2 + 1)T_{1,in} + (R_1 R_2 - 1)T_{2,in}}{R_1 R_2(R+1)} \tag{2.90}$$

$$T_{2,out} = \frac{R(R_1 R_2 - 1)T_{1,in} + (R + R_1 R_2)T_{2,in}}{R_1 R_2(R+1)} \tag{2.91}$$

$$q_1 = \frac{Cp_1 W_1(1 - R_1 R_2)(T_{1,in} - T_{2,in})}{R_1 R_2(R+1)} \tag{2.92}$$

$$q_2 = \frac{Cp_1 W_1(R_1 R_2 - 1)(T_{1,in} - T_{2,in})}{R_1 R_2(R+1)} \tag{2.93}$$

where

$$R = \frac{Cp_1 W_1}{Cp_2 W_2} \tag{2.94}$$

$$R_1 = e^{\frac{AU}{Cp_1 W_1}} \qquad (2.95)$$

$$R_2 = e^{\frac{AU}{Cp_2 W_2}} \qquad (2.96)$$

Worked Example 2.8

In a double-pipe heat exchanger with 10 m² of effective area, the hot fluid enters at 100 °C with a flow rate of 1.97 kg/s. The entrance temperature of the cold fluid is 15 °C and its flow rate is 0.6 kg/s. The average overall heat transfer coefficient can be taken as $U = 500$ W/m²K. The specific heats of the hot and cold fluids are 1.88 and 4.19 J/gK, respectively. Determine the heat load and exit temperatures if the streams are arranged in counterflow and parallel flow [18].

Solution

Setting the hot and cold fluids as fluids 1 and 2, respectively, the data of the heat exchanger streams is summarized as follows:

Fluid 1 (hot)	Fluid 2 (cold)
$T_{1,in} = 100.0$ °C	$T_{2,in} = 15.0$ °C
$W_1 = 1.97$ kg/s	$W_2 = 0.6$ kg/s
$Cp_1 = 1,880.0$ J/kg·K	$Cp_2 = 4,190.0$ J/kg·K

With eqs. (2.85)–(2.87) we can evaluate the terms R, R_1, and R_2 as follows:

$$R = \frac{Cp_1 W_1}{Cp_2 W_2} = \frac{1.9700 \cdot 1,880.0}{0.60000 \cdot 4,190.0} = 1.4732$$

$$R_1 = e^{\frac{AU}{Cp_1 W_1}} = e^{\frac{10.000 \cdot 500.00}{1.9700 \cdot 1,880.0}} = 3.8576$$

$$R_2 = e^{\frac{AU}{Cp_2 W_2}} = e^{\frac{10.000 \cdot 500.00}{0.60000 \cdot 4,190.0}} = 7.3072$$

For the counterflow arrangement, using eqs. (2.81) and (2.82), the exit temperatures of the hot and cold fluids are

$$T_{1,out} = \frac{R_2 T_{1,in}(R-1) + T_{2,in}(-R_1 + R_2)}{RR_2 - R_1}$$

$$= \frac{100.00 \cdot 7.3072(1.4732 - 1) + 15.000(-3.8576 + 7.3072)}{1.4732 \cdot 7.3072 - 3.8576} = 57.550 \text{ °C}$$

$$T_{2,out} = \frac{RT_{1,in}(-R_1 + R_2) + R_1 T_{2,in}(R-1)}{RR_2 - R_1}$$

$$= \frac{1.4732 \cdot 100.00(-3.8576 + 7.3072) + 15.000 \cdot 3.8576(1.4732 - 1)}{1.4732 \cdot 7.3072 - 3.8576} = 77.537 \ °C$$

Equation (2.92) allows the calculation of the heat transfer rate from the hot fluid as follows:

$$q_1 = \frac{Cp_1 W_1 (R_1 - R_2)(T_{1,in} - T_{2,in})}{RR_2 - R_1}$$

$$= \frac{1.9700 \cdot 1,880.0(100.00 - 15.000)(3.8576 - 7.3072)}{1.4732 \cdot 7.3072 - 3.8576} = (-1.5722e+5) \ W$$

Since fluid 1 is the hot stream, the negative sign just indicates that it loses energy.

For the parallel flow arrangement, R, R_1, and R_2 are the same as evaluated previously, and then the exit temperatures are obtained from eqs. (2.90) and (2.91):

$$T_{1,out} = \frac{T_{1,in}(RR_1 R_2 + 1) + T_{2,in}(R_1 R_2 - 1)}{R_1 R_2 (R+1)}$$

$$= \frac{100.00(1.4732 \cdot 3.8576 \cdot 7.3072 + 1) + 15.000(3.8576 \cdot 7.3072 - 1)}{3.8576 \cdot 7.3072(1.4732 + 1)} = 66.851 \ °C$$

$$T_{2,out} = \frac{RT_{1,in}(R_1 R_2 - 1) + T_{2,in}(R + R_1 R_2)}{R_1 R_2 (R+1)}$$

$$= \frac{1.4732 \cdot 100.00(3.8576 \cdot 7.3072 - 1) + 15.000(1.4732 + 3.8576 \cdot 7.3072)}{3.8576 \cdot 7.3072(1.4732 + 1)} = 63.835 \ °C$$

In addition, eq. (2.92) gives the heat load transferred from the hot fluid in the exchanger:

$$q_1 = \frac{Cp_1 W_1 (T_{1,in} - T_{2,in})(-R_1 R_2 + 1)}{R_1 R_2 (R+1)}$$

$$= \frac{1.9700 \cdot 1,880.0(100.00 - 15.000)(-3.8576 \cdot 7.3072 + 1)}{3.8576 \cdot 7.3072(1.4732 + 1)} = (-1.2277e+5) \ W$$

Alternatively, using heat balances of the fluids (eqs. (2.77) and (2.78)), along with the calculated exit temperatures, the heat transfer rates for both streams are checked below:

$$q_1 = Cp_1 W_1 (T_{1,out} - T_{1,in}) = 1.9700 \cdot 1,880.0(66.851 - 100.00) = (-1.2277e+5) \ W$$

$$q_2 = Cp_2 W_2 (T_{2,out} - T_{2,in}) = 0.60000 \cdot 4,190.0(63.835 - 15.000) = (1.2277e+5) \ W$$

Conclusion

From the previous calculations, we verify that by simply choosing the counterflow arrangement, the heat transfer rate increases in about 28%.

Worked Example 2.9

A 3,000 kg/h stream of hot water is cooled from 50 to 30 °C in the tube of a double-pipe heat exchanger, using 6,000 kg/h of water at a temperature of 10 °C. The fluids are contacted in parallel flow. The heat transfer coefficients for both sides were estimated at 5,000 W/m²K. The fouling and tube wall conductive resistances are negligible. (a) What is the required heat transfer area to perform this service? Assuming the heat transfer coefficients obey the Sieder–Tate equation for turbulent flow, determine (b) the exit temperatures if the hot water flow rate doubles, and (c) if the flow rate of both fluids is doubled [18].

Solution

a) Case 1: original flow rates.

Let the hot water be named fluid 1, then its average bulk temperature is

$$T_{1m} = \frac{T_{1,in}}{2} + \frac{T_{1,out}}{2} = \frac{50.000}{2} + \frac{30.000}{2} = 40.000 \text{ °C}$$

From Appendix C.13, its specific heat at 40 °C is estimated as $Cp_1 = 4{,}080.0$ J/(kg · K). Therefore, we have the heat transfer rate of the hot fluid as follows:

$$q_1 = W_1 Cp_1 (T_{1,out} - T_{1,in}) = 0.83333 \cdot 4{,}080.0(30.000 - 50.000) = -68{,}000 \text{ W}$$

As a first estimate, the specific heat of the cold water (fluid 2) is evaluated at the entrance temperature as $Cp_2 = 4070.0$ J/((kg · K)), then the exit temperature can be calculated from the heat balance equation:

$$T_{2,out} = T_{2,in} + \frac{q_2}{Cp_2 W_2} = 10.000 + \frac{68{,}000}{1.6667 \cdot 4{,}070.0} = 20.020 \text{ °C}$$

The log mean temperature difference for parallel flow is

$$\Delta T_1 = T_{1,in} - T_{2,in} = -10.000 + 50.000 = 40.000 \text{ °C}$$

$$\Delta T_2 = T_{1,out} - T_{2,out} = -20.020 + 30.000 = 9.9800 \text{ °C}$$

$$\Delta T_{lm,p} = \frac{\Delta T_1 - \Delta T_2}{\ln\left(\dfrac{\Delta T_1}{\Delta T_2}\right)} = \frac{40.000 - 9.9800}{\ln\left(\dfrac{40.000}{9.9800}\right)} = 21.624 \text{ °C}$$

With the initial values of the heat transfer coefficients, the overall heat transfer coefficient for the whole exchanger area is

$$U = \left(\frac{1}{h_o} + \frac{1}{h_i}\right)^{-1} = \frac{5,000.0}{2} = 2,500.0 \ \text{W}/(\text{m}^2 \cdot \text{K})$$

And the heat transfer surface can be calculated from eq. (2.8):

$$A = \frac{q}{U \Delta T_{\text{lm,p}}} = \frac{68,000}{21.624 \cdot 2,500.0} = 1.2579 \ \text{m}^2$$

b) Case 2: hot water flow rate doubled.
The fluid flow rates are

$$W_1 = W_{\text{new}} = 2 \cdot W_{\text{old}} = 2 \cdot 0.83333 = 1.6667 \ \text{kg/s}$$

$$W_2 = 1.6667 \ \text{kg/s}$$

The Sieder–Tate correlation for turbulent flow gives that $h \propto \text{Re}^{0.8} \propto W^{0.8}$; therefore, the internal heat transfer coefficient (h_i) for the new flow rate is

$$h_i = h_{\text{new}} = h_{\text{old}} \left(\frac{W_{\text{new}}}{W_{\text{old}}}\right)^{0.8} = 5,000.0 \left(\frac{1.6667}{0.83333}\right)^{0.8} = 8,705.7 \ \text{W}/(\text{m}^2 \cdot \text{K})$$

The overall heat transfer coefficient is

$$U = \left(\frac{1}{h_o} + \frac{1}{h_i}\right)^{-1} = \left(\frac{1}{8,705.7} + \frac{1}{5,000.0}\right)^{-1} = 3,175.9 \ \text{W}/(\text{m}^2 \cdot \text{K})$$

From eqs. (2.94) to (2.96), the parameters R, R_1, and R_2 are evaluated:

$$R = \frac{Cp_1 W_1}{Cp_2 W_2} = \frac{4,080.0}{4,070.0} = 1.0025$$

$$R_1 = e^{\frac{AU}{Cp_1 W_1}} = e^{\frac{1.2579 \cdot 3,175.9}{1.6667 \cdot 4,080.0}} = 1.7995$$

$$R_2 = e^{\frac{AU}{Cp_2 W_2}} = e^{\frac{1.2579 \cdot 3,175.9}{1.6667 \cdot 4,070.0}} = 1.8021$$

Equations (2.90) and (2.91) give the exit temperatures for parallel flow:

$$T_{1,\text{out}} = \frac{T_{1,\text{in}}(RR_1R_2 + 1) + T_{2,\text{in}}(R_1R_2 - 1)}{R_1R_2(R+1)}$$

$$= \frac{10.000(1.7995 \cdot 1.8021 - 1) + 50.000(1.0025 \cdot 1.7995 \cdot 1.8021 + 1)}{1.7995 \cdot 1.8021(1.0025 + 1)}$$

$$= 36.185 \ °\text{C}$$

$$T_{2,\text{out}} = \frac{RT_{1,\text{in}}(R_1R_2 - 1) + T_{2,\text{in}}(R + R_1R_2)}{R_1R_2(R + 1)}$$

$$= \frac{1.0025 \cdot 50.000(1.7995 \cdot 1.8021 - 1) + 10.000(1.0025 + 1.7995 \cdot 1.8021)}{1.7995 \cdot 1.8021(1.0025 + 1)}$$

$$= 23.850 \,°C$$

Using the entrance temperatures of fluids $T_{1,\text{in}}$ and $T_{2,\text{in}}$, the rate of heat loss from the hot fluid can be evaluated from eq. (2.92):

$$q_1 = \frac{Cp_1 W_1(T_{1,\text{in}} - T_{2,\text{in}})(-R_1R_2 + 1)}{R_1R_2(R + 1)}$$

$$= \frac{1.6667 \cdot 4,080.0(-10.000 + 50.000)(-1.7995 \cdot 1.8021 + 1)}{1.7995 \cdot 1.8021(1.0025 + 1)} = -93,946 \,W$$

Moreover, the heat gained by the cold (eq. (2.93)) fluid is

$$q_2 = \frac{Cp_1 W_1(T_{1,\text{in}} - T_{2,\text{in}})(R_1R_2 - 1)}{R_1R_2(R + 1)}$$

$$= \frac{1.6667 \cdot 4,080.0(-10.000 + 50.000)(1.7995 \cdot 1.8021 - 1)}{1.7995 \cdot 1.8021(1.0025 + 1)} = 93,946 \,W$$

We can double-check the streams' heat loads by making the calculation using the four process temperatures in the heat balance equations (2.77) and (2.78):

$$q_1 = Cp_1 W_1(T_{1,\text{out}} - T_{1,\text{in}}) = 1.6667 \cdot 4,080.0(36.185 - 50.000) = -93,944 \,W$$

$$q_2 = Cp_2 W_2(T_{2,\text{out}} - T_{2,\text{in}}) = 1.6667 \cdot 4,070.0(-10.000 + 23.850) = 93,951 \,W$$

c) Case 3: both streams' flow rates doubled.
In the present case, both streams have their flow rates doubled:

$$W_1 = W_{\text{new}} = 2 \cdot W_{\text{old}} = 2 \cdot 0.83333 = 1.6667 \,\text{kg/s}$$

$$W_2 = W_{\text{new}} = 2 \cdot W_{\text{old}} = 2 \cdot 1.6667 = 3.3334 \,\text{kg/s}$$

Updating the heat transfer coefficients with the flow rate changes, we have

$$h_i = h_{\text{new}} = h_{\text{old}}\left(\frac{W_{\text{new}}}{W_{\text{old}}}\right)^{0.8} = 5,000.0\left(\frac{1.6667}{0.83333}\right)^{0.8} = 8,705.7 \,W/(m^2 \cdot K)$$

$$h_o = h_{\text{new}} = h_{\text{old}}\left(\frac{W_{\text{new}}}{W_{\text{old}}}\right)^{0.8} = 5,000.0\left(\frac{3.3334}{1.6667}\right)^{0.8} = 8,705.5 \,W/(m^2 \cdot K)$$

The overall heat transfer coefficient is

$$U = \left(\frac{1}{h_o} + \frac{1}{h_i}\right)^{-1} = \left(\frac{1}{8,705.7} + \frac{1}{8,705.5}\right)^{-1} = 4,352.8 \,W/(m^2 \cdot K)$$

Then, the new exit temperatures for this overall heat transfer coefficient are calculated:

$$R = \frac{Cp_1 W_1}{Cp_2 W_2} = \frac{1.6667 \cdot 4,080.0}{3.3334 \cdot 4,070.0} = 0.50123$$

$$R_1 = e^{\frac{AU}{Cp_1 W_1}} = e^{\frac{1.2579 \cdot 4,352.8}{1.6667 \cdot 4,080.0}} = 2.2371$$

$$R_2 = e^{\frac{AU}{Cp_2 W_2}} = e^{\frac{1.2579 \cdot 4,352.8}{3.3334 \cdot 4,070.0}} = 1.4972$$

$$T_{1,out} = \frac{T_{1,in}(RR_1 R_2 + 1) + T_{2,in}(R_1 R_2 - 1)}{R_1 R_2 (R + 1)}$$

$$= \frac{10.000(1.4972 \cdot 2.2371 - 1) + 50.000(0.50123 \cdot 1.4972 \cdot 2.2371 + 1)}{1.4972 \cdot 2.2371(0.50123 + 1)}$$

$$= 31.310 \ ^\circ C$$

$$T_{2,out} = \frac{RT_{1,in}(R_1 R_2 - 1) + T_{2,in}(R + R_1 R_2)}{R_1 R_2 (R + 1)}$$

$$= \frac{0.50123 \cdot 50.000(1.4972 \cdot 2.2371 - 1) + 10.000(0.50123 + 1.4972 \cdot 2.2371)}{1.4972 \cdot 2.2371(0.50123 + 1)}$$

$$= 19.368 \ ^\circ C$$

The heat loads evaluated with eqs. (2.92) or (2.93) are

$$q_1 = \frac{Cp_1 W_1 (T_{1,in} - T_{2,in})(-R_1 R_2 + 1)}{R_1 R_2 (R + 1)}$$

$$= \frac{1.6667 \cdot 4,080.0(50.000 - 10.000)(-1.4972 \cdot 2.2371 + 1)}{1.4972 \cdot 2.2371(0.50123 + 1)} = (-1.2709e + 5) \ W$$

$$q_2 = \frac{Cp_1 W_1 (T_{1,in} - T_{2,in})(R_1 R_2 - 1)}{R_1 R_2 (R + 1)}$$

$$= \frac{1.6667 \cdot 4,080.0(50.000 - 10.000)(1.4972 \cdot 2.2371 - 1)}{1.4972 \cdot 2.2371(0.50123 + 1)} = (1.2709e + 5) \ W$$

On the other hand, using the evaluated exit temperatures, from the heat balance equations, we have

$$q_1 = Cp_1 W_1 (T_{1,out} - T_{1,in}) = 1.6667 \cdot 4,080.0(31.310 - 50.000) = (-1.2709e + 5) \ W$$

$$q_2 = Cp_2 W_2 (T_{2,out} - T_{2,in}) = 3.3334 \cdot 4,070.0(19.368 - 10.000) = (1.2710e + 5) \ W$$

Conclusion

The following table summarizes the obtained results:

	Case 1	Case 2	Case 3
$T_{1,in}$	50 °C	50 °C	50 °C
$T_{2,in}$	10 °C	10 °C	10 °C
W_1	0.83333 kg/s	1.6667 kg/s	1.6667 kg/s
W_2	1.6667 kg/s	1.6667 kg/s	3.3334 kg/s
$T_{1,out}$	30 °C	36.185 °C	31.310 °C
$T_{2,out}$	20.020 °C	23.850 °C	19.368 °C
q	68,000 W	93,946 W	(1.2709e + 5) W

Comparing to the initial operating conditions (case 1), the heat transfer rate increases after doubling the flow rate of any of the fluids. Notice that in case 2, despite the increased heat load of the heat exchanger, there is a counterintuitive increase in the exit temperature of fluid 1 (hot), which is a consequence of the greater flow rate.

Problems

(2.1) Consider the carbon steel heat exchanger tube shown in Fig. 2.3. Given the convective heat transfer coefficients of $h_i = 1{,}841.8$ BTU/(ft$^2 \cdot$ h $\cdot$ °F) and $h_o = 94.298$ BTU/(ft$^2 \cdot$ h $\cdot$ °F), the pipe dimensions ID = 1 in and OD = 4 in, and the material heat conductivity of $k = 29.467$ BTU/(ft h °F) determine the:
a) Clean overall heat transfer coefficient
b) Simplified clean overall heat transfer coefficient ignoring the wall conductive resistance
c) Design overall heat transfer coefficient
d) Error introduced if the simplified clean overall heat transfer coefficient is used to design the exchanger

Answer:
(a) $U_c = 48.503$ BTU/(ft^2 h °F)
(b) U_c (ignoring wall) = 78.269 BTU/(ft^2 h °F)
(c) the same as U_c
(d) error: 38%

(2.2) A fluid is heated from t_1 to t_2 by cooling another fluid from T_1 to T_2. Their heat capacities are wc and WC, respectively. If the overall heat transfer coefficient can be assumed approximately constant, derive the expressions for the log mean temperature difference for (a) parallel and (b) countercurrent flow.

Answer:

(a) $\Delta T_{lm} = \dfrac{T_1 - t_1 + t_2 - T_2}{\ln\left(\frac{T_1 - t_1}{T_2 - t_2}\right)}$

(b) $\Delta T_{lm} = \dfrac{T_1 - t_2 + t_1 - T_2}{\ln\left(\frac{T_1 - t_2}{T_2 - t_1}\right)}$

(2.3) Crude oil is heated from 20 to 60 °C with saturated vapor condensing at 100 °C. Evaluate the LMTD for (a) parallel flow and (b) counterflow.

Answer:
(a) LMTD = 57.7 °C
(b) LMTD = 57.7 °C

(2.4) A battery of counterflow double-pipe exchangers of effective area 14 m² is used to heat up a 2.5 kg/s stream of water from 30 to 80 °C with a 4 kg/s flue gas discharge cooling from 280 to 157 °C. Evaluate the overall heat transfer for this heat exchanger. The physical properties of the flue gas were found to be very similar to carbon dioxide at the same operating conditions.

Answer:
$U = 226.85$ W/m²·K

(2.5) A stream of ethanol with 5.2 kg/s flow rate is heated from 30 to 65 °C using 4 kg/s of liquid hot water at 90 °C. The service is performed in a counterflow carbon steel heat exchanger with 26.928 m² effective area. Determine:
a) Heat transfer rate
b) Exit temperature of the hot water
c) Log mean temperature difference
d) Overall heat transfer coefficient

Answer:
(a) $q = 5.0960e + 5$ W
(b) $T_{out} = 59.5$ °C
(c) LMTD = 27.2 K
(d) $U_d = 695.87$ W/m² K

(2.6) Consider the service from Problem 2.5. Assuming the ethanol is controlling the heat transfer rate in turbulent flow, the overall heat transfer of the exchanger can be taken as approximately proportional to its mass flow rate (i.e., $U \propto w^{0.8}$). Estimate the heat transfer rate and ethanol exit temperature if its flow rate changes to (a) 3 kg/s and (b) 7 kg/s. As a first approximation, assume that the water exit temperature remains unchanged.

Answer:
(a) $q = 3.1893e + 5$ W, $T_{out} = 67.9$ °C
(b) $q = 6.5925e + 5$ W, $T_{out} = 63.6$ °C

(2.7) Considering the same changes in the ethanol flow rate to (a) 3 kg/s and (b) 7 kg/s (from Problem 2.6), determine the "exact" heat transfer rate and the error introduced in the heat transfer rate by the assumed simplification of constant exit water temperature.

Answer:
(a) $q = 3.4131e + 5$ W, error = −6.5%
(b) $q = 6.1451e + 5$ W, error = 7.3%

(2.8) In a chemical plant, 3.3 kg/s of liquid benzene at 90 °C is cooled to 60 °C by heating a phenol stream from 30 to 50 °C. Assume a clean overall heat transfer coefficient of $U_c = 190$ W/(m² K). A fouling factor of $R_{di} = 0.00053$ m²K/W is provided for the phenol stream flowing in the inner tube. Calculate the required heat transfer areas if the flow arrangement is (a) parallel and (b) counterflow. (c) What is the most cost-effective flow arrangement in this case?

Answer:
(a) 35.9 m²
(b) 28.8 m²
(c) Countercurrent flow

(2.9) An effluent of 31,747 lb/h hot wastewater must be cooled from 176 to 95 °F, before being discharged to a watercourse. A process stream of aniline is available at 68 °F, and the outlet temperature is designed to be 104 °F. Fouling factors are specified for both streams as $R_{di} = 0.002$ ft² h °F/BTU and $R_{do} = 0.001$ ft² h °F/BTU. Carbon steel double-pipe exchangers are used, which have the following piping data:
Inner pipe: ID = 2.067 in, OD = 2.375 in
Outer pipe: ID = 7.981 in, OD = 8.625 in
Heat conductivity: 29.467 BTU/(ft h °F)

The convective heat transfer coefficient for the inner tube and annulus are 1841.8 BTU/(ft² h °F) and 94.3 BTU/(ft² h °F), respectively. Determine (a) the design overall heat transfer coefficient and (b) required area to perform this service.

Answer:
(a) $U_d = 66.697$ BTU/(ft² h °F)
(b) $A = 823.55$ ft²

(2.10) Given the process conditions and exchanger specifications, determine the expressions for the temperature profiles of hot and cold fluids in a parallel flow double-pipe heat exchanger by solving the differential equations (2.34) and (2.36).

(2.11) Calculate the temperature distributions for the fluids in a counterflow double-pipe heat exchanger. Determine the actual mean temperature difference by integrating the difference of the temperature distribution curves.

3 Wall temperatures

The Sieder-Tate correlation and most modern equations for the prediction of the heat transfer coefficient, such as Petukhov-Popov, Gnielinski and Notter-Sleicher, among others, require the evaluation of fluid physical properties in the temperature of the solid surface heating or cooling the fluid, the so-called wall temperature.

Rigorously, there are two "wall temperatures," one for the inner surface (t_{wi}) and another on the outer surface (t_{wo}), which depend on the convective coefficient of both streams exchanging heat, on the thermal conductivity of the solid sheet separating the fluids, and on its thickness. Typically, a cylindrical metallic wall separates the fluids; however, there are heat exchangers made with more uncommon materials, for instance, plastic or ceramics.

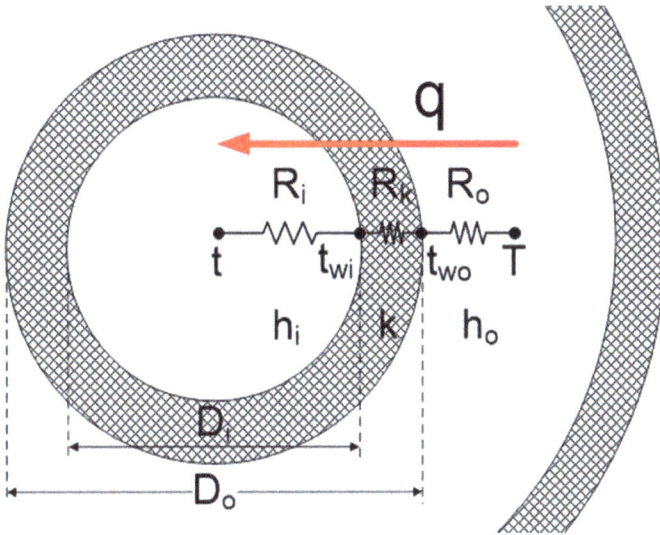

Fig. 3.1: Schematic of the resistances in the thermal circuit in a double-pipe exchanger with the cold fluid in the inner tube.

In this section, firstly, we are going to develop general equations for calculating the internal and external wall temperatures in a tubular heat exchanger. Then, the general equations can be simplified to the most common cases of heat exchangers, such as where the conductive resistance of the pipe wall is negligible in comparison to the convective resistances of the fluids.

https://doi.org/10.1515/9783110585872-003

3.1 Cold fluid in the inner pipe

When two fluids exchange energy with a solid wall through the heat path, there are three resistances opposing this heat flux: a convective resistance of the inner fluid, the wall conductive resistance, and the convective resistance in the outer fluid. This circuit is represented in Fig. 3.1 for the case where the cold fluid flows within the inner pipe passage.

The heat transfer rate in the individual resistances may be written as follows:

$$q = A_i h_i (t_{wi} - t) \qquad \text{(a)}$$

$$q = \frac{2A_o k (t_{wo} - t_{wi})}{D_o \ln\left(\frac{D_o}{D_i}\right)} \qquad \text{(b)}$$

$$q = A_o h_o (T - t_{wo}) \qquad \text{(c)}$$ (3.1)

$$q = \frac{T - t}{\frac{D_o \ln\left(\frac{D_o}{D_i}\right)}{2A_o K} + \frac{1}{A_o h_o} + \frac{1}{A_i h_i}} \qquad \text{(d)}$$

Equations (3.1) (a–c) can be combined to form the system of equations:

$$A_i h_i (t_{wi} - t) = \frac{2A_o k (t_{wo} - t_{wi})}{D_o \ln\left(\frac{D_o}{D_i}\right)} \qquad \text{(a)}$$

(3.2)

$$A_o h_o (T - t_{wo}) = \frac{2A_o k (t_{wo} - t_{wi})}{D_o \ln\left(\frac{D_o}{D_i}\right)} \qquad \text{(b)}$$

where the internal and external heat transfer surfaces of the inner tube may be expressed as follows:

$$A_i = \pi D_i L \qquad\qquad (3.3)$$

$$A_o = \pi D_o L$$

Solving eqs. (3.2) (a, b) for the inner wall (t_{wi}) and outer wall (t_{wo}), temperatures, and using eq. (3.3), we have

$$t_{wi} = \frac{D_i h_i \left(D_o h_o \ln\left(\frac{D_o}{D_i}\right) + 2k\right) t + 2D_o h_o kT}{D_i h_i \left(D_o h_o \ln\left(\frac{D_o}{D_i}\right) + 2k\right) + 2D_o h_o k} \qquad (3.4)$$

$$t_{wo} = \frac{D_o h_o \left(D_i h_i \ln\left(\frac{D_o}{D_i}\right) + 2k\right) T + 2D_i h_i kt}{D_i h_i \left(D_o h_o \ln\left(\frac{D_o}{D_i}\right) + 2k\right) + 2D_o h_o k} \qquad (3.5)$$

In the limiting case of a very thin pipe wall, or highly conductive material, or where the conductive thermal resistance is small compared with the convective resistances, the inner and outer wall temperatures become nearly the same, i.e., $t_{wi} = t_{wo} = t_w$.

Noticeably, these equations may be quite cumbersome for convenient use; however, we may verify that the terms in parenthesis have the same units of the tube's thermal conductivity, assuming the meaning of a corrected thermal conductivity for the effects of pipe curvature. Then, let us define the factors K_i, K_o, and K_{io} as follows:

$$K_i = D_i h_i \ln\left(\frac{D_o}{D_i}\right) + 2k \tag{3.6}$$

$$K_o = D_o h_o \ln\left(\frac{D_o}{D_i}\right) + 2k \tag{3.7}$$

$$K_{io} = D_i h_i K_o + 2 D_o h_o k \tag{3.8}$$

With eqs. (3.6)–(3.8), the wall temperature equations can be reduced to the more compact forms of the following equations:

$$t_{wi} = \frac{D_i h_i K_o t + 2 D_o h_o k T}{K_{io}} \tag{3.9}$$

$$t_{wo} = \frac{D_o h_o K_i T + 2 D_i h_i k t}{K_{io}} \tag{3.10}$$

3.1.1 Negligible conductive resistance

If the wall conductive resistance is very small when compared with the convective resistances of the fluids, eqs. (3.4) and (3.5) reduce to the following form:

$$t_w = \frac{D_i h_i t + D_o h_o T}{D_i h_i + D_o h_o} \tag{3.11}$$

Notice that, although the wall conductive resistance vanished in eq. (3.11), the effect of pipe curvature on the heat flux is still there, being taken account by the presence of the inner and outer diameters.

3.1.2 Thin tube wall

For a relatively thin tube wall, we have that $D_i \approx D_o$; therefore, eqs. (3.4) and (3.5) take the following form:

$$t_w = \frac{h_o T + h_i t}{h_i + h_o} \tag{3.12}$$

A comparison of eq. (3.11) with eq. (3.12) reveals that the thin wall approximation is a stronger simplification than small conductive wall resistance, since that later result ignores more system descriptive information, namely, the internal and external diameters of the inner pipe of the heat exchanger.

3.2 Hot fluid in the inner pipe

With the hot fluid passing through the inner tube, the heat transfer equations for the thermal circuit are similar, as follows:

$$
\begin{align}
&q = A_i h_i (T - t_{wi}) && \text{(a)} \\[4pt]
&q = \frac{2A_o k (t_{wi} - t_{wo})}{D_o \ln\left(\frac{D_o}{D_i}\right)} && \text{(b)} \\[4pt]
&q = A_o h_o (t_{wo} - t) && \text{(c)} \\[4pt]
&q = \frac{T - t}{\frac{D_o \ln\left(\frac{D_o}{D_i}\right)}{2A_o K} + \frac{1}{A_o h_o} + \frac{1}{A_i h_i}} && \text{(d)}
\end{align}
\tag{3.13}
$$

Then, the system of equations for obtaining the wall temperatures is given by

$$
\begin{align}
&A_i h_i (T - t_{wi}) = \frac{2A_o k (t_{wi} - t_{wo})}{D_o \ln\left(\frac{D_o}{D_i}\right)} && \text{(a)} \\[6pt]
&A_o h_o (t_{wo} - t) = \frac{2A_o k (t_{wi} - t_{wo})}{D_o \ln\left(\frac{D_o}{D_i}\right)} && \text{(b)}
\end{align}
\tag{3.14}
$$

Accordingly, the solutions of eqs. (3.14) (a, b) are

$$
t_{wi} = \frac{D_i h_i \left(D_o h_o \ln\left(\frac{D_o}{D_i}\right) + 2k\right) T + 2D_o h_o k t}{D_i h_i \left(D_o h_o \ln\left(\frac{D_o}{D_i}\right) + 2k\right) + 2D_o h_o k}
\tag{3.15}
$$

$$
t_{wo} = \frac{D_o h_o \left(D_i h_i \ln\left(\frac{D_o}{D_i}\right) + 2k\right) t + 2D_i h_i k T}{D_i h_i \left(D_o h_o \ln\left(\frac{D_o}{D_i}\right) + 2k\right) + 2D_o h_o k}
\tag{3.16}
$$

And using the K factors from eqs. (3.6) to (3.8), we have

$$
t_{wi} = \frac{D_i h_i K_o T + 2D_o h_o k t}{K_{io}}
\tag{3.17}
$$

$$
t_{wo} = \frac{D_o h_o K_i t + 2D_i h_i k T}{K_{io}}
\tag{3.18}
$$

3.2.1 Negligible conductive resistance

For a high thermally conductive material, the convective resistances dominate the heat transfer process, and both eqs. (3.15) and (3.16) take the same form of the following equation:

$$t_w = \frac{D_i h_i T + D_o h_o t}{D_i h_i + D_o h_o} \qquad (3.19)$$

3.2.2 Thin tube wall

A simplified version of eq. (3.19) is obtained for the approximation of a tube wall with small thickness, i.e., $D_i \approx D_o$:

$$t_w = \frac{h_i T + h_o t}{h_i + h_o} \qquad (3.20)$$

3.3 General form of wall temperature equations

Defining a notation wherein the temperatures of the fluids in the inner tube and annulus are t_i and t_o, respectively, disregarding which is the hot or cold fluid, a single set of equations may be written as described below.

3.3.1 Significant conductive and convective resistances

The full equations including the effects of curvature on the conductive and convective thermal resistances are

$$t_{wi} = \frac{D_i h_i \left(D_o h_o \ln\left(\frac{D_o}{D_i}\right) + 2k \right) t_i + 2 D_o h_o k t_o}{D_i h_i \left(D_o h_o \ln\left(\frac{D_o}{D_i}\right) + 2k \right) + 2 D_o h_o k} \qquad (3.21)$$

$$t_{wo} = \frac{D_o h_o \left(D_i h_i \ln\left(\frac{D_o}{D_i}\right) + 2k \right) t_o + 2 D_i h_i k t_i}{D_i h_i \left(D_o h_o \ln\left(\frac{D_o}{D_i}\right) + 2k \right) + 2 D_o h_o k} \qquad (3.22)$$

Or in the reduced forms of the following equations:

$$t_{wi} = \frac{D_i h_i K_o t_i + 2 D_o h_o k t_o}{K_{io}} \qquad (3.23)$$

$$t_{wo} = \frac{D_o h_o K_i t_o + 2 D_i h_i k t_i}{K_{io}}$$ (3.24)

where

$$K_i = D_i h_i \ln\left(\frac{D_o}{D_i}\right) + 2k$$ (3.25)

$$K_o = D_o h_o \ln\left(\frac{D_o}{D_i}\right) + 2k$$ (3.26)

$$K_{io} = D_i h_i K_o + 2 D_o h_o k$$ (3.27)

3.3.2 Negligible conductive wall resistance

$$t_w = \frac{D_i h_i t_i + D_o h_o t_o}{D_i h_i + D_o h_o}$$ (3.28)

3.3.3 Thin tube wall

$$t_w = \frac{h_i t_i + h_o t_o}{h_i + h_o}$$ (3.29)

It is important to notice that the formulas given in eqs. (3.21) and (3.22) or (3.23) and (3.24) are recommended for more accurate estimations of the internal and external wall temperatures. However, in ordinary situations, the incurred error from using the simplified forms (eqs. (3.28) and (3.29)) are likely to be admissible, and, indeed, these are the expressions typically offered by some reliable technical sources (e.g., [19, 28, 44, 47]).

Worked Example 3.1

Consider a double-pipe exchanger with the hot fluid allocated in the inner tube with an average bulk temperature of 82 °C and convective heat transfer coefficient of 500 W/m² K. The cold fluid flows in the annulus and has an average temperature of 55 °C and an estimated heat transfer coefficient of 300 W/m²K. The high-pressure inner pipe (schedule XXS) has internal and external diameters of 22.8 and 42.2 mm, respectively. (a) Estimate the inner tube wall temperature assuming a negligible conductive resistance. (b) What is the wall temperature for the thin wall approximation?

Solution

(a) If the conductive resistance of the wall is disregarded, eq. (3.28) may be used with inner fluid temperature $t_i = 82$ °C and outer temperature $t_o = 55$ °C. Substituting the numeric values, we have

$$t_w = \frac{D_i h_i t_i + D_o h_o t_o}{D_i h_i + D_o h_o} = \frac{0.02280 \cdot 500.0 \cdot 82.00 + 0.04220 \cdot 300.0 \cdot 55.00}{0.02280 \cdot 500.0 + 0.04220 \cdot 300.0} = 67.79\ ^\circ\text{C}$$

(b) The thin wall approximation is a stronger simplification, under which the wall temperature is given by eq. (3.29) as follows:

$$t_w = \frac{h_i t_i + h_o t_o}{h_i + h_o} = \frac{(300.0 \cdot 55.00 + 500.0 \cdot 82.00)}{300.0 + 500.0} = 71.88\ ^\circ\text{C}$$

Notice that the consideration of the pipe curvature brings the average wall temperature closer to the outer fluid temperature, which, in this case, is the cold fluid. The difference between both t_w values is about 4.09 °C.

Worked Example 3.2
The exchanger analyzed in Worked Example 3.1 has the inner pipe built in stainless steel 430, with thermal conductivity $k = 8.1$ W/mK. How do the "exact" inner and outer wall temperatures compare with the one estimated with the thin wall simplification?

Solution
Using eqs. (3.25)–(3.27) the K's can be evaluated as follows:

$$K_i = D_i h_i \ln\left(\frac{D_o}{D_i}\right) + 2k = 0.02280 \cdot 500.0 \ln\left(\frac{0.04220}{0.02280}\right) + 2 \cdot 8.100 = 23.22\ \text{W}/(\text{m}\cdot\text{K})$$

$$K_o = D_o h_o \ln\left(\frac{D_o}{D_i}\right) + 2k = 0.04220 \cdot 300.0 \ln\left(\frac{0.04220}{0.02280}\right) + 2 \cdot 8.100 = 23.99\ \text{W}/(\text{m}\cdot\text{K})$$

$$K_{io} = D_i K_o h_i + 2 D_o h_o k = 0.02280 \cdot 23.99 \cdot 500.0 + 2 \cdot 0.04220 \cdot 300.0 \cdot 8.100$$
$$= 478.6\ \text{W}^2/(\text{m}^2 \cdot \text{K}^2)$$

Then, the internal and external wall temperatures are obtained from eqs. (3.23) and (3.24):

$$t_{wi} = \frac{1}{K_{io}}(D_i K_o h_i t_i + 2 D_o h_o k t_o)$$
$$= \frac{1}{478.6}(0.02280 \cdot 23.99 \cdot 500.0 \cdot 82.00 + 2 \cdot 0.04220 \cdot 300.0 \cdot 55.00 \cdot 8.100) = 70.43\ ^\circ\text{C}$$

$$t_{wo} = \frac{1}{K_{io}}(2 D_i h_i k t_i + D_o K_i h_o t_o)$$
$$= \frac{1}{478.6}(2 \cdot 0.02280 \cdot 500.0 \cdot 8.100 \cdot 82.00 + 0.04220 \cdot 23.22 \cdot 300.0 \cdot 55.00) = 65.42\ ^\circ\text{C}$$

In this case, the use of the common assumption of thin wall estimates the tube temperature as $t_w = 71.88$ °C. This result would outcome a wall temperature error of
$t_w - t_{wi} = 1.45$ °C for the pipe (hot) fluid
$t_w - t_{wo} = 6.46$ °C for the annulus (cold) fluid

Problems

(3.1) The heat transfer coefficients for the cold and hot fluids in a double-pipe heat exchanger are $h_i = 2{,}667$ W/m^{2}K and $h_o = 940$ W/m^{2}K, respectively. The bulk mean temperature for the cold fluid is 47.5 °C and for the hot fluid is 81.2 °C. What is the average wall temperature in this heat exchanger?

Answer:

$t_w = 56.3$ °C

(3.2) Triethylene glycol is heated from 30 to 65 °C, using hot water varying from 90 to 80 °C in a counterflow heat exchanger. The overall heat transfer coefficient was evaluated as $U = 160$ W/m^{2}K. In this service, the water flows through the annulus and develops a convective heat transfer coefficient of $h_o = 777$ W/m^2 K. Estimate (a) the convective heat transfer coefficient of the triethylene glycol, (b) the wall temperature at the cold terminal, (c) at the hot terminal, and (d) the mean wall temperature.

Answer:

(a) $h_i = 201.49$ W/m^2 K
(b) $t_{w(cold)} = 69.7$ °C
(c) $t_{w(hot)} = 84.8$ °C
(d) $t_{w(mean)} = 77.3$ °C

(3.3) The heat exchanger used in Problem 3.2 has an internal pipe of DN 50 mm schedule 160. What are the wall temperatures if the effect of the pipe curvature is considered (a) at the cold terminal, (b) at the hot terminal, and (c) the average wall temperature?

Answer:

(a) $t_{w(cold)} = 72.2$ °C
(b) $t_{w(hot)} = 86.1$ °C
(c) $t_{w(mean)} = 79.2$ °C

(3.4) A high-pressure carbon steel heat exchanger with an NPS 2-1/2 – sch XXS internal pipe cools an ethylene glycol stream using regular cooling water. The bulk temperatures are 149 and 78 °F for ethylene glycol and water, respectively. Ethylene glycol flows inside the inner pipe with convective heat transfer coefficient $h_i = 136$ BTU/(ft^2 h °F). The water convective heat transfer coefficient is $h_o = 974$ BTU/ (ft^2 h °F). Calculate the wall temperature assuming the approximations of (a) thin wall and (b) negligible wall conductive resistance. Additionally, determine

(c) the temperature difference between both sides of the inner tube wall and the temperature deviations between the common thin wall approximation and the more exact estimations given by (d) eq. (3.23) and (e) eq. (3.24).

Answer:
(a) $t_w = 86.7\ ^\circ F$
(b) $t_w = 83.6\ ^\circ F$
(c) $(t_{wi} - t_{wo}) = 9.4\ ^\circ F$
(d) $(t_w - t_{wi}) = -5.5\ ^\circ F$
(e) $(t_w - t_{wo}) = 3.8\ ^\circ F$

(3.5) A chemical plant uses 39683.0 lb/h of cold water from 77.0 to 94.7 °F to cool down an explosive stream of 15873.0 lb/h of 2,4,6-trinitrotoluene from 336.2 to 212.0 °F in a carbon steel ($k = 29.467$ BTU/(h ft°F)) double-tube heat exchanger. Its inner pipe is NPS 2 in – schedule STD and carries the 2,4,6-trinitrotoluene stream. The internal and external heat transfer coefficients are $h_i = 47$ BTU/ (h ft²°F) and $h_o = 427$ BTU/(h ft²°F), respectively. The trinitrotoluene normal melting point is about 179.3 °F.

a) What is the wall thickness of the inner tube?
b) Do you anticipate the fouling formation during the operation of this exchanger? Why?
c) Calculate the average wall temperature in this exchanger assuming thin wall.
d) Do you think that the approximation of thin wall yields a significant error? Why?
e) What is the temperature difference between t_w from the negligible conductive resistance and internal and external wall temperatures t_{wi} and t_{wo}?

Answer:
(a) 0.154 in
(b) Yes (justify!)
(c) $t_{w(\text{thin wall})} = 104.5\ ^\circ F$
(d) No (justify!)
(e) $(t_w - t_{wi}) = -2.94\ ^\circ F$ and $(t_w - t_{wo}) = 0.28\ ^\circ F$

(3.6) Consider that the trinitrotoluene/water exchanger from Problem 3.5 has an effective length of 192 ft.
a) Evaluate the heat fluxes and
b) Evaluate the heat transfer rates for the internal and external fluids using the wall temperature estimated through the thin wall approximation (Problem 3.5, item c).

c) Does the previously calculated heat transfer rate apparently violate the principle of conservation of energy? Explain this result.
d) Determine the heat fluxes and the heat transfer rates over the inner and outer surfaces of the internal pipe using their specific wall temperatures t_{wi} and t_{wo}, respectively.

Answer:
(a) $(q/A_i) = -7970.3$ BTU/(ft$^2 \cdot$h), $(q/A_o) = 7972.1$ BTU/(ft$^2 \cdot$h)
(b) $q_i = -8.28e+5$ BTU/h, $q_o = 9.52e+5$ BTU/h
(c) yes (justify!)
(d) $(q/A_i) = -7936.0$ BTU/(ft$^2 \cdot$h), $(q/A_o) = 6908.9$ BTU/(ft$^2 \cdot$h)
(e) $q_i = -8.24e+5$ BTU/h, $q_o = 8.24e+5$ BTU/h

4 Heat transfer coefficients

The calculation of the overall heat transfer coefficient depends on the determination of the pipe wall conductive resistance and on the convective heat transfer coefficients for both surfaces of the internal pipe. The thermal conductive resistance of the pipe is not such a problem because it can be straightforwardly calculated from the geometric properties, such as the inner diameter and wall thickness, along with the materials selected to build the heat exchanger. On the other hand, the convective heat transfer coefficients for the fluids are key information for evaluating the heat exchanger performance or size, and normally impose greater uncertainty on the design than the one embedded in the conductive resistance.

4.1 Methods of estimation

The evaluation of the heat transfer coefficients for the fluids in the heat exchanger can be performed by several approaches:
1. Analytical solutions of the conservation equations
2. Numerical solutions of the conservation equations
3. Correlations:
   a. Empirical: derived from measured data
   b. Theoretical: derived from the analytical or numerical solution of a mathematical model
   c. Semiempirical: the fitted solution of the mathematical model is achieved using experimental data

Relatively simple geometries such as flat plates, spheres, and cylindrical ducts admit analytical treatment for the determination of the convective heat transfer coefficients. In these cases, simplifying hypotheses (e.g., one-dimensional, and laminar flow) are applied to reduce the mathematical complexity of the equations to a level achievable by the known analytical methods of solution of differential equations. Unfortunately, current real-world problems are commonly characterized by elaborated geometric features and turbulent flow, making the problem solution usually unattainable by analytical methods.

In a few cases, the heat transfer coefficients may be obtained with some accuracy from direct computation, however, they are typically estimated using empirical correlations involving nondimensional groups. Nowadays, convection coefficients may be calculated numerically by solving the differential equations expressing the conservation of mass, momentum, and energy. The tools and methods involved in these solutions comprise a whole branch of engineering named computational fluid dynamics (CFD), which are out of the scope of this text. The main difficulty encountered with

https://doi.org/10.1515/9783110585872-004

numerical methods is still the phenomenon of turbulence, which is particularly hard to simulate with reasonable accuracy and within a practical time frame necessary to design heat exchangers. Another challenge faced by CFD is related to the geometry of a heat exchanger, which may be complex for various types of industrial importance. Intricate geometries require refined discretization meshes, significantly raising the time span required to solve the equations. As a result, even with the great computational power at rather low cost available today, the predominant approach for estimating the heat transfer coefficients is derived from experimental data posed in the form of mathematical correlations.

4.1.1 Types of correlations

4.1.1.1 Empirical correlations

In practice, empirical correlations are the predominant source for calculation of the convective heat transfer coefficients used in heat exchanger design and analysis. Empirical correlations are essentially mathematical relations among sets of nondimensional numbers obtained from experimental data. Therefore, an empirical correlation is as good and reliable as is the accuracy of the original measured data, provided the conditions of use are respected. The engineer must select an adequate correlation for the studied system. A very accurate correlation may give poor results if applied to a wrong geometry, for which it was not developed.

4.1.1.2 Theoretical correlations

Several heat transfer correlations are not simply the statistical fitting of experimental data points. In some cases, the correlation represents the results obtained by means of a mathematical model solved analytically or numerically, for example, the solution of the differential equations of conservation of mass, energy, and momentum. In such situation, the correlation can be named a "theoretical correlation," because its data source has no component of experimental data.

The validity of theoretical correlations must be confirmed by experimental data. Although a theoretical correlation is based entirely on calculated (not measured) results, it should always be validated against reliable experimental data before using in practical situations. The reason for this is very straightforward: there is no such thing as a "perfect" mathematical model. Nature is complex and every single model has assumptions and simplifications to be solvable. The manner to confirm if a model produces good results is "asking the Nature" if the mathematical model predictions are correct, and we ask questions directly to Nature by doing experiments.

At this point, you may be questioning right now: why someone would need a correlation if the exact solution were at hand? Well, the answer has to do with practicality, appropriateness, and convenience. Engineers like practical things, methods, procedures, and so on. Practicality reverts to efficiency, saved time, and resources after all. Typically, the data source of a theoretical correlation is the fluid temperature distribution in the form of a complicated function, involving summation, integrals, or even differential equations for extraction of eigenvalues.[2] Therefore, the translation of the temperature profile into the heat transfer coefficient required for equipment design may not be a trivial task. On the other hand, a simple and straightforward equation, with the necessary accuracy, is probably the adequate tool for the designer.

4.1.1.3 Semiempirical correlations
There is also what is commonly called semiempirical correlations. It is a blend of theory and experimentation. A semiempirical correlation fits the solution of the mathematical model, however, to solve accurately the model, some piece of experimental data may be required. This is usual for turbulent flows, for example, while solving the energy equation for a flowing fluid, we must know the turbulent velocity distribution. There are two ways to get this information: (1) solving the momentum equation coupled with the energy equation, along with some turbulence model; (2) make use of an experimental measured velocity profile. The second option was the choice for several heat transfer correlations available in the technical literature.

4.2 Using correlations

Correlations for convective heat transfer give the value of the Nusselt (Nu) or Stanton (St) numbers as a function of Reynolds (Re), Prandtl (Pr), Grashof (Gr), and other relevant nondimensional numbers. With the value of the Nusselt or Stanton numbers, we can evaluate the heat transfer coefficient for the studied problem. Correlations for prediction of forced convection heat transfer coefficients usually involve only the Reynolds number; however, there are exceptions. For example, heat transfer correlations combining the effects of forced and natural convection relate the Nusselt or Stanton numbers as a function of both Reynolds and Grashof numbers.

Keep in mind that every correlation has its "user manual" and you must stick to its rules! In general, an empirical correlation is tied to a specific geometry and domain of validity, usually defined by the minimum and maximum values of the nondimensional groups measured in the experiments. The domain of validity for an empirical correlation may be given as ranges for the nondimensional numbers involved, or sometimes, as ranges for products of primary nondimensional numbers. If the value of the nondimensional numbers for the case studied are not within the

---

2 The Notter–Sleicher equations presented in Section 4.5.2 match this case (see [69]).

valid ranges of a given heat transfer correlation, you should look for another more adequate correlation.

Never put all the eggs (estimations) in a single basket (correlation). Every correlation you find in a reference book or scientific paper is susceptible to errors, be they experimental, statistical, or typos. Therefore, if you are designing a heat exchanger and evaluate heat transfer coefficients from a mistaken correlation, you may be in serious trouble. I have already found typos in widely used and referenced books, which generate errors of some orders of magnitude. It is recommended to evaluate heat transfer coefficients from a few sources before picking the final value to apply in a project. Use at least two or three sources for a given task.

Considering a set of correlations claiming comparable accuracies, usually, the older and more reproduced is a heat transfer correlation, the better. From the time of its publication, the fact that a heat transfer correlation is often reprinted and referenced in textbooks, handbooks, technical manuals, etc., is an indication of higher reliability. An early correlation is more probable to have been used in several projects, providing a greater chance for problems to be found. Avoid the use of a recent or obscure heat transfer correlation as the single source.

4.3 Flow regimes

Convective heat transfer in internal flows is largely dependent on the flow regime, which can be laminar, turbulent, or assume some transition behavior in between the two. The heat transfer rate for a fluid in contact with a heated or cooled solid surface is significantly increased by the macroscopic mixing present in turbulent flow, which renews continuously the fluid particles adjacent to the heat transfer surface. The replacement of the fluid near the wall increases the local temperature difference, preventing the boundary layer from reaching the thermal equilibrium with the heat transfer surface.

The fundamental dimensionless number for classifying the flow regime is the Reynolds number. The shift from laminar to turbulent flow depends on boundary factors affecting the fluid motion, such as the wall roughness, mechanical vibration, or sudden deviations in the flow direction. For internal flow in circular pipes, the Reynolds number below 2,000 indicates a sustained laminar regime in most practical situations; however, this laminar threshold is often stretched to 2,300 [47]. In the range ~ 2,000 < Re < 4,000, the flow behavior swings unsteadily between laminar and turbulent, corresponding to the transition regime.

In nonisothermal flows, a significant viscous layer may persist near the wall even at 4,000 < Re < 10,000, impairing the convective heat transfer.[3] Because of

---

3 For a detailed explanation of this phenomenon, please refer to Ref. [50].

this, when heat transfer is involved, the transition regime may be considered to occur in the range of ~ 2,000 < Re < 10,000. A Reynolds number above 10,000 characterizes the fully turbulent regime, where only a very thin viscous sublayer remains adjacent to the solid wall.

4.4 Effects of fluid flow development

The flow of a fluid inside a pipe may be classified as developing or fully developed, hydrodynamically and/or thermally. In developing flows, the velocities and temperature profiles of the fluid change for each cross section of the pipe. This implies that the conditions for momentum and heat transfer inside the fluid vary from the entrance to a certain length of the pipe, which is commonly called "entrance length" or "entrance region."

The entrance length of laminar flow is significantly greater than for turbulent flow. For this reason, the common practice is to use results for fully developed flows in design calculations when the flow regime is turbulent, ignoring completely the hydrodynamic and thermal entrance lengths [47]. On the other hand, the entrance length should be taken into consideration if the flow field is laminar, accordingly the more accurate correlations for forced laminar convection evaluates the Nusselt or Stanton numbers as a function of the dimensionless number (D/L), where L is the pipe length.

In any case, to achieve higher accuracy, the engineer is recommended to identify if the flow is fully developed hydrodynamically and thermally for the approached design, to apply the proper methods for evaluating the friction factors and heat transfer coefficients.

For the laminar regime of a Newtonian fluid, the entrance length – i.e., the pipe length counted from the fluid inlet, in which the velocity profile is still deforming – may be estimated by the following equation [48]:

$$\frac{L_{fd}}{D_h} = 0.05 Re \qquad\qquad Re \leq 2{,}100 \ (laminar) \qquad\qquad (4.1)$$

D_h is the hydraulic diameter, which is the internal diameter D of a circular tube, Re is based on D_h, and L_{fd} is the fully developed entrance length.

Some texts recommend the fine-tuning of the coefficient to 0.04 (see [49]) and other suggests increasing it to 0.056 as in Ref. [18]. Since the transition from undeveloped to fully developed flow with the position along the pipe is smooth, these modifications of eq. (4.1) are immaterial for practical purposes.

The thermal entrance length for laminar flow is given by eq. (4.2) [50], where $L_{fd,T}$ is the fully developed entrance length for the thermal boundary layer:

$$\frac{L_{fd,T}}{D_h} = 0.05 Re Pr \qquad\qquad Re \leq 2100 \ (laminar) \qquad\qquad (4.2)$$

In very highly turbulent flow, the eddy diffusivity dominates the transport of energy and momentum, and the thermal entrance length is essentially not dependent on the Prandtl number. For practical situations, the turbulent thermal and hydrodynamic entrance lengths may be considered about the same, varying from 10 to 20 tube diameters. For more precise assessments, the hydrodynamic entrance length for turbulent flow can be evaluated as follows [18]:

$$\frac{L_{fd}}{D_h} = 1.359 Re^{0.25} \qquad\qquad Re > 10000 \ (turbulent) \qquad (4.3)$$

For turbulent flow, the thermal entrance length is estimated from the following equation [48]:

$$\frac{L_{fd,T}}{D_h} = 1.359 Re^{0.25} Pr \qquad\qquad Re > 10000 \ (turbulent) \qquad (4.4)$$

4.4.1 Fully developed laminar flow in pipes

Can we apply fully developed flow heat transfer coefficients for heat exchangers operating in the laminar regime? Well, the estimation of the heat transfer coefficient assuming fully developed flow is conservative for the design, i.e., the estimate is smaller than the real value. Therefore, it is a safer way to go, especially in situations where some degree of overdesign is allowed or even recommended. However, you should not forget that overdesign also means additional cost, which is generally undesirable.

Worked Example 4.1
A stream of $W = 0.58$ kg/s of 1-butanol at a mean bulk temperature of 93 °C is heated in a pipe with internal diameter ID = 0.0301m. What is the pipe length necessary to achieve fluid dynamic and thermal fully developed flows?

Solution
From Appendix C.9 the physical properties for the fluid at the mean bulk temperature are

Physical properties

1-Butanol, 366.15 K, 101,325 Pa

$\mu = 0.00063397$ s Pa
$\rho = 761.9$ kg/m^3
$Cp = 2,529.0$ J/(kg K)
$k = 0.1392$ W/(m K)
$Pr = 11.518$

The cross-flow area is

$$S = \frac{\pi D^2}{4} = \frac{\pi 0.0301^2}{4} = 0.00071158 \; m^2$$

The mass flux and Reynolds number are

$$G = \frac{W}{S} = \frac{0.58}{0.00071158} = 815.09 \frac{kg}{s m^2}$$

$$Re = \frac{DG}{\mu} = \frac{0.0301}{0.00063397} 815.09 = 38699.32 \; (turbulent \; flow)$$

For turbulent flow, the hydrodynamic entrance length is evaluated from eq. (4.3):

$$L_{fd} = 1.359D \sqrt[4]{Re} = 1.359 \cdot 0.0301 \sqrt[4]{38699.32} = 0.57m$$

And the thermal entrance length for turbulent regime is given by eq. (4.4) as follows:

$$L_{fd,T} = 1.359DPr \sqrt[4]{Re} = 1.359 \cdot 0.0301 \cdot 11.518 \sqrt[4]{38699.32} = 6.61m$$

Considering the previous results, we may observe that the high value of the Pr number causes a significantly slower development of the thermal boundary layer in comparison with the hydrodynamic boundary layer.

4.5 Heat transfer coefficients for circular pipes

The inner pipe of a double-pipe heat exchanger is just a circular duct. There is a great variety of correlations for predicting the heat transfer in forced convection for such geometry. We will discuss here the most used and proven correlations, considering their domain of validity, for the involved nondimensional numbers. The Reynolds number is commonly a reliable indicator of the fluid flow regime, which can be laminar, transitional, or turbulent. There is no single correlation able to predict with high accuracy the heat transfer coefficients for all fluid dynamic regimes; therefore, those correlations are typically bound to a fluid flow regime, delimited by a Reynolds number range. In some cases, the same mathematical formula is applied; however, different sets of parameters are used for each fluid dynamic region.

4.5.1 Laminar flow

4.5.1.1 Exact solutions for laminar fully developed flow
The convective heat transfer in laminar flow is suitable for certain analytical treatment and some important results may be derived, which provide insight for the development of correlations for other flow regimes, such as transitional and turbulent.

In laminar flow, the way the heat transfer occurs at the surface of the pipe, i.e., the thermal boundary conditions, is a major issue, having great impact on the value of the mean heat transfer coefficient. The most common boundary conditions we can find are:
1. Uniform heat transfer rate at the pipe wall
2. Uniform temperature of the pipe wall

Uniform wall temperature of the pipe
In this situation, the heat transfer surface of the pipe is maintained at a constant temperature, meaning that every cross section along the pipe axis has the same temperature on the wall. This boundary condition is appropriate, in good approximation, for heat exchangers where one of the fluids performs a phase change, such as evaporators and condensers. Similarly, liquid-to-gas heat exchangers [18], where the liquid flow rate is high, are well represented by this condition, since the gas typically is the controlling fluid and has a relatively low heat capacity, causing the wall temperature to stay very close to the liquid temperature.

Assuming fully developed laminar flow, negligible axial conduction, and that the fluid properties can be considered constant with a uniform wall temperature boundary condition, the governing differential equations of momentum and energy for the fluid can be solved directly[4] to yield a constant value for the Nusselt number based on the pipe's internal diameter D:

$$Nu = 3.658 \tag{4.5}$$

Fact sheet	
Fluids	Gases and liquids
Pr	>0.6
Properties	Evaluated using the local fluid bulk temperature
Conditions	– Laminar – Fully developed flow – Uniform wall temperature

The Péclet number (Pe) can be used to evaluate the importance of the fluid's axial conduction. For Pe > 10, the axial conduction is negligible [47]; however, for smaller values, a more accurate Nu is obtained from [51]

$$Nu = \begin{cases} 4.1807(1 - 0.0439Pe) & Pe < 1.5 \\ 3.6568\left(1 + 1.227/Pe^2\right) & Pe > 5 \end{cases} \tag{4.6}$$

---

4 See Ref. [124] for a detailed derivation of the solution.

Uniform heat flux at the pipe wall

A boundary condition of constant heat transfer rate in the wall holds when the local heat transfer flux (heat transfer rate per unit of peripheral pipe surface) is approximately constant along the axial direction of the pipe. Therefore, the lateral area corresponding to each cross section transfers the same amount of heat per unit time. A direct consequence of such thermal condition is that the pipe wall temperature does not remain constant axially. As the fluid flows in contact with the heat transfer surface, it gains or loses internal energy, changing its temperature, then the wall temperature must necessarily increase or decrease, respectively, to keep the heat flux at the same value. For this case, the Nusselt number is estimated by:

$$Nu = 4.363 \qquad (4.7)$$

Fact sheet	
Fluids	Gases and liquids
Pr	>0.6
Properties	Evaluated using the local fluid bulk temperature
Conditions	– Laminar – Fully developed flow – Uniform heat flux

4.5.1.2 Sieder–Tate

The empirical correlations proposed by Sieder and Tate are still among the most reliable available in the open technical literature and cover both laminar and turbulent flow regimes. They are based on over a hundred points of experimental data, measured from several fluids while heating and cooling in a double-pipe heat exchanger with an internal smooth copper tube, positioned in both horizontal and vertical. The authors also included experimental data available from literature and other sources at the time. The fluids were mostly hydrocarbon mixtures (crude oil cuts) but also included others, such as glycerol, benzene, kerosene, spindle oil, and water [62].

The correlation was proven to be accurate for gases and liquids in general; however, the heat transfer coefficient predictions for water suffer the larger deviations, once this fluid was a minor part of the experimental data set. The mean deviations of the correlation from experimental points for water were one order of magnitude greater than the deviations for oils, reaching +63% in the transition range (2,100 < Re < 10,000). For turbulent flow (Re > 10,000), the water deviations remained in a more acceptable value of ±10% [62]. In addition, apparently the equation overestimated consistently the water heat transfer coefficient for Re > 60,000, which leads to a tangible risk of nonconservative heat exchanger designs.

Despite the recommendations against the use of Sieder–Tate for water [44, 62], a more recent experimental study [59] measured the heating of water in smooth tubes for $10^4 < Re < 10^5$ and $6.0 < Pr < 11.6$, and performed an assessment of several popular correlations, including Sieder–Tate. It has been found that the Sieder and Tate turbulent correlation (eq. (4.10)) regularly underestimated their data in about – 5% to – 15%. Such result endorses the use of Sieder–Tate for water, by placing its predictions for water in the safe side for design. After all, if the designer avoids the flow transition range, the equation seems acceptable, even for water, since the deviation for higher Reynolds numbers is not large and might be accounted for by the provision of a proper safety factor.

Probably one major contribution of Sieder and Tate, apart from the correlations themselves, was the introduction of the dimensionless group $(\mu/\mu_w)^{0.14}$, a viscosity correction factor, to compensate the significant deformations of the fluid velocity profile when the heat transfer rate is large, and the physical properties of the fluid cannot be regarded as constants along the pipe. The viscosity correction factor allowed fitting Nusselt's experimental data for cooling and heating with a single unified curve. Earlier correlations, e.g., the Dittus–Boelter equation (4.9), were offered as two separated equations for cooling and heating, since the heat transfer data could not be adequately represented by a single functional form.

Sieder and Tate pointed out that their correlations are able to estimate local heat transfer coefficients, provided the physical properties are evaluated using the bulk and the wall temperature of the corresponding tube's cross section [62]. It is worth noting that the correlations presented herein (eqs. (4.8) and (4.10)) are almost invariably attributed to Sieder and Tate; nevertheless, they are not presented in the currently used forms in their original work [62].

The application of Sieder–Tate for gases deserves some caution. Despite the fact that there are important references which do not point out any warnings on its use for gases (e.g., [42, 44, 47, 63], among others), there are indications of poor predictions of this equation, for gases, see for example [56]. In addition, it is noticed that the original work of Sieder and Tate applied strictly experimental data from liquids (tab. 2 from Ref. [62]) in the correlation development.

In the laminar flow regime, the Sieder–Tate equation is the following equation, where L is the effective pipe length performing the heat transfer. All the properties should be evaluated at the fluid bulk temperature, except for the viscosity μ_w, which is based on the temperature of the heating or cooling surface T_w:

$$Nu = 1.86 \left(RePr \frac{D}{L} \right)^{1/3} \left(\frac{\mu}{\mu_w} \right)^{0.14} \tag{4.8}$$

Fact sheet	
Fluids	Gases, organic liquids, and aqueous solutions [44] and water [28, 59]
Re	<2,100
Pr	0.48–16,700 [62]
μ/μ_w	0.0044–9.75 [62]
(Re Pr $D/L)^{1/3}(\mu/\mu_w)^{0.14}$	>2
Properties	All properties were evaluated using the fluid bulk temperature T_b, except for μ_w, which is calculated at the wall temperature T_w. The average bulk temperature $(T_b = (T_{b,in} + T_{b,out})/2)$ may be used to calculate the mean heat transfer coefficient for the whole surface
Error	±12% in the range 100 < Re < 2,100 [44] ±20% for Re Pr D/L > 10 [43]
Conditions	– Heat transfer coefficients are estimated not conservatively for water [44]; however, there remains some controversy about this statement – Applicable for smooth and rough pipes – Variable fluid properties – Isothermal wall boundary condition – Ref. [63] asserts the applicability for uniform heat flux along pipe, although no supportive study is mentioned

Among the conditions of use of eq. (4.8) is that (Re Pr $D/L)^{1/3}(\mu/\mu_w)^{0.14}$ > 2. This is necessary, because the Nusselt number given in this correlation will approach zero for very long pipes ($D/L \rightarrow 0$), which is physically incorrect. In such case, the predicted Nusselt number is expected to approach Nu = 3.658, as estimated by eq. (4.5) for a fully developed laminar flow inside a pipe with uniform wall temperature.

Note that eq. (4.8) accounts for the entrance length of the tube, where the heat transfer coefficients reach higher values. In basic theory, the Nusselt number is infinite at the beginning of the tube and diminishes asymptotically along the tube axis to the fully developed flow value.

4.5.2 Turbulent flow

4.5.2.1 Dittus–Boelter
Dittus–Boelter [52] gave the precursor form of his famous equation in 1930s. It is one of the earliest equations for the prediction of convective heat transfer coefficients in pipe flow, being originally developed for the design of car radiators in the automotive industry. Because it was published a long time ago, this equation is

presented in almost all text and reference books, being widely known by engineers and practitioners, and probably in great use even these days.

Several versions of this correlation may be found in the literature. Usually, the structure of the equation is preserved, and the coefficients and powers are modified. Apparently, the most famous (and current) form of the Dittus–Boelter equation is due to McAdams [53], which recalibrated its coefficient from the original value of 0.0243–0.023 [54].

The modern formulation of the Dittus–Boelter correlation (with the McAdams' modifications [53]) is:

$$Nu = 0.023 Re^{0.8} Pr^n \qquad (4.9)$$

$$n = 0.4 \; (heating, T_w > T_b)$$
$$n = 0.3 \; (cooling, T_w < T_b)$$

Fact sheet[5]	
Fluids	Gases and liquids
Re	10^4–10^7
Pr	0.7–100
L/D	>60
Properties	Evaluated using the average of the fluid bulk temperature: $T_b = (T_{b,in} + T_{b,out})/2$
Error	±25% from correlated data −10% to −25% for gases and +40% for liquids, confirmed by later studies
Conditions	– Turbulent – Fully developed flow – Constant fluid properties (small-to-moderate temperature difference T_b–T_w and low heat transfer rate) – Both uniform wall temperature and uniform heat flux (with deviation of ±25% of correlated experimental data) – Smooth pipes

There is some disagreement about the validity domain of this relation, see, for example, [42, 47, 48, 50, 55–58]. Sometimes the discordance is on the Prandtl number, where the minimum value ranges from 0.5 to 0.7, and the maximum valid Prandtl is given as 100, 120, 160, or even 170, depending on the source. The lowest recommended

---

5 The Reynolds range of applicability varies significantly in the literature, for example, Ref. [50] recommends a wider range of $Re = 6 \times 10^3 - 10^7$, and Ref. [48] recommends a Prandtl range of $Pr = 0.7–120$. Therefore, we are going conservative here by recommending a narrower validity domain to reduce deviations and consequent design risk.

Reynolds number is also not settled, being reported as 2,500, 6,000, or 10,000. The later divergence is of special importance, since it endorses the use of this equation in the transition flow regime (roughly Re = 2,000 – 10,000 for pipes), which is strongly NOT recommended.

The correlation is appropriate for uniform or constant fluid properties, i.e., the temperature difference between the entrance and exit of the fluid should not be large. In other words, the rate of heating or cooling of the fluid must not be capable of deforming the velocity distribution from the isothermal conformation. All physical properties are to be evaluated at the fluid bulk temperature of the fluid T_b, whereas an average bulk temperature along the pipe ($T_b = (T_{b,in} + T_{b,out})/2$) may be used to find an average heat transfer coefficient for the whole heat transfer surface. For rough calculations, the exponent $n = 0.4$ can be used for both heating and cooling [54].

The application of Dittus–Boelter correlation (eq. (4.9)) over the years revealed some inaccuracies for certain ranges of Reynolds and Prandtl numbers. It was reported that this correlation may overestimate the heat transfer coefficient by 10–25% for gases, and may underestimate to about 40% the heat transfer coefficient for liquids [56]. Another rigorous study has confirmed underestimations of 5–15% for water flowing at several Prandtl values [59]. Additionally, experiments from Ref. [60] revealed underpredictions of 2–12% and 20% for water and oil, respectively. Reference [61] confirmed predictions with –20% deviations for water.

Such deviations, especially the overvaluing error, pose a critical risk while sizing or analyzing a heat exchanger. In compensation for the possibility of overestimation of the heat transfer coefficient, the designer should preventively apply a generous safety factor to his/her calculations. For that reason, nowadays it is advised against the use of this equation, except with the purpose of cross-checking along with other predictions of the heat transfer coefficient in question. Of course, in the absence of more reliable correlation or experimental data, the Dittus–Boelter equation is satisfactorily dependable for practical use, provided an appropriate safety factor is in place.

4.5.2.2 Sieder–Tate

For fully developed turbulent flow, use eq. (4.10), which follows the same application rules of the Sieder–Tate correlation for the laminar regime (eq. (4.8)) [62]:

$$Nu = 0.027 Re^{0.8} Pr^{1/3} \left(\frac{\mu}{\mu_w} \right)^{0.14} \tag{4.10}$$

From the reasoning that eq. (4.8) is based mostly on viscous liquids, and also considering additional experimental evidence not available at the publication time of

Fact sheet	
Fluids	Gases,[6] organic liquids, and aqueous solutions [44] and water ([28, 59])
Re	$>10^4$
Pr	0.48–16,700 [62]
L/D	>10 [43]
μ/μ_w	0.0044–9.75 [62]
Properties	All properties were evaluated using the fluid bulk temperature T_b, except for μ_w, which is calculated at the wall temperature T_w. The average bulk temperature ($T_b = (T_{b,in} + T_{b,out})/2$) may be used to calculate the mean heat transfer coefficient for the whole surface
Error	−10% to +15% for Re > 10^4 [44] ±20% in the range 10^4 < Re < 10^6 and 0.6 < Pr < 100 [43]
Conditions	– Heat transfer coefficients are estimated not conservatively for water [44]; however, there remains some controversy about this statement – Applicable for smooth and rough pipes – Fully developed flow – Variable fluid properties – Isothermal wall boundary condition – Reference [63] asserts the applicability for uniform heat flux along the pipe, although no supportive study is mentioned

the original Sieder and Tate equations, fine-tuning of the coefficient is proposed [19] in the following manner:
– 0.027 for more viscous liquids,
– 0.023 for less viscous liquids, and
– 0.021 for gases.

Although the change of the coefficient according to the fluid category may yield some accuracy improvement in the prediction of the Nusselt number, to some engineers, this gain seems not worth for normal practical applications. It is important to warn, however, that the small absolute difference between these coefficients is quite misleading. In fact, considering the larger range from 0.021 to 0.027, we are talking about a variation of nearly 30% in the heat transfer coefficient!

---

6 There are a few contraindications against the use of this correlation with gases in the technical literature; see, for example, [56].

4.5.2.3 Petukhov–Popov

The Petukhov–Popov correlation is a remarkable instance of successful theoretical development in the turbulent heat transfer arena. Based on an elaborated analytical and numerical procedure, Petukhov and Popov [64] proposed a solution for the turbulent heat transfer in fully developed flow in a smooth tube, assuming constant physical properties and uniform heat flux at the wall. Their correlation is in fact an interpolation formula; however, the correlated database was entirely produced by their calculations and validated later with selected experimental data.

The assumed hypotheses limited the results to the range of $10^4 < Re < 5 \times 10^6$ and $0.5 < Pr < 2{,}000$, covering gases, and nonviscous and viscous liquids, except liquid metals flowing in fully turbulent regime. Although the calculations presumed the uniform heat flux wall condition, it was shown previously [65–67] that within the assumed Pr and Re ranges, the difference between the results for uniform heat flux and uniform temperature at the pipe wall is immaterial.

For the case in which the fluid properties can be taken as approximately constants along the pipe, they proposed an equation resembling the original Prandtl correlation [20] in the form:

$$Nu = \frac{(f_F/2)RePr}{K_1 + K_2(f_F/2)^{1/2}(Pr^{2/3} - 1)} \tag{4.11}$$

$$K_1 = 1 + 13.6 f_F$$

$$K_2 = 11.7 + 1.8/Pr^{1/3}$$

where f_F is the Fanning friction factor for constant fluid properties in smooth pipes given by Filonenko's equation [68]:

$$f_F = \frac{1}{4}(0.79 \ln(Re) - 1.64)^{-2} \tag{4.12}$$

Fact sheet	
Fluids	Gases and liquids, except liquid metals
Re	$10^4 - 5 \times 10^6$
Pr	0.5–2,000
Properties	All properties were evaluated using the fluid bulk temperature T_b. The average bulk temperature ($T_b = (T_{b,in} + T_{b,out})/2$) may be used to calculate the mean heat transfer coefficient for the whole surface
Conditions	– Applicable for smooth and rough pipes – Fully developed flow – Constant fluid properties – Isothermal wall boundary condition – Uniform heat flux at the wall

(continued)

Fact sheet	
Error[7]	Under ±1% from theoretical predictions for $10^4 < Re < 5 \times 10^5$ and $0.5 < Pr < 200$
	±1–2% from theoretical predictions in the range $5 \times 10^5 < Re < 5 \times 10^6$ and $200 < Pr < 2{,}000$
	±5–6% from experimental data tested in the range $10^4 < Re < 4 \times 10^5$ and $1 < Pr < 100$

Any correlation other than eq. (4.12) may be used for the calculation of the friction factor coefficient; thus, eq. (4.11) is not restricted to smooth tubes. If an appropriate friction factor for rough tubes is used, it remains valid for estimating the heat transfer coefficient.

Liquids and gases with constant properties (compact form)

$$Nu = \frac{(f_F/2)RePr}{1.07 + 12.7(f_F/2)^{1/2}(Pr^{2/3} - 1)} \qquad (4.13)$$

In eq. (4.13), f_F is the Fanning friction factor for constant fluid properties in smooth pipes given by Filonenko's equation [68]:

$$f_F = \frac{1}{4}(0.79 \ln(Re) - 1.64)^{-2} \qquad (4.14)$$

Fact sheet	
Fluids	Gases and liquids, except liquid metals
Re	$10^4 - 5 \times 10^6$
Pr	0.5–2,000
Properties	All properties were evaluated using the fluid bulk temperature T_b. The average bulk temperature ($T_b = (T_{b,in} + T_{b,out})/2$) may be used to calculate the mean heat transfer coefficient for the whole surface
Conditions	– Applicable for smooth and rough pipes
	– Fully developed flow
	– Constant fluid properties
	– Isothermal wall boundary condition
	– Uniform heat flux at the wall

---

7 From Ref. [72].

(continued)

Fact sheet	
Error[8]	$\pm 5\text{–}6\%$ from experimental data tested in the range $10^4 < Re < 5 \times 10^6$ and $0.5 < Pr < 200$
	$\pm 10\%$ from experimental data tested in the range $10^4 < Re < 5 \times 10^6$ and $0.5 < Pr < 2{,}000$

Alternative correlations for estimating the friction factor instead of eq. (4.14) may be used with eq. (4.13); therefore, it is not restricted to smooth tubes. It remains valid for estimating the heat transfer coefficient using an appropriate friction factor equation for rough tubes.

Variable physical properties

Heat transfer coefficient in liquids with variable properties
The Nusselt numbers given by eqs. (4.11) and (4.13) can be corrected to consider the effect of varying physical properties using the following factor:

$$Nu = Nu_0 \left(\frac{\mu}{\mu_w}\right)^n \tag{4.15}$$

$$n = \begin{cases} 0.11 & (\textit{heating}) \\ 0.25 & (\textit{cooling}) \end{cases}$$

where Nu_0 is the Nusselt number for the condition of constant fluid properties, μ and μ_w are the fluid viscosities at the bulk and wall temperatures, respectively.

Fact sheet[9]	
Fluids	Liquids, except liquid metals
Re	$10^4 \text{–} 1.25 \times 10^5$
Pr	$2\text{–}140$
μ_w/μ	$0.08\text{–}40$
μ/μ_w	$0.025\text{–}12.5$

---

8 From Ref. [72].
9 From Ref. [72].

(continued)

Fact sheet[9]	
Properties	μ is evaluated at bulk temperature T_b and μ_w at the wall temperature T_w. The average temperatures $T_b = (T_{b,in} + T_{b,out})/2$ and $T_w = (T_{w,in} + T_{w,out})/2$ may be used to calculate the mean heat transfer coefficient for the whole surface of the tube.
Conditions	– Applicable for smooth and rough pipes – Fully developed flow – Variable liquid properties – Isothermal wall boundary condition – Uniform heat flux at the wall

Heat transfer coefficient in gases with variable properties (compact form)

$$Nu = Nu_0 \left(\frac{T}{T_w}\right)^n \tag{4.16}$$

$$n = \begin{cases} 0.11 \ (heating) \\ 0.25 \ (cooling) \end{cases}$$

where Nu_0 is the Nusselt number for the condition of constant fluid properties, T and T_w are the bulk and wall temperatures, respectively.

Fact sheet[10]	
Fluids	Gases
Re	10^4–10^6
Pr	2–140
T_w/T	0.37–3.1
T/T_w	0.32–2.70
Conditions	– Smooth and rough pipes – Fully developed flow – Variable physical properties – Isothermal wall boundary condition – Uniform heat flux at the wall
Error	±6% from analytical results

---

10 From Ref. [72].

Friction factor for liquids with variable properties

$$f_F = f_{FO} \; \frac{1}{6}\left(7 - \frac{\mu}{\mu_w}\right) \qquad \frac{\mu}{\mu_w} > 1 \; (heating)$$

(4.17)

$$f_F = f_{FO} \; \left(\frac{\mu}{\mu_w}\right)^{-0.24} \qquad \frac{\mu}{\mu_w} < 1 \; (cooling)$$

where f_{FO} is the Fanning friction factor for the condition of constant fluid properties.

Fact sheet[11]	
Fluids	Liquids, except liquid metals
Re	$10^4 – 2.3 \times 10^5$
Pr	1.3–10
μ/μ_w	0.5–2.86
Properties	μ is evaluated at bulk temperature T_b and μ_w at the wall temperature T_w. The average temperatures $T_b = (T_{b,in} + T_{b,out})/2$ and $T_w = (T_{w,in} + T_{w,out})/2$ may be used to calculate the mean heat transfer coefficient for the whole surface of the tube.
Conditions	– Smooth and rough pipes – Fully developed flow – Variable liquid properties – Isothermal wall boundary condition – Uniform heat flux at the wall

Friction factor for gases with variable properties (compact form)

$$f_F = f_{FO} \left(\frac{T}{T_w}\right)^n$$

(4.18)

$$n = \begin{cases} 0.52 \; (heating) \\ 0.38 \; (cooling) \end{cases}$$

---

11 From Ref. [72].

Fact sheet[12]	
Fluids	Gases
T_w/T	0.37–3.7
T/T_w	0.27–2.7
Conditions	− Smooth and rough pipes − Fully developed flow − Variable physical properties − Isothermal wall boundary condition − Uniform heat flux at the wall
Error	±7% from analytical results for heating ±4% from analytical results for cooling

4.5.2.4 Notter–Sleicher

Notter–Sleicher [56, 69] solved numerically the energy equation for the nonisothermal fluid flow in a circular pipe with boundary conditions of uniform temperature and uniform heat flux at the wall. The determination of the two-dimensional temperature distribution of the flow field required the assumption of constant fluid properties and the use of empirical data for the turbulent velocity profile and eddy diffusivity distribution in the energy differential equation. The obtained numerical results were correlated within ±10% of deviation in the range $10^4 < Re < 10^6$ and 0.1 $< Pr < 10^4$. The authors also performed validation with available experimental data. The Notter–Sleicher fitted equation was given as follows:

$$Nu = 5 + 0.015 Re^a Pr^b$$
$$a = 0.88 - \frac{0.24}{(4 + Pr)}$$
$$b = \frac{1}{3} + 0.5e^{-0.6Pr}$$

(4.19)

Fact sheet	
Fluids	Gases and liquids
Re	10^4–10^6
Pr	0.1–10^4
Properties	Evaluated at bulk temperature T_b. The average bulk temperature $T_b = (T_{b,in} + T_{b,out})/2$ may be used to calculate the mean heat transfer coefficient for the whole surface of the tube.

---

12 From Ref. [72].

(continued)

Fact sheet	
Conditions	– Constant fluid properties – Smooth pipes – Fully developed flow – Isothermal wall boundary condition – Uniform heat flux at the wall
Error	±10% from analytical results ±10% from experimental data tested for high Prandtl numbers Negligible deviation from tested experimental data for gases

Notice that eq. (4.19) has a Pr range able to account for liquid metals. A more accurate and simplified version of eq. (4.19) specifically for gases takes the following form, which correlates the source data within only 4% of deviation:

$$Nu = 5 + 0.012Re^{0.83}(Pr + 0.29) \tag{4.20}$$

Fact sheet	
Fluids	Gases
Re	10^4–10^6
Pr	0.6–0.9
Properties	Evaluated at bulk temperature T_b. The average bulk temperature $T_b = (T_{b,in} + T_{b,out})/2$ may be used to calculate the mean heat transfer coefficient for the whole surface of the tube.
Conditions	– Constant fluid properties – Smooth pipes – Fully developed flow – Isothermal wall boundary condition – Uniform heat flux at the wall
Error	±4% from analytical results

Variable physical properties

A few years after the publication of Notter–Sleicher equation, the Sleicher–Rouse correlation was proposed as an adaptation for use with nonuniform properties fluid flows [56]. The Sleicher–Rouse equation (4.21) is essentially the same as the Notter–Sleicher's, but with a key difference: the evaluation of physical properties at several temperatures, namely, the bulk, film, and wall temperatures.

Notice that eq. (4.21) can be applied to uniform fluid properties and for variable fluid properties; however, the domain of validity for Re_f and Pr_w is further restricted when the fluid properties change significantly along the pipe. The authors reported a maximum deviation of ±20% and an average deviation of about ±7% from the experimental data set. The correlated heat transfer experimental data was limited to measurements of local heat transfer coefficients; therefore, the correlation is supposed to be adequate for local heat transfer evaluations:

$$Nu_b = 5 + 0.015Re_f^a Pr_w^b \qquad (4.21)$$

$$a = 0.88 - \frac{0.24}{(4 + Pr_w)}$$

$$b = \frac{1}{3} + 0.5e^{-0.6Pr_w}$$

where Nu_b is the Nusselt number with fluid properties at bulk temperature, Re_f is the Reynolds number with properties at the film temperature, and Pr_w is the Prandtl number evaluated at the wall temperature.

Fact sheet	
Fluids	Liquids and gases
Re_f	10^4–10^6 for constant properties fluids 10^4–5×10^5 for variable properties fluids[13]
Pr_w	0.1–10^4 for constant properties fluids 0.7–75 for variable properties fluids[13]
Properties	Nu_b, Re_f, and Pr_w are evaluated at bulk, film, and wall temperatures, respectively. The average temperatures for the whole tube length may be used to calculate the mean heat transfer coefficient for the whole surface of the tube.
Conditions	– Constant fluid properties – Variable fluid properties – Smooth pipes – Fully developed flow – Isothermal wall boundary condition – Uniform heat flux at the wall
Error	±20% maximum deviation in some points from correlated experimental data ±6.9% average deviation from correlated experimental data

---

[13] This range of validity seems to be restricted by the experimental data for varying fluid properties available to the authors at that time.

The modified correlation for gas flow with nonuniform properties is

$$Nu_b = 5 + 0.012 Re_f^{0.83}(Pr_w + 0.29) \tag{4.22}$$

Fact sheet	
Fluids	Gases
Re_f	10^4–10^6 for constant properties of gases 10^4–5×10^5 for variable properties of gases13
Pr_w	0.6–0.9
Properties	Nu_b, Re_f, and Pr_w are evaluated at bulk, film, and wall temperatures, respectively. The average temperatures for the whole tube length may be used to calculate the mean heat transfer coefficient for the whole surface of the tube.
Conditions	– Constant fluid properties – Variable fluid properties – Smooth pipes – Fully developed flow – Isothermal wall boundary condition – Uniform heat flux at the wall
Error	±6.9% average deviation from correlated experimental data

4.5.2.5 Gnielinski

Gnielinski ([70] in [20]) modified the theoretical equation from Petukhov–Kirillov by adding the Hausen corrector ([71] in [45]) to consider the effect of the entrance length on the heat transfer coefficient.

The deformation of the fluid velocity and temperature profiles due to nonisothermal variations of the physical properties were taken into account with a factor in the same fashion proposed by Petukhov [72]; however, this factor was recalibrated in order to correlate simultaneously the heat transfer data for heating and cooling.

The Gnielinski correlation (eq. (4.23)) seems to be increasing in popularity. There is not complete consensus, but important references [48] consider it the most accurate correlation at the present time, while others like Ref. [47] consider the Petukhov's equations the best, due to its solid theoretical grounds.

$$Nu = \frac{0.5 f_F (Re - 1000)\, Pr}{1 + 8.98 f_F^{1/2}(Pr^{2/3} - 1)} \left(1 + \left(\frac{D}{L}\right)^{2/3}\right) \phi \tag{4.23}$$

where f_F is the Fanning friction factor for constant fluid properties in smooth pipes given by Filonenko's equation [68]:

$$f_F = \frac{1}{4}(0.79 \ln(Re) - 1.64)^{-2} \tag{4.24}$$

Variable physical properties

The recommended correction factor for variable properties flow is distinct for liquids or gases, given as:

$$\phi = \begin{cases} (Pr/Pr_w)^{0.11} & (liquids) \\ (T/T_w)^{0.45} & (gases) \end{cases} \tag{4.25}$$

where the Prandtl numbers Pr and Pr_w are evaluated at the fluid bulk temperature (T) and wall temperature (T_w), respectively.

Fact sheet	
Fluids	Gases and liquids
Re	2,300–10^6 [20]
Pr	0.6–2,000 (for constant fluid properties) 0.6–200 (for variable fluid properties) [50]
D/L	<1
Pr/Pr_w	0.1–10 [20]
T/T_w	0–0.5 [20]
Properties	All properties were evaluated using the fluid bulk temperature T_b. The average bulk temperature ($T_b = (T_{b,in} + T_{b,out})/2$) may be used to calculate the mean heat transfer coefficient for the whole surface [45]
Conditions	– Applicable for smooth and rough pipes – Fully developed and developing flow – Constant and variable fluid properties – Isothermal wall boundary condition – Uniform heat flux at the wall
Error[14]	±20% from tested experimental data [20]

This correlation was extensively verified against experimental data and presented a very good agreement with the Prandtl number varying from 0.6 to 1,000 [70].

The Sieder–Tate correction factor for variable physical properties may also be used with eq. (4.23) [48]:

$$\phi = (\mu/\mu_w)^{0.14} \tag{4.26}$$

---

14 From Ref. [20].

Fact sheet	
Fluids	Gases and liquids
μ/μ_w	1–40 [48]

In practice, the difference between the Prandtl-based and viscosity-based correction factor is usually small; therefore, if both factors are within the range of validity, the choice is merely a matter of taste.

Worked Example 4.2

A stream of 1.7 m/s of heptanol is heated from 30 to 50 °C in the inner pipe of a heat exchanger with 5 cm ID and 3 m long. The pipe internal surface is maintained at $T_w = 90$ °C. Estimate the Nusselt number using eq. (4.23) and evaluate the error introduced if the flow is assumed to be fully developed.

Solution

From Appendix C.9, the physical properties of 1-heptanol at the mean bulk ($T = 40$ °C) and wall temperatures are:

Physical properties: 1-heptanol, 101,325 Pa
$T = 313.15$ K
$\rho = 805.72$ kg/m^3 $k = 0.15563$ W/(m · K) $\mu = 0.0036201$ s · Pa   Pr = 47.706
$T_w = 363.15$ K
$\mu_w = 0.0010822$ s · Pa   Pr$_w$ = 16.867

The Reynolds number can be calculated as:

$$Re = \frac{D\rho}{\mu}v = \frac{0.050000}{0.0036201}1.7000 \cdot 805.72 = 18918. \rightarrow \textit{fully turbulent}$$

This result indicates that the flow regime is fully turbulent; therefore, eq. (4.23) is within its domain of validity and can be used for the Nusselt estimation.

Using Filonenko's correlation (eq. (4.24)), the Fanning friction factor is

$$f_F = \frac{0.25}{(0.79 \ln(Re) - 1.64)^2} = \frac{0.25}{(0.79 \ln(18918.) - 1.64)^2} = 0.0066229$$

And the correction factor accounting for variation of physical properties in liquids is given by eq. (4.25):

$$\phi = \left(\frac{Pr}{Pr_w}\right)^{0.11} = \left(\frac{47.706}{16.867}\right)^{0.11} = 1.1212$$

Combining the last results into eq. (4.23), the Nusselt for a noncompletely developed flow is

$$Nu = \frac{0.5 Pr f_F (Re - 1000.0)}{8.98 f_F^{0.5} \left(Pr^{\frac{2}{3}} - 1.0\right) + 1.0} \left(\left(\frac{D}{L}\right)^{\frac{2}{3}} + 1.0\right) \phi$$

$$= \frac{0.5 \cdot 0.0066229 \cdot 1.1212 \cdot 47.706 (18918. - 1000.0) \left(\left(\frac{0.050000}{3.0000}\right)^{\frac{2}{3}} + 1.0\right)}{8.98 \cdot 0.0066229^{0.5} \left(47.706^{\frac{2}{3}} - 1.0\right) + 1.0} = 342.11$$

The Nusselt number for fully developed flow (Nu_{fd}) may be obtained simply by setting $D/L = 0$ in eq. (4.23), then

$$Nu_{fd} = \frac{0.5 Pr f_F (Re - 1000.0)\phi}{8.98 f_F^{0.5} \left(Pr^{\frac{2}{3}} - 1.0\right) + 1.0} = \frac{0.5 \cdot 0.0066229 \cdot 47.706 (18918. - 1000.0) \cdot 1.1212}{8.98 \cdot 0.0066229^{0.5} \left(47.706^{\frac{2}{3}} - 1.0\right) + 1.0} = 321.16$$

The relative deviation from the Nusselt number accounting for the effect of the entrance length on the heat transfer is

$$Error \ (\%) = \frac{100}{Nu} \left(Nu_{fd} - Nu\right) = \frac{100}{342.11} (321.16 - 342.11) = -6.1238\%$$

Conclusion

By ignoring the flow development length, the Nusselt number is underestimated in about 6%.

4.5.3 Transition flow

The design of a heat exchanger operating in the transition region between laminar and turbulent flow is not recommended. Do that only if it is strictly necessary. The fluctuations of the friction factor and Nusselt number in the transitional flow with very small changes in the Reynolds number are relatively higher than those for laminar or turbulent flow. Such amplified fluctuations may cause operational instabilities resulting from minor variations of the flow rate in the process stream, which of course is undesirable and poses an additional difficulty for the equipment control.

The critical Reynolds number for fluid flow in rounded pipes is Re = 2,300. The transition regime for circular tubes in normal process conditions is usually defined with the Reynolds number in the range $2,300 < Re < 10^4$. For $Re > 10^4$, it is generally safe to assume that the flow regime is fully turbulent. Although with mechanical vibrations on the pipeline, it is possible to find transitional flow at $2,000 < Re < 2,300$, if $Re < 2,000$, the flow field returns to laminar, even under strong intermittent vibrations [73].

It is of major importance for the engineer to correctly identify the flow regime under which each stream of the heat exchanger will operate, to calculate appropriately the heat transfer coefficients. For example, if the flow regime used in the design is turbulent and the actual flow regime when the equipment is in production is laminar, then certainly the heat exchanger will underperform and possibly not match the heat transfer duty. The behavior of a turbulent flow, and consequently the heat transfer coefficient, depends on some aspects such as the configuration of the tube entrance,[15] the roughness of the tube's internal surface, and disturbances in the fluid velocity. These elements are of minor importance if the flow regime is laminar.

There are several methods of heat transfer coefficient estimation for the transition flow regime, and some of them are presented further.

4.5.3.1 Gnielinski correlation
The Gnielinski equation (4.23) combined with eqs. (4.24) and (4.25) may be used for estimating Nusselt numbers in the transition range within ±20% of accuracy.

4.5.3.2 Nusselt number interpolation
The Nusselt number for transition flow may be approximated by simple interpolation between the limiting Nusselt values for laminar and turbulent regimes. This method was verified to agree very well with experimental data [45].

The Nusselt number for any transition Reynolds is calculated from

$$Nu = (1 - y) Nu_{(2300)} + y Nu_{(10^4)} \qquad (4.27)$$

with y being given by

$$y = \frac{Re - 2300}{10^4 - 2300}, 0 \le y \le 1 \qquad (4.28)$$

$Nu_{(2,300)}$ and $Nu_{(10^4)}$ are the average Nusselt numbers for $Re = 2,300$ and $Re = 10^4$, respectively, evaluated with any appropriate equation. If the pipe is assumed at constant wall temperature, the average Nusselt number of laminar flow with $Re = 2,300$ in a tube of length L can be calculated from equation (4.29) [45].

---

15 Methods of estimation of Nusselt number for various entrance shapes may be found in Ref. [47].

$$Nu_{(2300)} = 53.8 \left(\left(\frac{DPr}{L} \right)^{3/2} \left(21.3 \left(\frac{DPr}{L} \right)^{1/3} - 0.7 \right)^3 \left(\frac{1}{22Pr+1} \right)^{1/2} \right)^{1/3} \tag{4.29}$$

On its turn, for turbulent flow at either constant wall temperature or constant heat flux, $Nu_{(10^4)}$ may be taken, for example, from eqs. (4.11) or (4.23).

4.5.3.3 Hausen correlation

The Hausen equation (4.30) is frequently recommended for the transition flow regime for Reynolds in the range $2{,}100 < Re < 10^4$, given by [74].

$$Nu = 0.116 \left(Re^{2/3} - 125 \right) Pr^{1/3} \left(\frac{\mu}{\mu_w} \right)^{0.14} \left(1 + \left(\frac{D}{L} \right)^{2/3} \right) \tag{4.30}$$

Fact sheet	
Fluids	Liquids and gases
Re	$2{,}100\text{--}10^4$
Properties	All properties were evaluated using the fluid bulk temperature T_b, except for μ_w, which is calculated at the wall temperature T_w. The average bulk temperature $(T_b = (T_{b,in} + T_{b,out})/2)$ may be used to calculate the mean heat transfer coefficient for the whole surface
Conditions	– Transitional flow – Smooth pipe

4.5.4 Heat transfer coefficients for water

The heat transfer coefficient of water flow in tubes can be estimated within practical accurateness from general equations like Petukhov–Popov (eqs. (4.11) and (4.13)) or Gnielinski (eq. (4.23)). However, a more accurate estimation is attainable through specific data for water, such as provided by Eagle and Ferguson [75]. Empirical correlations based on Eagle–Ferguson measurements are presented in eqs. (4.31) or (4.32).[16] Notice that those equations do not use nondimensional groups, as is usual. On the contrary, they are tied to a set of units, and the convective heat transfer coefficient (h) is a function of the water velocity (v) and temperature (T).

---

16 From Ref. [58].

In the SI system, we have:

$$h = 1057(1.352 + 0.02T)\frac{v^{0.8}}{D^{0.2}} \qquad (4.31)$$

$$[h] = W/m^2{}^\circ C$$
$$[v] = m/s$$
$$[D] = m$$
$$[T] = {}^\circ C$$

Alternatively, in US customary units:

$$h = 0.13(1 + 0.011T)\frac{v^{0.8}}{D^{0.2}} \qquad (4.32)$$

$$[h] = Btu/hft^2{}^\circ F$$
$$[v] = ft/s$$
$$[D] = ft$$
$$[T] = {}^\circ F$$

Fact sheet

Fluids	Liquid water
T	5–104 °C (40–220 °F)

4.5.5 Liquid metals

For some applications, liquid metals may be the most suitable option as the cooling fluid. The low vapor pressure at high temperatures combined with the greater heat capacity usually leads to more cost-effective heat exchanger, with smaller size and operating pressure. The use of liquid metals as a heat transfer fluid is possible for most types of heat exchangers, including the double-pipe type; however, some caution should be taken with welded and brazed heat exchangers, since increased chemical corrosion could take place when the liquid metal contacts some alloys [76].

Axial thermal diffusion becomes an important mechanism of heat transfer in the forced convection of liquid metals. The thermal conductivity of liquid metals is distinctly high in comparison with other fluids. In accordance, liquid metals are characterized by very low Prandtl numbers (Pr), being typical values in the range of Pr = 0.001–0.1.

The convective heat transfer coefficient in liquid metals is strongly dependent on the Péclet number (Pe). Most correlations for forced convection in internal flows give the Nusselt number as a function of the Reynolds number (Re) and Prandtl number (Pr) separately, i.e., Nu = f(Re,Pr). For liquid metals, the Péclet number can accurately explain Nu experimental data in many cases; therefore, it is also common to correlate the Nusselt number as a sole function of the Péclet number (i.e., Nu = f(Pe)), which is a combination of both Re and Pr, defined as Pe = RePr. The Péclet number provides a measure of the relative importance of the heat transported in the form of internal energy while the fluid moves, compared to the heat transferred by thermal diffusion in the flow direction. The smaller the Péclet number, the greater is the role of thermal heat conduction for the convective heat transfer rate.

4.5.5.1 Notter–Sleicher

Semiempirical equations based on a combination of experimental data and numerical analysis were proposed by Notter and Sleicher [69] for the boundary conditions of uniform heat flux and uniform temperature at the pipe wall.

Uniform wall heat flux

$$Nu = 6.3 + 0.0167Re^{0.85}Pr^{0.93} \tag{4.33}$$

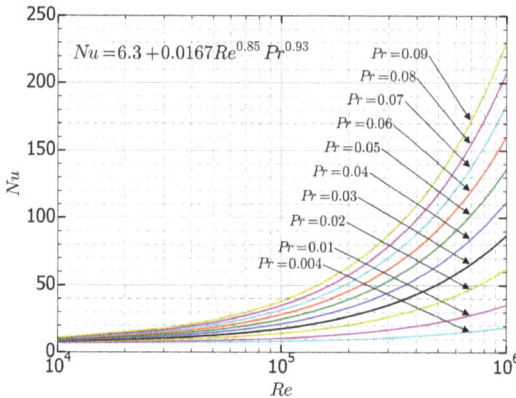

Fig. 4.1: Nusselt numbers for liquid metals in fully developed flow and uniform heat flux boundary condition (eq. (4.33)).

Uniform wall temperature

$$Nu = 4.8 + 0.0156Re^{0.85}Pr^{0.93} \tag{4.34}$$

Fact sheet	
Fluids	Liquid metals in general
Re	10^4–10^6
Pr	$0 - 0.1$
L/D	>10
Properties	Evaluated using the average of the fluid bulk temperature: $T_b = (T_{b,in} + T_{b,out})/2$
Error	±5% in the range $0.004 < \text{Pr} < 0.1$ and $10^4 < \text{Re} < 5 \times 10^5$ ±10% in the range $0 < \text{Pr} < 0.1$ and $10^4 < \text{Re} < 10^6$
Conditions	– Turbulent – Fully developed flow – Small value of the difference $T_b - T_w$ – Smooth pipe

The accuracy of correlations (4.33) and (4.34) can be improved to a maximum deviation of 5% when applied within a narrower range of $0.004 < \text{Pr} < 0.1$ and $10^4 < \text{Re} < 5 \times 10^5$.

For convenient reading of Nusselt values, a graphic representation of eqs. (4.33) and (4.34) is given in Figs. 4.1 and 4.2, respectively.

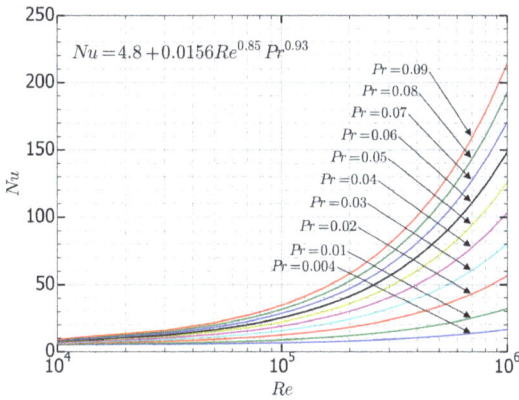

Fig. 4.2: Nusselt numbers for liquid metals in fully developed flow and uniform wall temperature boundary condition (eq. (4.34)).

4.5.5.2 Skupinski, Tortel, and Vautrey

This is an empirical correlation based on experiments performed with an alloy of 44% sodium and 56% potassium (NaK) under fully developed turbulent flow with uniform heat flux at the wall. The Nusselt number evaluated is local with the properties evaluated at the local bulk temperature. An average Nusselt number corresponding to the entire heat transfer surface may be estimated using the mean bulk temperature for evaluating the fluid properties [77]:

$$Nu = 4.82 + 0.0185(RePr)^{0.827}$$

or   (4.35)

$$Nu = 4.82 + 0.0185(Pe)^{0.827}$$

where Pe is the Péclet number using the pipe diameter as the characteristic length.

Fact sheet	
Fluids	Liquid metals in general
Re	$3.6 \times 10^3 - 9.05 \times 10^5$
Pr	0.003–0.05
Pe	$10^2 - 10^4$
Conditions	− Turbulent − Fully developed flow − Constant heat flux on the wall − Smooth pipe
Properties	Evaluated using the average of the fluid bulk temperature
Error	±15%

4.5.5.3 Seban and Shimazaki

The correlation of Seban and Shimazaki [65] is based on experiments using lead–bismuth as thermal fluid. It evaluates the local Nusselt number of fully developed turbulent flows under the boundary condition of constant wall temperature:

$$Nu = 5.0 + 0.025(RePr)^{0.8}$$

or   (4.36)

$$Nu = 5.0 + 0.025(Pe)^{0.8}$$

where Pe is the Péclet number using the pipe diameter as the characteristic length.

Fact sheet	
Fluids	Liquid metals in general
Re	3.6×10^3–9.05×10^5
Pr	0.003–0.05
Pe	>100
L/D	>30
Properties	Evaluated using the local fluid bulk temperature
Error	±15%
Conditions	− Turbulent − Fully developed flow − Constant wall temperature − Smooth pipe

4.5.5.4 Effect of impurities

Correlations (4.33), (4.34), (4.35), and (4.36) are valid for cooling or heating of the liquid metal. However, care should be taken to the concentration of impurity content, e.g. oxygen, nitrogen, etc., in the liquid metal, which is required to be below the oxide solubility limit at the operating temperature of the heat exchanger [20]. When there is oxide deposition in the wall, the heat transfer coefficient may be significantly decreased by the oxide layer. An estimation of the reduced Nusselt number may be obtained from:

$$Nu = 4.3 + 0.0021(RePr)$$

or (4.37)

$$Nu = 4.3 + 0.0021(Pe)$$

Fact sheet	
Fluids	Liquid metals in general with high concentration of impurities (occurrence of oxide deposition)
Pe	10^2–10^4
L/D	>60
Properties	Evaluated using the local fluid bulk temperature
Conditions	− Fully developed flow − Smooth pipe

4.6 Heat transfer coefficients for the annulus

The annulus region formed by the assembly of two concentric tubes in a double-pipe heat exchanger is characterized geometrically by the external diameter of the internal pipe (D_o) and the internal diameter of the shell (D_s), depicted in Fig. 4.3.

The annular channel of two concentric tubes exhibits a distinct hydrodynamic and thermal behavior from a circular duct. The fluid flowing in the annulus may exchange heat in three ways:
1. From/to the inner surface with the outer surface insulated
2. From/to the outer surface with the inner surface insulated
3. From/to both inner and outer surfaces simultaneously

Considering the possible combinations of the heat transfer surface (inner, outer, or both) and the heat transfer mode (heating or cooling), there are eight arrangements for the fluid exchanging heat, each one with slightly different velocity and thermal profiles, and consequently distinct heat transfer behavior.

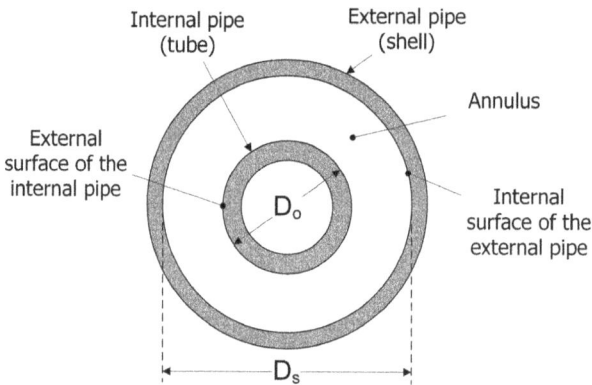

Fig. 4.3: Schematic representation of the annulus channel of two concentric pipes.

Typically, the outer pipe of a double-pipe heat exchanger is insulated (or the heat loss is ignored); therefore, the most encountered boundary conditions for this equipment are:
1. fluid heated by the inner surface and
2. fluid cooled by the inner surface.

There are several methods for evaluation of the heat transfer coefficient for these situations. Let us mention some of them further.

4.6.1 Annulus equivalent diameter

Perhaps, the most used technique to estimate the annulus heat transfer coefficient is using the hydraulic diameter (D_h) as a kind of equivalent diameter (D_e) to adapt the Nusselt number and friction factor data originally tailored for circular tubes. Such approximation works satisfactorily for several noncircular ducts,[17] including the annulus of a double-pipe exchanger, within 15% accuracy [47], being a commonplace recommendation in several technical sources [18, 19, 47, 50, 57, 58].

The hydraulic diameter for a noncircular channel is defined in the following form:

$$D_h = 4 \times \frac{\textit{flow cross section}}{\textit{wetted perimeter}} \tag{4.38}$$

Setting D_o as the outer diameter of the internal tube and the inner diameter of the outer tube (shell) as D_s (Fig. 4.3), the flow's cross section and wetted perimeter are given by

$$\textit{flow cross section} = \frac{\pi}{4}\left(D_s^2 - D_o^2\right) \tag{4.39}$$

$$\textit{wetted perimeter} = \pi(D_s + D_o) \tag{4.40}$$

Substituting in the hydraulic diameter expression and simplifying, we have the equivalent diameter as follows:

$$D_e = D_h = \frac{D_s^2 - D_o^2}{D_s + D_o} = D_s - D_o \tag{4.41}$$

In general, throughout this text, unless stated otherwise, the equivalent diameter (D_e) of noncircular ducts is going to be taken as equal to the hydraulic diameter (D_h).

4.6.2 Petukhov–Roizen corrector

A method [20, 78] for diminishing the deviation from using a circular pipe equation to estimate the annulus heat transfer coefficient consists in the application of a correction factor on the Nusselt number for a circular pipe, as follows:

$$Nu_a = F_{a(inner)}Nu_t \qquad (\textit{inner tube heated or cooled})$$

$$\tag{4.42}$$

$$F_{a(inner)} = 0.86\left(\frac{D_o}{D_s}\right)^{-0.16}$$

---

[17] Despite the great convenience, the use of the hydraulic diameter as the characteristic length in circular pipe correlations for estimating the friction factor and Nusselt number for noncircular ducts with sharp corners produces unacceptable deviations [47].

$$Nu_a = F_{a(outer)} \, Nu_t \qquad \text{(outer tube heated or cooled)}$$

$$F_{a(outer)} = 1 - 0.14 \left(\frac{D_o}{D_s}\right)^{0.6} \tag{4.43}$$

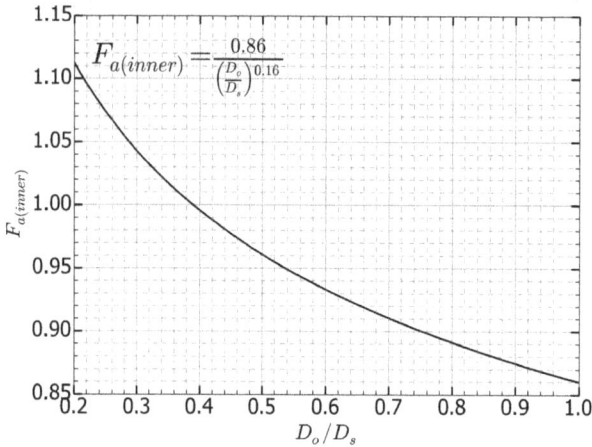

Fig. 4.4: Nusselt number correction factor for an annulus passage with the inner pipe heated or cooled (eq. (4.42)).

The Nusselt number in eqs. (4.42) and (4.43) is evaluated from any correlation targeted to circular tubes, such as eqs. (4.11), (4.19), or (4.23), with the characteristic length as the pipe's internal diameter. Nu_a is the Nusselt number for the annulus and Nu_t is the Nusselt number calculated from equations tailored for circular tubes. The Nusselt numbers Nu_a and Nu_t and any nondimensional numbers used in their determination must be evaluated using the hydraulic diameter $D_h = D_s - D_o$.

In the case of a double-pipe heat exchanger, the equation to be used is eq. (4.42), since typically the outer pipe wall is approximately adiabatic in practical situations. For convenience of calculation, Fig. 4.4 presents the curve corresponding to the Nusselt correction factor for the annulus with the inner tube heated or cooled.

4.6.3 Gnielinski annulus equation

A relatively recent method for predicting the heat transfer coefficient for annular ducts is a modification of the Petukhov–Kirillov equation proposed by Gnielinski after correlating a significant amount of experimental data [45, 79]. The suggested equation is eq. (4.44), in which Nu and Re are based on the annulus hydraulic diameter (D_h) [79]:

$$Nu = \frac{(f_F/2)RePr}{k_1 + 12.7(f_F/2)^{1/2}(Pr^{2/3} - 1)}\left(1 + \left(\frac{D_h}{L}\right)^{2/3}\right)F_a \tag{4.44}$$

$$k_1 = 1.07 + \frac{900}{Re} - \frac{0.63}{(1 + 10Pr)} \tag{4.45}$$

The annulus Fanning friction factor f_F used in eq. (4.44) is also adapted from Filonenko's equation for the annular geometry, using a modified Reynolds number in the form:

$$f_F = \frac{1}{4}\left(0.79 \ln\left(Re^*\right) - 1.64\right)^{-2} \tag{4.46}$$

$$Re^* = Re + \frac{(1 + a^2)\ln(a) + (1 - a^2)}{(1 - a)^2 \ln(a)} \tag{4.47}$$

$$a = \frac{D_o}{D_s} \tag{4.48}$$

The factor F_a is recalibrated from the Petukhov–Roizen corrector (eq. (4.42)), and accounts for the thermal boundary condition under which the heat is transferred to or from the fluid in the annulus. When the heat is transferred through the inner pipe wall and the outer pipe is insulated, this factor becomes

$$F_{a(inner)} = 0.75a^{-0.17} \tag{4.49}$$

If the heat is transferred through the outer pipe wall and the inner pipe is adiabatic, the boundary corrector is given as follows:

$$F_{a(outer)} = 0.9 - 0.15a^{0.6} \tag{4.50}$$

Fact sheet	
Fluids	Liquids and gases
Re	10^4–10^6
Pr	0.6–1,000
D/L	0–1
Properties	Evaluated using fluid's mean bulk temperature: $T_b = (T_{b,in} + T_{b,out})/2$
Conditions	– Turbulent – Fully developed flow – Smooth pipe

Problems

(4.1) About 2 kg/s of benzene flows in a 10 m long pipe of ID = 52.502 mm. The benzene inlet and outlet temperatures are 52 and 30 °C, respectively. The pipe's absolute roughness is about 0.035 mm. Determine:
a) Reynolds number
b) Fluid dynamic and thermal entrance lengths
c) The mean convective heat transfer coefficient estimated by the Sieder–Tate equation
d) The mean convective heat transfer coefficient estimated by the Petukhov–Popov equation, using the friction factor from the Filonenko's equation
e) The mean convective heat transfer coefficient estimated by the Petukhov–Popov equation with the friction factor given by the Haaland's equation

Answer:
(a) Re = 98,784
(b) L_{fd} = 1.2649 m, $L_{fd,T}$ = 6.9065 m
(c) h = 1,246.5 W/(m²·K)
(d) h = 1,380.4 W/(m²·K)
(e) h = 1,523.4 W/(m²·K)

(4.2) A 3.46 kg/s stream of liquid acetone is heated from 20.0 to 30.0 °C in the annulus of a 5 m long double-pipe hairpin. The annulus inner and outer diameters are 60.325 mm and 128.19 mm, respectively; and can be considered smooth with its inner surface maintained at 34.5 °C. Determine:
a) The hydraulic diameter and Reynolds number
b) Fluid dynamic and thermal entrance lengths
c) The relative deviation between the Petukhov's and Sieder–Tate's heat transfer coefficients, assuming constant physical properties. Take Petukhov's estimation as the reference
d) The viscosity and Prandtl-based correction factors for variable physical properties of the fluid
e) The convective heat transfer coefficient given by Petukhov's equation, considering the variation of fluid properties with the Prandtl correction factor

Answer:
(a) D_h = 0.067865 m, Re = 76,164
(b) L_{fd} = 1.5322 m, $L_{fd,T}$ = 5.7915 m
(c) Error: – 3.3%
(d) $(\mu/\mu_w)^{0.14}$ = 1.0120, $(Pr/Pr_w)^{0.11}$ = 1.0048
(e) h = 828.77 W/(m²·K)

(4.3) Cooling water at a flow rate of 2.23 kg/s and 20 °C is used to chill 3 kg/s of diethanolamine from 110 to 90 °C in a double-pipe exchanger with tubes DN 50 mm/40S and DN 90 mm/40S. The diethanolamine is allocated in the inner tube. Evaluate (a) the heat transfer coefficients assuming constant fluid properties, (b) the wall temperature, (c) the Prandtl-based correction factors for variable fluid properties in the tube and annulus, (d) the actual (corrected) heat transfer coefficients, (e) the overall heat transfer coefficient for this heat exchanger, (f) the overall heat transfer coefficient if noncorrected convective heat transfer coefficients were used, and (g) the error introduced on its value.

Answer:
(a) $h_i = 470.15$ W/(m²·K) (transition), $h_o = 3321.5$ W/(m²·K) (turbulent)
(b) $t_w = 41.0$ °C
(c) $\phi_i = 0.74693$, $\phi_o = 1.0142$
(d) $h_i = 351.17$ W/(m²·K), $h_o = 3,368.7$ W/(m²·K)
(e) $U_c = 273.90$ W/(m²·K)
(f) $U_c = 364.30$ W/(m²·K)
(g) Error: +30%

(4.4) In an experiment to measure heat transfer coefficients, a 32.8 ft pipe, ID = 2.0661 in, has its internal surface at 68.9 °F. An air stream of 2,381 lb/h is fed at 230.0 °F and exits at 140.0 °F. Determine:
a) Total heat transfer rate of the air stream
b) Mean temperature difference between the fluid and the pipe
c) Convective heat transfer coefficient
d) The pipe fraction corresponding to the hydrodynamic entrance length
e) The pipe fraction corresponding to the thermal entrance length
f) The relative deviation of the calculated convective heat transfer coefficient from the value predicted by the Sieder–Tate equation

Answer:
(a) $q = 51,859$ BTU/h
(b) 110.03 °F
(c) $h = 47.182$ BTU/(ft²·h·°F)
(d) 17%
(e) 12%
(f) $h_{\text{(Sieder–Tate)}}$ is +41% greater from the experimental value

(4.5) As an alternative process, the air stream referred on Problem 4.4 is cooled in a double-pipe heat exchanger using 1,587.3 lb/h of water passing through the annulus with inlet temperature of 68.0 °F. The exchanger piping is carbon steel NPS 2 in/40S and 2–1/2 in/40S. Evaluate:

a) Outlet temperature of the water
b) Pipe wall temperature, assuming negligible wall thermal conductive resistance
c) Annulus heat transfer coefficient
d) The heat exchanger overall heat transfer coefficient

Answer:
(a) $T_{water,out} = 101.65\ °F$
(b) $t_w = 91.6\ °F$
(c) $h_o = 571.07\ BTU/(ft^2 \cdot h \cdot °F)$
(d) $U = 37.62\ BTU/(ft^2 \cdot h \cdot °F)$

5 Pressure drop

When both fluids exchanging heat pass through a hairpin heat exchanger, the initial pressures decrease along the flow path, from the inlet to the outlet nozzles, respectively. The total pressure drop is a combination of twofold causes:

1. Distributed pressure losses along the linear pipe length of inner tube and annulus
2. Localized pressure losses at the fittings, connecting the pipes and annuli to build the hairpin assembly

Then, the total pressure drop for the whole fluid path may be given by eq. (5.1), where ΔP_{dist} and ΔP_{loc} are the distributed and localized pressure losses, respectively:

$$\Delta P = \Delta P_{dist} + \Delta P_{loc} \tag{5.1}$$

The pipe distributed losses along ducts are evaluated with the aid of the friction factor. There are several definitions of friction factor; however, the most common are that of Fanning (f_F) and Darcy (f_D), related by $f_F = f_D/4$. Equation (5.2) evaluates the duct-distributed pressure loss based on the Fanning friction factor as follows:

$$\Delta P_{dist} = 2 f_F \frac{L}{D} \rho v^2 \tag{5.2}$$

Or in terms of the mass flux G (kg/m² s), with eq. (5.3):

$$\Delta P_{dist} = 2 f_F \frac{L}{D} \frac{G^2}{\rho} \tag{5.3}$$

Formerly, the localized pressure drop in fittings and flow path "events" was taken as proportional to the number of velocity heads of the stream, being the proportionality constant the loss coefficient K. A velocity head expressed in pressure units is the quantity $\rho v^2/2$, i.e., the kinetic energy per unit volume of the fluid. Therefore, local losses in a pipe circuit are calculated from the following equation:

$$\Delta P_{loc} = K \frac{\rho v^2}{2} \tag{5.4}$$

Or in terms of the mass flux G (kg/m²s), with the following equation:

$$\Delta P_{loc} = K \frac{G^2}{2\rho} \tag{5.5}$$

The current knowledge is that the relationship between localized pressure drop and the fluid velocity is not that simple. Accumulated experimental evidence has shown that the loss coefficient K is a function of Reynolds number, pipe diameter and

https://doi.org/10.1515/9783110585872-005

roughness. Nonetheless, the relations given by eqs. (5.4) and (5.5) are still valid if we keep in mind that $K = K(\text{Re},D)$, and evaluate the loss coefficient accordingly.

The (constant) K-factor method is still in wide use. As the uncertainties involved in Reynolds number and diameter variations are covered by generous safety factors, more advanced methods such as the New Crane K [80], the Hooper's 2-K [81], and the Darby's 3-K [82, 83] methods are in increasing acceptance.

5.1 Friction factor in circular tubes

We need friction factors to calculate pressure losses. When a fluid flows inside a pipe, provided the nonslip condition holds, a force will be exerted on the internal surface of the wall with the same direction of the flow. This force per unit area creates a tangential stress on the wall τ_w, which can be made dimensionless by dividing by the volumetric specific kinetic energy of the flowing fluid $\rho v^2/2$ to give:

$$f_F = \frac{\tau_w}{\frac{\rho v^2}{2}} \tag{5.6}$$

The left-hand side of eq. (5.6) is the Fanning friction factor, a dimensionless number. Notice that the denominator 2 in this equation is arbitrary. Any other number could be used, and the equation would remain a dimensionless definition for a friction factor. In fact, there are other types of friction factors. For example, the Darcy friction factor is in common use by civil and mechanical engineers, which is defined as $f_D = 4f_F$.

So, how we can evaluate pressure drop from the friction factor? A straightforward force balance for a fluid flowing with bulk velocity v in a tubular section of length L and inner diameter D shows that:

$$\pi DL\tau_w = \frac{\pi D^2}{4}\Delta P \tag{5.7}$$

Solving for ΔP, we get

$$\Delta P = 4\frac{L}{D}\tau_w \tag{5.8}$$

But, from eq. (5.6), τ_w is

$$\tau_w = f_F \frac{\rho v^2}{2} \tag{5.9}$$

Therefore, combining eqs. (5.9) into (5.8), the relation between the pressure drop and the Fanning friction factor is obtained as follows:

$$\Delta P = 2 f_F \frac{L}{D} \rho v^2 \tag{5.10}$$

Dimensional analysis shows that the friction factor depends on the Reynolds number of the flow and the relative roughness of the duct wall in contact with the fluid, i.e., $f_F = f_F(\text{Re}, \varepsilon/D)$.

5.1.1 Isothermal laminar flow

The laminar regime admits accurate analytical treatment and the friction factor for $\text{Re} < 2{,}100$ is calculated from the following equation:

$$f_F = \frac{16}{\text{Re}} \tag{5.11}$$

Fact sheet	
Fluids	Liquids and gases
Re	Re < 2,100
Properties	Evaluated using fluid mean bulk temperature: $T_b = (T_{b,in} + T_{b,out})/2$
Conditions	– Laminar – Fully developed flow – Smooth and rough pipes

5.1.2 Isothermal turbulent flow

5.1.2.1 Smooth tubes

Possibly, the simplest friction factor correlation for turbulent flow inside a smooth tube is the Blasius' equation [45]:

$$f_F = \frac{0.0791}{\text{Re}^{0.25}} \tag{5.12}$$

Fact sheet	
Fluids	Liquids and gases
Re	$4 \times 10^3 < Re < 10^5$
Properties	Evaluated using fluid mean bulk temperature: $T_b = (T_{b,in} + T_{b,out})/2$
Conditions	– Turbulent – Fully developed flow – Smooth pipes

5.1.2.2 Rough tubes

For rough pipes in the turbulent flow range, the Fanning friction factor f_F may be evaluated from the Haaland's equation (5.13) [84]. This correlation agrees within about +1.21% [47] to the Colebrook–White [85] implicit equation, considered the golden standard for friction factors in circular pipes:

$$f_F = 0.41 \left(\ln \left(0.23 \left(\frac{\epsilon}{D} \right)^{\frac{10}{9}} + \frac{6.9}{Re} \right) \right)^{-2} \tag{5.13}$$

Fact sheet	
Fluids	Liquids and gases
Re_D	$4 \times 10^3 - 10^8$ [43, 47]
ϵ/D	0–0.05 [43] 2×10^{-8}–0.1 [47]
Properties	Evaluated using fluid mean bulk temperature: $T_b = (T_{b,in} + T_{b,out})/2$
Conditions	– Turbulent – Fully developed flow – Smooth and rough pipes

5.1.3 All flow regimes

5.1.3.1 Churchill equation

A quite general friction factor equation was developed by Churchill [86], throughout an interpolated [87] blend of results for various flow regimes. The correlation in the form of Fanning friction factor is shown in eq. (5.14). The author recommended the evaluation of the Reynolds number at the wall temperature. Equation (5.14) is not appropriate for very viscous flows (with $Re < 7$), in which case the friction factor should be calculated from the analytical result (5.11).

$$f_F = 2 \left(\left(\left(\frac{8}{Re_w} \right)^{10} + \left(\frac{Re_w}{36,500} \right)^{20} \right)^{-0.5} + \left(2.21 \ln \left(\frac{Re_w}{7} \right) \right)^{10} \right)^{-1/5} \tag{5.14}$$

Fact sheet	
Fluids	Liquids and gases
Re_w	>7
Properties	Re_w is calculated at the wall temperature T_w
Conditions	- Smooth circular ducts - Fully developed flow

5.1.3.2 Bhatti and Shah equation

The Bhatti–Shah [88] equation is targeted to all Reynolds regimes and can be used to evaluate the friction factor for smooth circular ducts with fully developed flow, as follows:

$$f_F = A + \frac{B}{Re^{\frac{1}{m}}} \tag{5.15}$$

$A = 0$	$B = 16$	$m = 1$	$Re < 2,100$
$A = 0.0054$	$B = 2.3 \times 10^{-8}$	$m = -\frac{2}{3}$	$2,100 < Re < 4,000$
$A = 1.28 \times 10^{-3}$	$B = 0.1143$	$m = 3.2154$	$Re > 4,000$

Fact sheet	
Fluids	Liquids and gases
Re_w	All
Properties	Evaluated using fluid mean bulk temperature: $T_b = (T_{b,in} + T_{b,out})/2$
Conditions	- Smooth circular ducts - Fully developed flow

5.1.4 Effect of variable physical properties

Sieder and Tate used a fair amount of experimental data to develop a corrector factor for the effect of large variations of fluid properties due to the heating or cooling of the fluid flowing in circular pipes [62]:

$$f_F = f_{FO}\phi \tag{5.16}$$

$$\phi = \begin{cases} 1.1\left(\dfrac{\mu}{\mu_w}\right)^{-0.25} & (\text{Re} < 2,100) \\[2mm] 1.02\left(\dfrac{\mu}{\mu_w}\right)^{-0.14} & (2,100 < \text{Re} < 10^5) \end{cases} \tag{5.17}$$

Fact sheet	
Fluids	Gases, organic liquids, and aqueous solutions and water [62]
Re[18]	$2,100–10^5$
μ/μ_w	0.0044–9.75 (laminar) 0.1–7 (turbulent)
Properties	All properties evaluated using the fluid bulk temperature T_b, except for μ_w, which is calculated at the wall temperature T_w. The average bulk temperature ($T_b = (T_{b,in} + T_{b,out})/2$) may be used to calculate the mean heat transfer coefficient for the whole surface
Conditions	– Smooth pipes – Variable fluid properties – Isothermal wall boundary condition – Fully developed flow

The friction factor for isothermal flow f_{FO} used in eq. (5.16) by Sieder and Tate is the Fanning friction factor, though the Darcy friction factor may also be corrected for nonisothermal flow with this approach.

5.2 Pressure drop calculation with the 2-K method

A more comprehensive technique for evaluating localized pressure drop is the 2-K method, which was developed by Hooper [81]. The 2-K method addresses some weak points of the conventional K-factor method, by incorporating the dependences of the

---

18 The actual experimental data for friction factor correlated by Ref. [62] spanned strictly to Re < 10,000; however, the extrapolation to greater values of Reynolds number is widely accepted.

head loss coefficient K on the Reynolds number and size of the fitting. The method is accurate for a wide range of diameters, even for large ones (ID = 0.2–80 in), and for low Reynolds number flows (Re = $10–10^6$). For further details about this method, please follow Refs. [16, 81, 89].

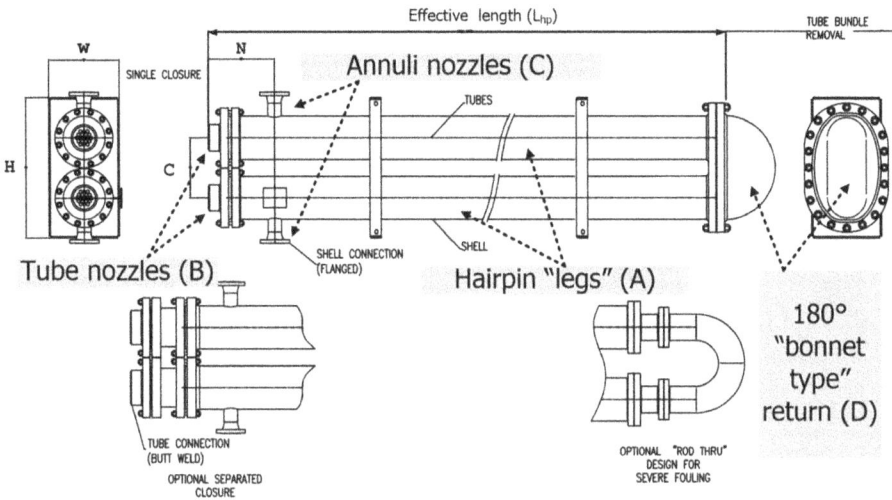

Fig. 5.1: Hairpin heat exchanger mechanical drawing. The annulus fluid returns through a bonnet (adapted from Ref. [15]. Courtesy of Heat Exchanger Design, Inc., Indianapolis-USA).

Considering its accuracy advantages, in this section we are going to apply the 2-K method in the development of a procedure for the calculation of pressure losses in the heat exchanger pipe and the annulus. A remark is worth about not using the 3-K method here. Although the 3-K method claims greater accuracy for a wider range of fitting diameters, their constant K's were adjusted using nominal diameters, while the 2-K method is optimized for the actual internal pipe diameter [81]. Since the internal diameter is already used for several calculations, such as mass flux, Reynolds number, and friction factor, and because of its greater simplicity, the presented procedure will be based on the 2-K method.

Figure 5.1 identifies the sources of distributed (A) and localized (B, C, and D) pressure losses for the pipe and the annulus of a hairpin with bonnet return.

5.3 Inner tube pressure drop

The main component of pressure drop in the inner tube is the linear pipe length. Entrance and exit losses are typically negligible in comparison with linear pipe losses [44] because the nozzles are usually aligned with the process pipeline, and the cross-sectional variation between the nozzles and the inner pipe is not large.

5.3.1 Distributed pressure drop

The distributed pressure drop for the inner pipe channel may be calculated using eq. (5.18), where $\Delta P_{i,\text{dist}}$ is the pressure drop for a mass flux G_i of fluid with specific mass ρ_i flowing along a straight pipe of length L and internal diameter D_i. f_{Fi} is the pipe Fanning friction factor, evaluated from any valid correlation. This equation is not tied to a unit system; therefore, any consistent unit set could be used:

$$\Delta P_{i,\text{dist}} = \frac{2G_i^2 L f_{Fi}}{D_i \rho_i} \tag{5.18}$$

The cross-flow area of the inner pipe is given by eq. (5.19), and for a mass flow rate W_i, the tube mass flux is calculated with eq. (5.20):

$$S_i = \frac{\pi D_i^2}{4} \tag{5.19}$$

$$G_i = \frac{W_i}{S_i} \tag{5.20}$$

It should be noted that for an association of N_{hp} hairpins in series, L comprises the full pipe length, including all exchanger legs; however, the length of the return bend is ignored – it is later considered as a local source of pressure loss – i.e., if the effective length of the exchanger is L_{hp}, the entire fluid path is computed as $L = 2N_{hp}L_{hp}$, since a "U" hairpin exchanger has two legs:

$$\Delta P_{i,\text{dist}} = \frac{4 L_{hp} N_{hp} f_{Fi}}{D_i \rho_i} G_i^2 \tag{5.21}$$

The Fanning friction factor to be applied in eq. (5.21) is evaluated from a valid equation for the flow conditions of the process. Notice that eq. (5.21) is essentially isothermal; hence, for large variations of fluid temperature, the friction factor should be corrected with a Sieder–Tate (eq. (5.16)) or Petukhov factor (eqs. (4.17) and (4.18)).

5.3.2 Localized pressure drop

The pipe fluid performs a 180° turn at one end of the hairpin exchanger. Moreover, because of the small heat transfer area per unit length, double-pipe exchangers are industrially mounted in banks (see Figs. 1.4, 1.5, and 5.4), where the inner tube of one exchanger is connected to the next exchanger by another 180° bend, which should also be accounted as a source of pressure loss. Assuming these bends have approximately the same internal diameter of the inner tube, only a small fluid expansion or contraction takes place.

According to 2-K head loss coefficients [81], the short radius screwed 180° bend produces the greatest pressure loss among the available types, which is about twice the pressure loss of a single flanged/welded 180° bend, given in the following equations:

$$K_{(180°)} = 0.35 + \frac{1,000}{Re} + \frac{8.89}{D_{h(mm)}} \tag{5.22}$$

$$K_{(180°)} = 0.35 + \frac{1,000}{Re} + \frac{0.35}{D_{h(in)}} \tag{5.23}$$

To make a safer pressure drop estimation, allowing for slight expansion and contraction losses in the joints between the 180° bend and the heat exchanger body, it is conservatively assumed that the head loss coefficient for a "generic" 180° bend will be twice the head loss of the flanged/welded bend. Therefore, the total number of velocity heads for the inner tube is evaluated from eq. (5.24), where K_i is the friction loss coefficient for the inner pipe of one hairpin with internal diameter D_i expressed in millimeters:

$$K_i = 0.7 + \frac{2,000}{Re_i} + \frac{17.78}{D_{i(mm)}} \tag{5.24}$$

Or, for D_i measured in inches, K_i is calculated by

$$K_i = 0.7 + \frac{2,000}{Re_i} + \frac{0.7}{D_{i(in)}} \tag{5.25}$$

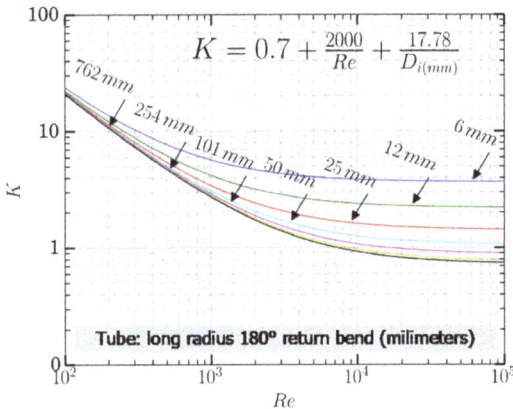

Fig. 5.2: K-factor head loss coefficient for long radius 180° return bend. D_i is the internal tube diameter expressed in millimeters (eq. (5.24)).

Equation (5.24) is represented in Fig. 5.2, while eq. (5.25) is depicted in Fig. 5.3, where the significant increase in the K value for laminar flows is noticeable. In this regime, the application of a constant number of velocity heads to account for local head losses may dangerously underestimate the exchanger pressure drop. As predicted by the classical K-factor method, the K value becomes practically constant for higher Reynolds number flows. The effect of the pipe diameter is also significant for smaller diameters. However, for greater pipe sizes, an increase in diameter affects the head loss coefficient to a lesser extent.

For a battery containing N_{hp} hairpin exchangers connected in series, with eqs. (5.5) and (5.24), there are $(N_{hp} - 1)$ additional 180° return bends connecting the tubes of adjacent hairpins, and then we write the total localized pressure loss in the inner pipe as follows:

$$\Delta P_{i,\,loc} = \frac{G_i^2 K_i}{2\rho_i} \left(2N_{hp} - 1\right) \tag{5.26}$$

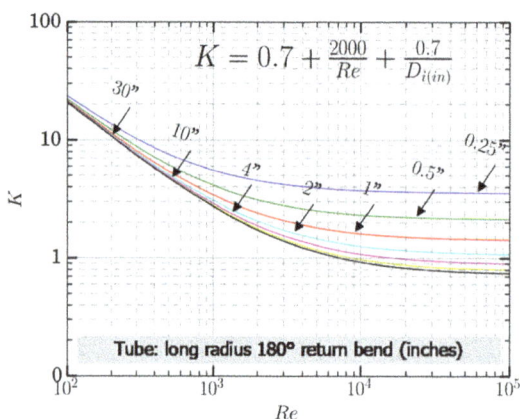

Fig. 5.3: K-factor head loss coefficient for long radius 180° return bend. D_i is the internal tube diameter expressed in inches (eq. (5.25)).

Hence, the total pressure drop for a bank of N_{hp} hairpin exchangers with length L_{hp} is given as follows:

$$\Delta P_i = \frac{G_i^2 K_i}{2\rho_i} \left(2N_{hp} - 1\right) + \frac{4L_{hp}N_{hp}f_{Fi}}{D_i\rho_i} G_i^2 \tag{5.27}$$

Worked Example 5.1
A process stream of 1.26 kg/s of ethylenediamine should be heated from 15 to 50 °C through the inner pipe in a serial bank of seven 5 m length hairpin double-pipe heat exchangers. The internal tube is carbon steel NPS 32 mm – schedule 40. Determine if the inner tube pressure drop is within an allowable limit of 150 kPa.

Solution

a) Fluid physical properties

The average temperature of ethylenediamine inside the exchanger is

$$T_m = \frac{T_{in}}{2} + \frac{T_{out}}{2} = \frac{288.15}{2} + \frac{323.15}{2} = 305.65 \text{ K}$$

From Appendix C.11, the estimated physical properties of the fluid are given in the following table:

Ethylenediamine: 305.65 K – 101, 325 Pa

$\mu = 0.00145 \text{ s} \cdot \text{Pa}$
$\mu = 1.45 \text{ cP}$
$\rho = 551.0 \text{ kg/m}^3$

---

b) Piping specifications

From Appendix A.1, the tube specifications are:

NPS 32 mm – schedule 40

$D_i = \text{ID} = 35.08 \text{ mm}$ (inner diameter)
$D_o = \text{OD} = 42.2 \text{ mm}$ (outer diameter)

---

c) Friction factor

Using the Haaland's equation (5.13) to calculate the friction factor for a smooth pipe, we have:

–   Flow cross section and mass flux

$$S_i = \frac{\pi D_i^2}{4} = \frac{\pi 0.035080^2}{4} = 0.00096652 \text{ m}^2$$

$$G_i = \frac{W_i}{S_i} = \frac{1.2600}{0.00096652} = 1,303.6 \text{ kg}/(\text{m}^2 \cdot \text{s})$$

–   Reynolds number

$$Re_i = \frac{D_i G_i}{\mu_i} = \frac{0.035080}{0.0014500} 1,303.6 = 31,538 \rightarrow \text{fully turbulent}$$

The absolute roughness of a new carbon steel pipe surface from Appendix B.1 is estimated as $\varepsilon = 0.035$ mm; therefore, the friction factor is given by:

$$f_{Fi} = \frac{0.41}{\ln^2\left(0.23\left(\frac{\varepsilon}{D_i}\right)^{\frac{10}{9}} + \frac{6.9}{Re_i}\right)} = \frac{0.41}{\ln^2\left(0.23\left(\frac{(3.5000e-5)}{0.035080}\right)^{\frac{10}{9}} + \frac{6.9}{31,538}\right)} = 0.0063571$$

d) Pressure drop

– Distributed pressure drop

Using eq. (5.21), the distributed pressure loss in the tube is:

$$\Delta P_{i,\text{dist}} = \frac{4L_{hp}N_{hp}f_{Fi}}{D_i\rho_i}G_i^2 = \frac{4 \cdot 0.0063571 \cdot 5.0000}{0.035080 \cdot 551.00}1,303.6^2 \cdot 7.0000$$

$$= 78,246 \text{ kg}/(\text{m}\cdot\text{s}^2) = 78.246 \text{ kPa}$$

– Localized pressure drop

With the internal diameter $D_{i(mm)} = 35.08$ mm, the friction loss coefficient can be evaluated from eq. (5.24):

$$K_i = 0.7 + \frac{2,000}{Re_i} + \frac{17.78}{D_{i(mm)}} = 0.7 + \frac{17.78}{35.080} + \frac{2,000}{31,538} = 1.2703$$

Therefore, the local pressure loss is given by eq. (5.26):

$$\Delta P_{i,\text{loc}} = \frac{G_i^2 K_i}{2\rho_i}(2N_{hp}-1) = \frac{1.2703 \cdot 1,303.6^2}{2 \cdot 551.00}(2 \cdot 7.0000 - 1) = 25,466 \text{ kg}/(\text{m}\cdot\text{s}^2)$$

$$= 25.466 \text{ kPa}$$

– Total pressure drop

The full pressure loss in the pipe is the summation of distributed and local losses, then:

$$\Delta P_i = \Delta P_{i,\text{dist}} + \Delta P_{i,\text{loc}} = 25,466 + 78,246 = (1.0371e + 5) \text{ kg}/(\text{m}\cdot\text{s}^2) = 103.71 \text{ kPa}$$

Considering the calculated pressure loss, we conclude that this bank of seven exchangers does not exceed the available pressure drop.

5.4 Annulus pressure drop

The fluid inside the annulus of a hairpin double-pipe exchanger experiences localized pressure drop at several places, namely, the entrance and exit nozzles and in the 180° return connecting the two hairpin legs (Fig. 5.1). Therefore, the total annulus pressure is a combination of linear or distributed friction losses along the annulus channel with the localized losses at these specified spots.

5.4.1 Distributed pressure drop

The distributed pressure loss is given by eq. (5.28), where $D_h = D_s - D_o$ is the hydraulic diameter of the annulus channel (see Section 4.6.1). The annulus Fanning friction factor f_{Fo} must be evaluated with the Reynolds number based on this hydraulic diameter:

$$\Delta P_{o,\,dist} = \frac{2G_o^2 L f_{Fo}}{D_h \rho_o} \tag{5.28}$$

The flow cross section and the mass flux through the annulus channel (G_o) are given by:

$$S_o = \frac{\pi}{4} \left(D_s^2 - D_o^2 \right) \tag{5.29}$$

$$G_o = \frac{W_o}{S_o} \tag{5.30}$$

5.4.2 Localized pressure drop

The equation for the local pressure loss in the annulus of a single hairpin, using the annulus friction loss coefficient K_o is:

$$\Delta P_{o,\,loc} = \frac{G_o^2 K_o}{2\rho_o} \tag{5.31}$$

Typically, the annulus channel of two adjacent hairpins composing a bank is joined with rather short pipe sections, which are taken as part of the internal friction losses of each hairpin; therefore, the localized pressure loss for a series association of N_{hp} exchangers may be expressed as:

$$\Delta P_{o,\,loc} = \frac{G_o^2 K_o N_{hp}}{2\rho_o} \tag{5.32}$$

And the total pressure loss for the annulus of a bank with N_{hp} exchangers of effective length L_{hp} comes from eqs. (5.28) and (5.32) in the form:

$$\Delta P_o = \frac{G_o^2 K_o N_{hp}}{2\rho_o} + \frac{4 L_{hp} N_{hp} f_{Fo}}{D_h \rho_o} G_o^2 \tag{5.33}$$

The way of estimating the annulus additional pressure drop in eq. (5.31) is not universally settled in the technical literature. Generally, the localized pressure drop is considered by adding a certain number of velocity heads regarding the fluid returns, and

entrance and exit losses; however, the recommended amount of velocity heads varies significantly depending on the referenced source. For example:
- Kern [44] and Coker [90] recommend the addition of one velocity head per hairpin exchanger to compensate for entrance and exit losses.
- Bejan and Kraus [48] account with four velocity heads for the return and localized losses in a single hairpin exchanger.
- Serth [28] suggests adding 1.2 (turbulent flow) to 1.5 (laminar flow) velocity head for a long radius 180° return bend, and the addition of another 1.5 velocity heads for entrance and exit losses if the flow is turbulent or 3 velocity heads for laminar flow, according to standard formulas for head losses in nozzles found in Ref. [20]. Therefore, this method adds a total of 2.7–4.5 velocity heads, almost quintuplicating the value placed by Refs. [44, 90].

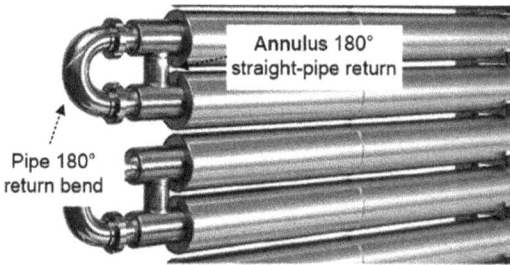

Fig. 5.4: Hairpin exchanger with annuli connected by a straight pipe (adapted from Ref. [34]).

The aforementioned methods for estimating the localized pressure losses in the annulus are all based on the standard friction loss coefficient concept, i.e., the conventional K-factor method [91, 92], which represents the number of velocity heads lost due to a fitting in a pipeline circuit. The friction loss coefficient K has a fixed value for a given pipe fitting, not varying with the fitting dimension or with the flow regime. It was verified that this approach to evaluate local pressure losses can have significant accuracy limitations, since the K value is not actually constant. It may increase expressively for viscous flows under Re < 1,000 and does not scale exactly with the size of the pipe fitting [16] because smaller diameter fittings are more sensitive to roughness and their K values tend to be higher in comparison with the same shape fitting in greater sizes, for example, the K for a ¼ in elbow does not equal its value for a 4 in elbow.

Currently, concerning the manner the annulus 180° return is assembled; heat exchanger manufactures may offer two distinct models of hairpins. This is illustrated in Figs. 5.1 and 5.4. In the first type (Fig. 5.1), the annulus fluid is reverted inside a chamber covered with a "bonnet" flanged to the end of the heat exchanger. In the second type shown in Fig. 5.4, the fluid return is simply a straight pipe connecting both annuli in each leg of the hairpin. Since such difference in construction

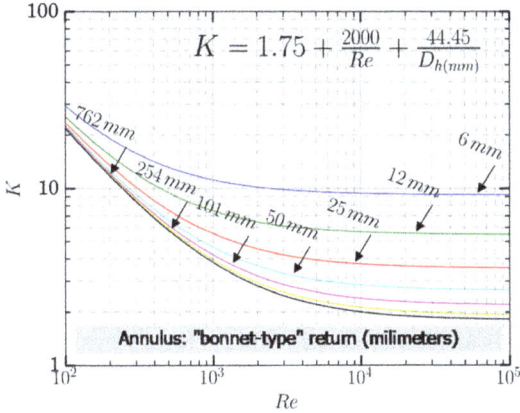

Fig. 5.5: *K*-factor head loss coefficient for "bonnet-type" annulus return. $D_h = D_s - D_o$ is the hydraulic diameter in millimeters (eq. (5.36)).

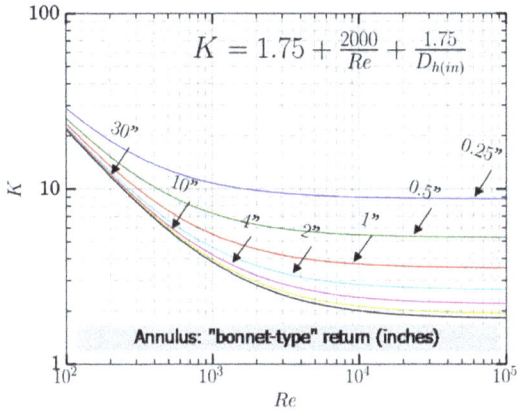

Fig. 5.6: *K*-factor head loss coefficient for "bonnet-type" annulus return. $D_h = D_s - D_o$ is the hydraulic diameter in inches (eq. (5.37)).

enables distinct spots of local friction losses, they are treated separately in the following Sections 5.4.3 and 5.4.4.

5.4.3 Annulus with bonnet-type return

If the hairpin exchanger has one or multiple "U"-shaped internal tubes, conventionally, the annulus fluid return is built with a flanged bonnet covering the inner pipes. Referring to Fig. 5.1, in the absence of the detailed nozzle specification, the entrance and exit regions of the annulus can be approximated as a branched tee, i.e., a tee with one leg blocked, forcing the fluid to perform a 90° curve. Based on the annulus hydraulic diameter, the 2-K head loss coefficient for a standard welded or flanged tee [81] is given as follows:

$$K_{(tee)} = 0.7 + \frac{500}{Re} + \frac{17.78}{D_{h(mm)}} \tag{5.34}$$

$$K_{(tee)} = 0.7 + \frac{500}{Re} + \frac{0.7}{D_{h(in)}} \tag{5.35}$$

Therefore, for this hairpin type, the flow path for the annulus fluid has its localized pressure losses approximated as two branched tees (mounted as 90° elbows) and a bonnet return of 180° (eqs. (5.22) and (5.23)), and the total number of velocity heads for such circuit is in eqs. (5.36) and (5.37), where D_h is the hydraulic diameter of the annulus measured in millimeters and inches, respectively. Figures 5.5 and 5.6 show a graphic representation of the aforementioned equations.

$$K_o = 1.75 + \frac{2,000}{Re_o} + \frac{44.45}{D_{h(mm)}} \tag{5.36}$$

$$K_o = 1.75 + \frac{2,000}{Re_o} + \frac{1.75}{D_{h(in)}} \tag{5.37}$$

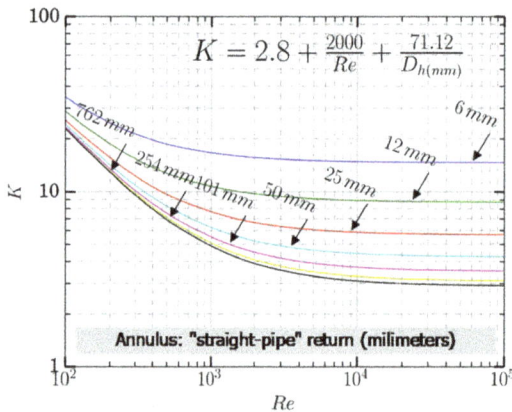

Fig. 5.7: K-factor head loss coefficient for "straight-pipe" annulus return. $D_h = D_s - D_o$ is the hydraulic diameter in millimeters (eq. (5.40)).

Using eq. (5.36), the complete equation for the exchanger pressure loss with a bonnet channel in the annulus is as follows:

$$\Delta P_o = \frac{G_o^2 N_{hp}}{2\rho_o} \left(1.75 + \frac{2,000}{Re_o} + \frac{44.45}{D_{h(mm)}} \right) + \frac{4 L_{hp} N_{hp} f_{Fo}}{D_h \rho_o} G_o^2 \tag{5.38}$$

$$\Delta P_o = \frac{G_o^2 N_{hp}}{2\rho_o} \left(1.75 + \frac{2,000}{Re_o} + \frac{1.75}{D_{h(in)}} \right) + \frac{4 L_{hp} N_{hp} f_{Fo}}{D_h \rho_o} G_o^2 \qquad (5.39)$$

5.4.4 Annulus with straight-pipe return

Manufacturers also offer a hairpin model in which the return of the annulus fluid is performed through a simple straight pipe connection between the two annuli. In this case, there is no bonnet. An example of this construction is shown in Fig. 5.4.

The direct annuli connection enforces the flow through a couple of abrupt 90° turns, which is approximated by the pressure losses of two branched tees, as given by eq. (5.34). Therefore, the localized total pressure loss for this hairpin type may be estimated with the following equations:

$$K_o = 2.8 + \frac{2,000}{Re_o} + \frac{71.12}{D_{h(mm)}} \qquad (5.40)$$

$$K_o = 2.8 + \frac{2,000}{Re_o} + \frac{2.8}{D_{h(in)}} \qquad (5.41)$$

For convenient quick calculations, the velocity head number charts based on eqs. (5.40) and (5.41) are plotted in Figs. 5.7 and 5.8, respectively.

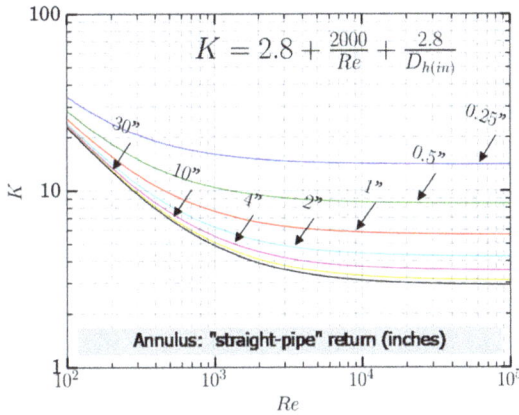

Fig. 5.8: K-factor head loss coefficient for "straight-pipe" annulus return. $D_h = D_s - D_o$ is the hydraulic diameter in inches (eq. (5.41)).

Worked Example 5.2

Determine the pressure drop for the annulus of a bank of eight NPS 32 mm (Sch 40) × 80 mm (Sch 40) × 6 m long carbon steel hairpin double-pipe heat exchangers. The processed fluid is a stream of 1.15 kg/s of heptanal, cooled from 65 to 38 °C.

Solution

a) Fluid physical properties

The average temperature of the hot fluid from 65 to 38 °C is:

$$T_m = \frac{T_{1,in}}{2} + \frac{T_{1,out}}{2} = \frac{311.15}{2} + \frac{338.15}{2} = 324.65 \text{ K}$$

Using Appendix C.11, the estimated physical properties of heptanal are:

Heptanal: 324.65 K – 101, 325 Pa

$\mu = 0.000642 \text{ s} \cdot \text{Pa}$
$\mu = 0.642 \text{ cP}$
$\rho = 689.0 \text{ kg/m}^3$

---

b) Piping specifications

The annulus channel of the heat exchanger is delimited by the external diameter of the inner pipe D_o (NPS 32 mm – schedule 40) and internal diameter of the outer pipe D_s (NPS 80 mm – schedule 40). From Appendix A.1, the tube specifications for both pipes are:

NPS 32 mm – schedule 40

$D_i = \text{ID} = 35.08 \text{ mm}$ (inner diameter)
$D_o = \text{OD} = 42.2 \text{ mm}$ (outer diameter)

NPS 80 mm – schedule 40

$D_s = \text{ID} = 77.92 \text{ mm}$ (inner diameter)
$\text{OD} = 88.9 \text{ mm}$ (outer diameter)

---

c) Friction factor

The annulus cross section and fluid mass flux are evaluated from eqs. (5.29) and (5.30) as follows:

– Annulus mass flux

$$S_o = \frac{\pi}{4} \left(D_s^2 - D_o^2 \right) = \frac{\pi}{4} \left(0.077920^2 - 0.042200^2 \right) = 0.0033699 \text{ m}^2$$

$$G_o = \frac{W_o}{S_o} = \frac{1.1500}{0.0033699} = 341.26 \text{ kg}/(\text{m}^2 \cdot \text{s})$$

– Hydraulic diameter (eq. (4.41)) and Reynolds number

$$D_h = D_s - D_o = 0.077920 - 0.042200 = 0.035720 \text{ m}$$

$$Re_o = \frac{D_h G_o}{\mu_o} = \frac{0.035720}{0.00064200} 341.26 = 18,987$$

The absolute roughness of the carbon steel (new) pipe surface from Appendix B.1 is estimated as $\varepsilon = 0.035$ mm; therefore, the friction factor is estimated by the Haaland's equation (5.13):

$$f_{Fo} = \frac{0.41}{\ln^2\left(0.23\left(\frac{\varepsilon}{D_h}\right)^{\frac{10}{9}} + \frac{6.9}{Re_o}\right)} = \frac{0.41}{\ln^2\left(0.23\left(\frac{(3.5000e-5)}{0.035720}\right)^{\frac{10}{9}} + \frac{6.9}{18,987}\right)} = 0.0069739$$

d) Distributed pressure drop

$$\Delta P_{o,dist} = \frac{4 L_{hp} N_{hp} f_{Fo}}{D_h \rho_o} G_o^2 = \frac{4 \cdot 0.0069739 \cdot 6.0000}{0.035720 \cdot 689.00} 341.26^2 \cdot 8.0000$$

$$= 6,336.0 \text{ kg}/(\text{m} \cdot \text{s}^2) = 6.336 \text{ kPa}$$

e) Localized pressure drop
From eq. (5.36), with the hydraulic diameter expressed in millimeters $D_{h(mm)} = 35.72$ mm:

$$K_o = 1.75 + \frac{2,000}{Re_o} + \frac{44.45}{D_{h(mm)}} = 1.75 + \frac{44.45}{35.720} + \frac{2,000}{18,987} = 3.0997$$

$$\Delta P_{o,loc} = \frac{G_o^2 K_o N_{hp}}{2 \rho_o} = \frac{3.0997 \cdot 341.26^2 \cdot 8.0000}{2 \cdot 689.00} = 2,095.7 \text{ kg}/(\text{m} \cdot \text{s}^2) = 2.0957 \text{ kPa}$$

f) Total annulus pressure drop
Combining the distributed and localized contributions, we get the total annulus pressure drop as follows:

$$\Delta P_o = \Delta P_{o,dist} + \Delta P_{o,loc} = 2,095.7 + 6,336.0 = 8,431.7 \text{ kg}/(\text{m} \cdot \text{s}^2) = 8.4317 \text{ kPa}$$

5.5 Allowable pressure drop

Before the installation of any piece of equipment in a process line, it should be assured that the line has enough head to support the new equipment in operation; otherwise, the stream will be obstructed. This premise applies to heat exchangers also. Therefore, right after the thermal design is done, the engineer is responsible for evaluating the pressure drop (or head loss) that the heat exchanger is going to produce during operation. If the process line has not enough pump (or compressor/blower) power available, then the heat exchanger must be redesigned to meet the allowable pressure drop requirements.

The allowance for pressure drop in a heat exchanger is a balance between the cost of the equipment itself and the cost of the pumping power for its operation. In basic terms, with given process conditions and equipment type, heat exchangers designed to produce higher pressure drop are almost always smaller (i.e., cheaper) than units with lower pressure drop. In compensation, the smaller exchanger with higher pressure drop will require more energy to run the pumps moving the fluids during the whole lifetime of the equipment. Therefore, clearly, we see a trade-off in the sense that there must be a pressure drop specification that minimizes the total cost (fixed and variable costs) associated with the heat exchanger.

In principle, there is no universal standard for the allowable pressure loss in a chemical plant. Each process and, more specifically, each process line may have a different pressure drop threshold. Commonly, the pressure drop for a process stream is determined by the nature of the process, or the process conditions. In vacuum lines, the pressure drop available may be quite restricted as a few Pa; in contrast, high-pressure systems may use dozens of kPa to drive the fluids. As a general rule for an initial reference value, for low-to-medium viscosity liquids, the allowable pressure drop for liquids should be in the range of 35–70 kPa (5–10 psi) [44]. Other possible pressure drop design limits are shown in Tab. 5.1.

Tab. 5.1: General reference design criteria for heat exchanger allowable pressure drop.

Allowable pressure drop (general rule)	
Liquids	
Low viscosity: <1 mPa s (1 cP)	35 kPa (5 psi) [19]
Medium viscosity: 1–10 mPa s (1–10 cP)	50–70 kPa (7–10 psi) [18, 19, 93, 94]
High viscosity	Possibly >70 kPa (10 psi), according to the pumping system
Gases and vapors, pressurized[19]	
1–2 bar	0.5× gauge pressure (50–100 kPa)
>10 bar	0.1× gauge pressure (>100 kPa) [19]

---

19 References [93, 94] recommend the range of 5–20 kPa (1–3 psi) as a rough rule.

Tab. 5.1 (continued)

Allowable pressure drop (general rule)	
Gases and vapors, vacuum	
High vacuum	0.4–0.8 kPa (0.06–0.12 psi)
Medium vacuum	0.1× absolute pressure [19]

Problems

(5.1) A fluid with a flow rate of 13,492.0 lb/h is cooled in a 32.8 ft long, ID = 1.049 in, straight pipe. The inlet and outlet temperatures are 140.0 and 104.0 °F, respectively. The pipe can be considered smooth, and its inner surface temperature is t_w = 95.1 °F. Determine:
a) Heat transfer coefficient, assuming constant physical properties
b) Heat transfer coefficient improved with the Prandtl correction factor
c) Fluid pressure drop

Fluid properties:

T = 122.0 °F (582.0°R)
μ = 3.27 lb/(ft · h)
μ = 1.35 cP
ρ = 52.3 lb/ft³
Cp = 0.597 BTU/(lb · °F)
k = 0.0868 BTU/(ft · h · °F)
Pr = 22.5

T = 95.1 °F (555.0°R)
μ = 4.36 lb/(ft · h)
μ = 1.8 cP
ρ = 53.4 lb/ft³
Cp = 0.581 BTU/(lb · °F)
k = 0.0886 BTU/(ft · h · °F)
Pr = 28.6

Answer:
(a) $h = 626.73$ BTU/(ft² h °F)
(b) $h = 610.41$ BTU/(ft² h °F)
(c) $\Delta P_{dist} = 6.39$ psi

(5.2) A cooling water stream is heated in the annulus of a double-pipe hairpin exchanger. The water flow rate, and entrance and exit temperatures are 16,574.0 lb/h, and 68.0 °F and 86.0 °F, respectively. The hairpin is $L_{hp} = 16.4$ ft long and has a straight pipe annulus return. The inner and outer annulus diameters are $D_o = 1.315$ in and $D_s = 2.469$ in. Calculate the (a) distributed, (b) localized, and (c) total annulus pressure drop.

Answer:
(a) $\Delta P_{o,dist} = 0.47431$ psi
(b) $\Delta P_{o,loc} = 0.31993$ psi
(c) $\Delta P_o = 0.79424$ psi

(5.3) A bank of six double-pipe exchangers is designed to heat 7 kg/s of benzene from 30 to 50 °C in the annulus, by cooling 3.94 kg/s of toluene in the tube from 85 to 53 °C. Each hairpin is built in stainless steel with 4 m length and piping of DN 50 mm and DN 90 mm, both schedule 40S. The annulus fluid return is a straight pipe. Determine the friction factor for (a) the tube and (b) the annulus, using the Haaland's equation; and (c) the localized pressure drop in the pipe. Assuming an allowable pressure drop of 70 kPa for both streams, (d) verify if this bank is suitable to perform the service regarding its developed pressure drop.

Answer:
(a) $f_F = 0.0044427$
(b) $f_F = 0.0051723$
(c) $\Delta P_{i,loc} = 23.225$ kPa
(d) No, it is not

(5.4) As an attempt to meet the service requirements indicated in Problem 5.3 with the already designed heat exchangers bank, the fluid allocation is inverted. Calculate (a) the tube and (b) the annulus total pressure drop.

Answer:
(a) $\Delta P_i = 167.6$ kPa
(b) $\Delta P_o = 49.734$ kPa

(5.5) In a biofuel production facility, a stream of 3 kg/s of ethanol need to be cooled before storage from 60 to 35 °C. Cooling water is available at an initial temperature of 25 °C. The supply of water is not an issue but its outlet temperature must not surpass 50 °C. The storehouse has in standby a battery of 11 carbon steel, DN 50 × 80 mm/10S, 4 m long double pipe hairpins. The annulus return is of bonnet type and the ethanol is allocated in the inner tube. Calculate the following:

a) Required heat transfer rate
b) Final overall heat transfer coefficient
c) Total pressure loss in the tube
d) Total pressure loss in the annulus

Answer:

(a) $2.1e + 5$ W
(b) $1,259.7$ W/$(m^2 \cdot K)$
(c) $\Delta P_i = 52.566$ kPa
(d) $\Delta P_o = 46.951$ kPa

6 Series–parallel arrangements

As an outcome of its simple construction and relatively small heat transfer area per volume ratio, quite rarely a single double-pipe heat exchanger is designed to match a real-world heat transfer service. Possibly, this would require a very long hairpin with large diameter nonstandardized pipes, which is not practical economically. The viable solution is to design an appropriately sized double-pipe exchanger and combine a number of them in a bank or battery by connecting in sequence their tubes and annulus, which we could designate as a "series–series" association for obvious reasons. With the fluids flowing in countercurrent mode, the maximum value of the log mean temperature difference (LMTD) is achieved in this arrangement, and the heat transfer area is minimized accordingly. This common series–series association is demonstrated in Fig. 6.1, in which two exchangers appear with their tubes and annulus connected in sequence.

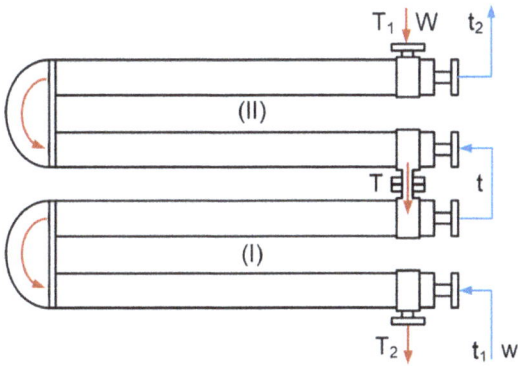

Fig. 6.1: Hairpin heat exchangers with both pipe and annulus associated in series.

If both process streams have comparable flow rates and temperature range between the inlet and outlet of the exchanger, the series–series arrangement will be the most probable to meet the service effectively; however, there are situations in which this arrangement is not the best design choice. Let us consider the integral energy balance for the fluids in the form:

$$CW(T_1 - T_2) = cw(t_2 - t_1) \qquad (6.1)$$

By rearranging, we have the mass flow rate ratio as follows:

$$\frac{W}{w} = \frac{c(t_2 - t_1)}{C(T_1 - T_2)} \qquad (6.2)$$

https://doi.org/10.1515/9783110585872-006

From eq. (6.2), we realize that the flow rate for a given stream is inversely proportional to the product of its specific heat with the temperature range. Therefore, if one of the fluids has a relatively small specific heat or a very short temperature range, its flow rate needs to be quite large to match the heat transfer duty. This situation can raise a difficulty in the specification of the heat exchanger because the fluid with the larger flow rate may exceed the available pressure drop in great extent, while the other fluid produces a relatively small pressure drop, impairing its heat transfer coefficient, and possibly the performance of the equipment as a whole. A conceivable solution is to divide both streams in halves and use two independent batteries of exchangers in parallel. While such design may solve the excess of pressure drop, the partition of the stream with smaller flow rate will reduce even more its heat transfer coefficient in both banks of exchangers, increasing the required total heat transfer area, the size of the individual hairpins, and the equipment cost as a whole.

Clearly, there are drawbacks with the previous approach, and a better design can be the sole division of the stream with larger flow rate, which is exceeding the allowable pressure loss. Such design is possible with the series–parallel association of double-pipe exchangers. As shown in Fig. 6.2, the inner fluid in the tube is divided with half flow rate entering the exchanger (I), and the other half entering the exchanger (II). As a result, the flow pipe length L and the mass flux G are reduced by a factor of ½. From eq. (5.18) for tube pressure loss, the factor $(L/2)(G/2)^2 = (1/8)(LG^2)$ cuts the pressure loss to 1/8 of the original value. Readily, we see that the division of the tube fluid in three parallel streams reduces the pressure loss by 1/27, and the division in N pipe parallel streams cuts the pressure loss by a factor of $1/N^3$, which is a significant reduction, permitting the accommodation of nearly arbitrarily large flow rates as the inner fluid.

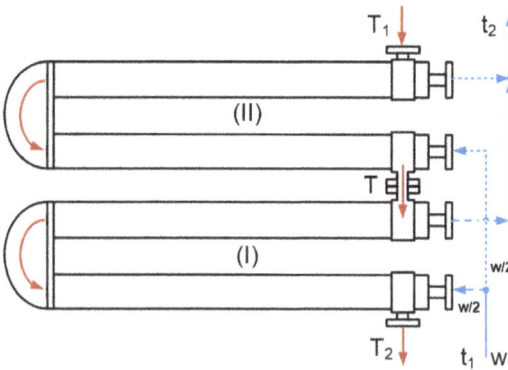

Fig. 6.2: Hairpin heat exchangers with annulus connected in series and the pipe connected in parallel.

Notice that if a service requires larger heat transfer surfaces, each hairpin depicted in Fig. 6.2 may in fact be replaced by a bank of exchangers with the pipe and annuli both connected in series inside the bank. A design in which two banks of hairpins are linked in series–parallel arrangement is presented in Fig. 6.3. In this case, the inner tube stream is divided into two parallel streams with each half fed in one of the banks. The annulus fluid flows in series through all the exchangers.

6.1 Mean temperature difference

Kern [44] demonstrated that the mean temperature difference is not the LMTD for the series–parallel association of double-pipe exchangers. In fact, the use of the LMTD for sizing could generate unacceptable deviations from the true mean temperature difference, and consequently for the calculated heat transfer area.

Consider the design equation (eq. (6.3)), where U_m is the mean overall heat transfer coefficient representing the entire heat transfer area A, and ΔT_m is the "exact" mean of all temperature differences along the whole exchanger:

$$q = U_m A \Delta T_m \tag{6.3}$$

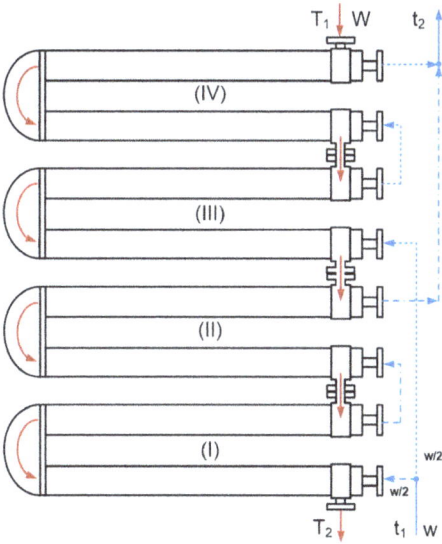

Fig. 6.3: Bank of four hairpin heat exchangers with annulus connected in series and the pipe fluid divided into two parallel streams.

Starting from eq. (6.3), Kern [44] showed that the actual mean temperature difference for the series–parallel association has the form given in eq. (6.4), i.e., a "series–parallel temperature factor" (F_{sp}), varying from zero to one, multiplied by the maximum temperature difference found in the equipment (T_1–t_1).

$$\Delta T_m = F_{sp}(T_1 - t_1) \tag{6.4}$$

6.1.1 Hot fluid in series and cold fluid in parallel

After extensive algebra, the F_{sp} factor for the hot fluid flowing in the annulus and the cold fluid divided into n_c parallel pipe streams was evaluated in the form of eq. (6.5). For a detailed derivation, please follow Ref. [44].

$$F_{sp} = \frac{P_1 + R_1(1 - P_1) - 1}{R_1 n_c \ln\left(\frac{1}{R_1}\left(R_1\left(\frac{1}{P_1}\right)\frac{1}{n_c} - \left(\frac{1}{P_1}\right)\frac{1}{n_c} + 1\right)\right)} \tag{6.5}$$

$$P_1 = \frac{T_2 - t_1}{T_1 - t_1} \tag{6.6}$$

$$R_1 = \frac{T_1 - T_2}{n_c(t_2 - t_1)} \tag{6.7}$$

Worked Example 6.1

In a service, a battery of double-pipe heat exchangers is associated in series–parallel, with the hot fluid through the annulus in series and temperature ranging from 250 to 100 °C. The cold fluid has a very large flow rate, being partitioned in four parallel streams in the tube of the exchangers, with temperature ranging from 83 to 120 °C. What is the mean temperature difference in this bank?

Solution

Since the cold fluid is in parallel, we should use eq. (6.5), with parameters given from eqs. (6.6) and (6.7) as follows:

$$P_1 = \frac{T_2 - t_1}{T_1 - t_1} = \frac{100 - 83}{250 - 83} = 0.1$$

$$R_1 = \frac{T_1 - T_2}{n_c(t_2 - t_1)} = \frac{250 - 100}{4(120 - 83)} = 1.01$$

The temperature difference factor is:

$$F_{sp} = \frac{0.10 + 1.01(-0.10 + 1) - 1}{1.01 \cdot 4 \ln\left(\frac{1}{1.01}\left(1.01\left(\frac{1}{0.10}\right)^{\frac{1}{4}} - \left(\frac{1}{0.10}\right)^{\frac{1}{4}} + 1\right)\right)} = 0.292$$

Substituting in eq. (6.4), the actual mean temperature difference for the entire bank is:

$$\Delta T_m = F_{sp}(T_1 - t_1) = 0.292(250 - 83) = 48.8\,^\circ C$$

Let us compare this result with the LMTD for countercurrent flow ($\Delta T_{lm.c}$), calculated as follows:

$$\Delta T_{lm,c} = \frac{T_1 - T_2 + t_1 - t_2}{\ln\left(\frac{T_1 - t_2}{T_2 - t_1}\right)} = \frac{-100 - 120 + 250 + 83}{\ln\left(\frac{-120 + 250}{100 - 83}\right)} = 55.5\,^\circ C$$

Considering that the logarithmic mean temperature difference is 55.5 °C and the "actual" reference value is 48.8 °C, the relative error between both results is about 13.8%. Since the actual mean temperature difference is lower, the use of the log mean temperature would lead to an underestimated heat transfer area.

6.1.2 Cold fluid in series and hot fluid in parallel

When the cold fluid goes through the annulus in series and the hot fluid is divided into n_h parallel tube streams, the proper equation for F_{sp} is as follows:

$$F_{sp} = \frac{P_2 R_2 - P_2 - R_2 + 1}{n_h \ln\left(-R_2\left(\frac{1}{P_2}\right)^{\frac{1}{n_h}} + R_2 + \left(\frac{1}{P_2}\right)^{\frac{1}{n_h}}\right)} \tag{6.8}$$

$$P_2 = \frac{T_1 - t_2}{T_1 - t_1} \tag{6.9}$$

$$R_2 = \frac{n_h(T_1 - T_2)}{t_2 - t_1} \tag{6.10}$$

Worked Example 6.2

Consider the same service presented in Worked Example 6.1. At this time, however, the cold fluid is the smaller stream and should be allocated in the annulus in series with its temperature ranging from 83 to 120 °C. The hot fluid will be divided into four parallel streams in the tube of the exchangers, with temperature ranging from 250 to 100 °C. What is the mean temperature difference in this bank?

Solution

With eqs. (6.9) and (6.10), we have:

$$P_2 = \frac{T_1 - t_2}{T_1 - t_1} = \frac{250 - 120}{250 - 83} = 0.78$$

$$R_2 = \frac{n_h(T_1 - T_2)}{t_2 - t_1} = \frac{4(250 - 100)}{120 - 83} = 16.2$$

The temperature factor comes from eq. (6.8):

$$F_{sp} = \frac{0.78 \cdot 16.2 - 0.78 - 16.2 + 1}{4 \ln\left(-16.2\left(\frac{1}{0.78}\right)^{\frac{1}{4}} + 16.2 + \left(\frac{1}{0.78}\right)^{\frac{1}{4}}\right)} = 0.214$$

Substituting in eq. (6.4), the actual mean temperature difference for the entire bank is:

$$\Delta T_m = F_{sp}(T_1 - t_1) = 0.214(250 - 83) = 35.8 \ ^\circ C$$

The LMTD in counterflow in this case is:

$$\Delta T_{lm,c} = \frac{T_1 - T_2 + t_1 - t_2}{\ln\left(\frac{T_1 - t_2}{T_2 - t_1}\right)} = \frac{-100 - 120 + 250 + 83}{\ln\left(\frac{-120 + 250}{100 - 83}\right)} = 55.5 \ ^\circ C$$

Considering that the counter current logarithmic mean temperature difference is 55.5 °C and the "actual" value is only 35.8 °C, the relative error between both results is about 55.2%. Such deviation is a dangerous risk in the design because if ΔT_{lm} were used to size this bank the heat transfer area would be seriously under-specified, and the energy duty would not be fulfilled at all.

6.1.3 Choice of the fluid division

Although a series–parallel association is commonly encountered with tube in parallel and the annulus in series, as in Figs. 6.2 and 6.5, this condition is not mandatory. The decision for splitting the tube or annulus stream is irrelevant for the mean temperature difference between the fluids; therefore, the designer may choose to set the annulus fluid in parallel if necessary.

That said, there is a reason for the division of the inner fluid being far more common industrially, and it is related to piping and fittings. Due to the very structure of a hairpin, with the annuli nozzles placed side by side in a bank, joining the annuli in series is easier and quicker, and requires less piping and man work.

6.2 Pressure drop calculation

6.2.1 Tube pressure drop

In a bank of N_{hp} hairpins of length L_{hp} connected in series–parallel arrangement, with the pipe fluid partitioned in n parallel streams, the pressure loss equation should be modified in the following ways:
1. The actual mass flux in each individual pipe stream is G_i/n.
2. The number of exchangers comprising each train is N_{hp}/n.

With these changes, the total pressure drop for the pipe may be calculated by rewriting eq. (5.27) in the form of eq. (6.11), where K_i is the friction loss coefficient for the inner pipe with internal diameter D_i expressed in millimeters, evaluated from eq. (5.24):

$$\Delta P_i = \frac{G_i^2 K_i}{2n^2 \rho_i} \left(\frac{2N_{hp}}{n} - 1 \right) + \frac{4G_i^2 L_{hp} N_{hp} f_{Fi}}{D_i n^3 \rho_i} \tag{6.11}$$

Notice that the inner tube friction factor (f_{Fi}) is to be evaluated with the actual mass flux of the divided stream G_i/n.

6.2.2 Annulus pressure drop

For a series–parallel association of double-pipe heat exchangers involving only the division of the tube stream, the pressure loss calculation for the annulus passage remains unaltered, as discussed in Section 5.4. Equation (5.33) may be used directly, which is reproduced below for convenience, where N_{hp} is the total number of exchangers with length L_{hp} and hydraulic diameter D_h. The friction loss factor K_o is obtained from eq. (5.36) for bonnet-type return or from eq. (5.40) if the exchanger has a direct return for the annular fluid:

$$\Delta P_o = \frac{G_o^2 K_o N_{hp}}{2\rho_o} + \frac{4G_o^2 L_{hp} N_{hp} f_{Fo}}{D_h \rho_o} \tag{6.12}$$

Worked Example 6.3
A stream of 1.8 kg/s of triethylene glycol is cooled with 2 kg/s water in a bank of 20 series–parallel carbon steel, 4 m long, double-pipe exchangers (hairpin with return bend). Triethylene glycol goes inside the inner tube is divided into two parallel streams, while the water flows through the annulus is in series in the whole bank. The mean bulk temperatures are 71 and 42.7 °C for triethylene glycol and water, respectively. The internal pipe is ID = 52.48 mm and OD = 60.3 mm. The outer tube is ID = 90.12 mm. Estimate the tube and annulus pressure drop for this bank.

Solution

1) Fluid properties

Using the mean bulk temperatures, the physical properties are estimated from Appendices C.8 and C.13 as follows:

Tube: triethylene glycol 344.15 K – 101,325 Pa	Annulus: water 305.15 K – 101,325 Pa
$\mu = 0.00732$ s Pa	$\mu = 0.000775$ s Pa
$\mu = 7.32$ cP	$\mu = 0.775$ cP
$\rho = 1,230.0$ kg/m^3	$\rho = 1,060.0$ kg/m^3
$Cp = 1,860.0$ J/(kg·K)	$Cp = 4,070.0$ J/(kg·K)
$k = 0.196$ W/(m·K)	$K = 0.62$ W/(m·K)
$Pr = 69.2$	$Pr = 5.09$

2) Tube

The inner fluid is divided into two streams; therefore, the actual mass flow rate is:

$$W_i = 0.9 \text{ kg/s}$$

For the internal diameter $D_i = ID = 52.48$ mm, the inner pipe cross section, mass flux, and Reynolds are:

$$S_i = \frac{\pi D_i^2}{4} = \frac{\pi 0.052480^2}{4} = 0.0021631 \text{ m}^2$$

$$G_i = \frac{W_i}{S_i} = \frac{0.90000}{0.0021631} = 416.07 \text{ kg/(m}^2 \cdot \text{s)}$$

$$Re_i = \frac{D_i G_i}{\mu_i} = \frac{0.052480}{0.0073200} 416.07 = 2,983.0 \rightarrow \text{transition}$$

From Appendix B.1, let us take the absolute roughness of a carbon steel pipe surface as $\varepsilon = 0.035$ mm. In this case, the friction factor accounting for the tube roughness can be evaluated conservatively from the Haaland's equation:

$$f_{Fi} = \frac{0.41}{\ln^2\left(0.23\left(\frac{\varepsilon}{D_i}\right)^{\frac{10}{9}} + \frac{6.9}{Re_i}\right)} = \frac{0.41}{\ln^2\left(0.23\left(\frac{(3.5000e-5)}{0.052480}\right)^{\frac{10}{9}} + \frac{6.9}{2,983.0}\right)} = 0.011238$$

We must also halve the total number of exchangers in the bank, then:

$$N_{hp} = \frac{N_{hp,\,total}}{n} = \frac{20.000}{2.0000} = 10.000$$

The distributed pressure drop is calculated from:

$$\Delta P_{i,\,dist} = \frac{4 L_{hp} N_{hp} f_{Fi}}{D_i \rho_i} G_i^2 = \frac{4 \cdot 0.011238 \cdot 10.000}{0.052480 \cdot 1,230.0} 4.0000 \cdot 416.07^2 = 4,822.2 \; kg/\left(m \cdot s^2\right)$$

$$= 4.8222 \; kPa$$

The localized pressure drop for the inner fluid is given by:

$$D_{i(mm)} = 52.48 \; mm$$

$$K_i = 0.7 + \frac{2,000}{Re_i} + \frac{17.78}{D_{i(mm)}} = 0.7 + \frac{17.78}{52.480} + \frac{2,000}{2,983.0} = 1.7093$$

$$\Delta P_{i,\,loc} = \frac{G_i^2 K_i}{2 \rho_i} \left(2 N_{hp} - 1\right) = \frac{1.7093 \cdot 416.07^2}{2 \cdot 1,230.0} \left(2 \cdot 10.000 - 1\right) = 2,285.4 \; kg/\left(m \cdot s^2\right)$$

$$= 2.2854 \; kPa$$

The total tube pressure drop, combining distributed and local losses, is:

$$\Delta P_i = \Delta P_{i,\,dist} + \Delta P_{i,\,loc} = 2,285.4 + 4,822.2 = 7,107.6 \; kg/\left(m \cdot s^2\right) = 7.1076 \; kPa$$

3) Annulus

The heat exchangers annuli are connected in series, then the total flow rate is used in the pressure drop calculation:

$$W_o = 2.0 \; kg/s$$

a) Reynolds number
    The cross-sectional area and mass flux for the annulus are:

$$S_o = \frac{\pi}{4}\left(D_s^2 - D_o^2\right) = \frac{\pi}{4}\left(0.090120^2 - 0.060300^2\right) = 0.0035229 \; m^2$$

$$G_o = \frac{W_o}{S_o} = \frac{2.0000}{0.0035229} = 567.71 \; kg/\left(m^2 \cdot s\right)$$

Using the equivalent diameter as the hydraulic diameter, we have:

$$D_e = D_h = D_s - D_o = 0.090120 - 0.060300 = 0.029820 \; m$$

And finally, the Reynolds number is:

$$Re_o = \frac{D_e G_o}{\mu_o} = \frac{0.029820}{0.00062400} 567.71 = 27,130 \rightarrow \text{fully turbulent}$$

b) Friction factor
    With the Haaland's equation, the Fanning friction factor is estimated as follows:

$$f_{Fo} = \cfrac{0.41}{\ln^2\left(0.23\left(\cfrac{\mathrm{\grave{o}}}{D_e}\right)^{\frac{10}{9}} + \cfrac{6.9}{Re_o}\right)} = \cfrac{0.41}{\ln^2\left(0.23\left(\cfrac{(3.5000e-5)}{0.029820}\right)^{\frac{10}{9}} + \cfrac{6.9}{27,130}\right)}$$

$$= 0.0066190$$

c) Pressure drop

The total number of hairpins in series is:

$$N_{hp} = N_{hp,\,total} = 20.000 = 20.000$$

From eq. (5.28), the distributed pressure loss can be calculated as:

$$\Delta P_{o,\,dist} = \frac{4 L_{hp} N_{hp} f_{Fo}}{D_h \rho_o} G_o^2 = \frac{4 \cdot 0.0066190 \cdot 20.000}{0.029820 \cdot 1,050.0} 4.0000 \cdot 567.71^2$$

$$= 21,802 \ \mathrm{kg}/\left(\mathrm{m} \cdot \mathrm{s}^2\right) = 21.802 \ \mathrm{kPa}$$

The friction loss coefficient for a straight pipe return from eq. (5.40) is:

$$D_{h(mm)} = 29.82 \ \mathrm{mm}$$

$$K_o = 2.8 + \frac{2,000}{Re_o} + \frac{71.12}{D_{h(mm)}} = 2.8 + \frac{71.12}{29.820} + \frac{2,000}{27,130} = 5.2587$$

The local pressure drop is then obtained as follows:

$$\Delta P_{o,\,loc} = \frac{G_o^2 K_o N_{hp}}{2 \rho_o} = \frac{20.000 \cdot 5.2587 \cdot 567.71^2}{2 \cdot 1,050.0} = 16,141 \ \mathrm{kg}/\left(\mathrm{m} \cdot \mathrm{s}^2\right) = 16.141 \ \mathrm{kPa}$$

And the total pressure loss in the annulus is the sum of distributed and local losses:

$$\Delta P_o = \Delta P_{o,\,dist} + \Delta P_{o,\,loc} = 16,141 + 21,802 = 37,943 \ \mathrm{kg}/\left(\mathrm{m} \cdot \mathrm{s}^2\right) = 37.943 \ \mathrm{kPa}$$

Problems

(6.1) In a battery of double-pipe heat exchangers, the hot fluid temperature varies from 50 to 30 °F, while the cold fluid temperature ranges from 22 to 41 °F. The heat load is $6.7e + 5$ BTU/h and the overall heat transfer coefficient was evaluated as $U = 103$ BTU/(ft$^2 \cdot$ h $\cdot$ °F). Assuming the overall heat transfer coefficient is nearly constant, determine the effective mean temperature difference and the required heat transfer area for this service when a series–parallel arrangement is used with (a) the cold fluid and (b) the hot fluid divided into three parallel streams.

Answer:
(a) $\Delta T_m = 3.8657\ °F$, $A = 1{,}682.7\ ft^2$
(b) $\Delta T_m = 2.8095\ °F$; $A = 2{,}315.3\ ft^2$

(6.2) The benzene/toluene heat transfer service examined in Problems 5.3 and 5.4 could not be accomplished by fluid relocation. Either way, the benzene stream exceeded the allowable pressure loss. Analyze an alternative design in which the bank of six exchangers is restructured with benzene allocated in the tube divided into two parallel streams, while the exchangers' annuli are kept connected in series.
a) Determine tube friction factor, using Haaland's equation
b) Determine annulus friction factor, using Haaland's equation
c) Determine distributed and localized pressure loss in the tube
d) Determine distributed and localized pressure loss in the annulus
e) Determine heat transfer coefficients for the internal and external fluids
f) Determine clean overall heat transfer coefficient (ignoring the wall conductive resistance)
g) Determine clean overall heat transfer coefficient
h) Does the issue of exceeding the allowable pressure drop was solved?
i) Can this series–parallel bank arrangement handle the required heat load?

Answer:
(a) $f_{Fi} = 0.004632$
(b) $f_{Fo} = 0.0053007$
(c) $\Delta P_{i,dist} = 12.927\ kPa$; $\Delta P_{i,loc} = 8.0132\ kPa$
(d) $\Delta P_{o,dist} = 25.967\ kPa$; $\Delta P_{o,loc} = 23.767\ kPa$
(e) $h_i = 2213.3\ W/(m^2 \cdot K)$; $h_o = 1{,}955.6\ W/(m^2 \cdot K)$
(f) $U_c = 970.41\ W/(m^2 \cdot K)$
(g) $U_c = 776.46\ W/(m^2 \cdot K)$
(h) Yes (justify!)
(i) No (justify!)

(6.3) A fluid is heated from 10 to 50 °C with 3.3 kg/s of another process stream varying from 55 to 40 °C. The heat is transferred by a battery of 33 hairpins, where the annuli are connected in series and the hot fluid flows inside the pipe is divided into three parallel streams, each one passing through 11 hairpins.
a) What is the mean temperature difference for this battery?
b) If the divided stream is the cold fluid, what is the mean temperature difference?

Answer:
(a) $\Delta T_m = 11.491\ °C$
(b) Not defined

(6.4) Acetic acid at a flow rate of 10.0 kg/s is cooled from 65 to 50 °C using cold water from 20 to 35 °C. The service is accomplished in a bank of four carbon steel double-pipe DN 50×90 mm/10S, 7 m long, exchangers with their annuli connected in series. Acetic acid flows in two parallel streams inside the tubes, while the water flows through the annulus. Determine:
a) Mean temperature difference for this battery
b) Tube and annulus convective heat transfer coefficient
c) Clean overall heat transfer coefficient
d) Pressure loss in the tube and annulus
e) Percentage of the over-surface

Answer:
(a) $\Delta T_m = 29.462\ °C$
(b) $h_i = 2846.4\ W/(m^2 \cdot K)$; $h_o = 4820.9\ W/(m^2 \cdot K)$
(c) $U_c = 1535.5\ W/(m^2 \cdot K)$
(d) $\Delta P_i = 35.188 kPa$; $\Delta P_o = 28.838 kPa$
(e) Over-surface $= 65\%$

(6.5) Consider the service specified in Problem 6.4. Test if a greater design margin can be attained by inverting the fluid allocation, associating the tubes in series, and dividing the annulus fluid into two parallel streams. Evaluate:
a) Mean temperature difference for this battery
b) Tube and annulus convective heat transfer coefficient
c) Clean overall heat transfer coefficient
d) Pressure loss in the tube and annulus
e) Percentage of over-surface

Answer:
(a) $\Delta T_m = 29.462\ °C$
(b) $h_i = 7,526.0\ W/(m^2 \cdot K)$; $h_o = 1,800.0\ W/(m^2 \cdot K)$
(c) $U_c = 1317.8\ W/(m^2 \cdot K)$
(d) $\Delta P_i = 53.159$ kPa; $\Delta P_o = 19.702$ kPa
(e) Over-surface $= 41.5\%$

7 Heat exchanger design

When dealing with heat exchanger design, the engineer's most common tasks are twofold:
1. Sizing a new heat exchanger.
2. Rating the performance of an existing heat exchanger.

The sizing procedure is detailed in Section 7.4, and the rating task is discussed in Section 7.5. In principle, sizing or rating a heat exchanger is a simple matter of following the calculations in a rather straightforward routine; however, the engineer needs to make important decisions in key steps, which, in the end, may determine the effectiveness of the design. Among these decisions are the fluid allocation and the acceptable margin embedded in the final design.

7.1 Fluid allocation

Commonly, one of the first decisions the engineer has to make is related to the fluid allocation, i.e., which fluid will be channeled through the inner tube or the annulus (shell) of the double-pipe heat exchanger. The allocation of fluids has a direct impact on the equipment capital and operating costs.

There is no strict rule for choosing if a fluid stream should flow internally or externally; therefore, each heat transfer service must be carefully considered under its own particularities. As a general guide, a more effective design can be achieved if the stream allocated in the inner tube has the following characteristics [94–96]:
1. More corrosive, erosive, or hazardous.
   The larger allowances for corrosion or erosion make the exchanger more expensive, once the required tube thickness may increase significantly. Placing the more aggressive fluid inside the tube prevents the selection of heavier schedules for both inner and outer tubes, since only the inner tube will be exposed to an accelerated damage.
   As it might seem obvious, the reason for allocating the hazardous (i.e., harmful or toxic) stream inside the tube is safety. If the leakage of one fluid involves great risk for people or to the environment, the allocation in the inner tube is strongly recommended. In this case, if the fluid leaks, it remains contained by the outer pipe.
2. More fouling.
   In general, the inner tube passage is more accessible for mechanical cleaning. Therefore, the allocation of the stream with an increased propensity for fouling in the tube results in lower maintenance costs and shorter downtime.

https://doi.org/10.1515/9783110585872-007

While the fouling formation may not be critical for fluid allocation in double-pipe exchangers, it might be a key factor to make this decision when designing multi-tube and finned heat exchangers, or even other types such as shell-and-tube exchangers.

3. Excessively high temperature and operating pressure.
   Tubular heat exchangers operating with streams in extreme conditions of temperature and pressure require thick tubes to sustain the physical integrity even under high mechanical stresses. Allocating the high-temperature or high-pressure stream in the inner tube usually leads to the most cost-effective design, since a thinner outer tube (shell) may be applied.

4. Lower viscosity.
   Usually, the fluid of lower viscosity produces the higher heat transfer coefficient. It is easier and less expensive to install external fins; therefore, having the controlling fluid in the annulus (shell) allows the placement of external fins, which may significantly increase the overall heat transfer coefficient of the exchanger.

5. Hotter fluid.
   Typically, the methods for designing heat exchangers presume that all energy lost by the hot fluid goes to the cold fluid, i.e., no energy is lost to the environment. Allocating the hot fluid in the annulus undermines this assumption, especially if the stream operates at relatively high temperature. To minimize the problem, the heat exchanger may be thermally insulated, implying in an additional cost.

7.2 Design margin

In broad terms, "design margin" in an engineering project refers to a performance excess attainable by the equipment, beyond what was primarily requested to the designer. The same concept appears with several denominations in the technical literature, such as "over-area," "excess area," "excess surface," "over-sizing," and "safety factor." The engineer must be aware of the exact definition stated in these sources.

For a heat exchanger, performance is usually related to its heat duty or heat transfer area. A deep discussion of design margins for heat exchangers is done in Refs. [6, 94, 97], which are recommended for the interested student; however, the key aspects are as follows:

1. Design margin can be good and can be bad, depending on its magnitude.
2. An appropriate design margin provides flexibility and operational robustness to the equipment against unadvised process disturbances, feedstock changes, environmental variations, etc.
3. The design margin provides room for plant revamps or increases of production without significant equipment changes.

4. While a design margin may increase the equipment response time, and consequently less sensitive to sudden variations, an excessive margin can make the control less manageable.
5. An excessive design margin increases the capital costs, produces larger and heavier exchangers (very undesirable in offshore applications), and decreases stream velocities, promoting fouling. This acceleration of the deposit formation is particularly a serious drawback.

When designing heat exchangers, there are three "kinds" of heat transfer areas considered for the equipment: the clean, design, and actual (final) heat transfer areas, denoted by A_c, A_d, and A, respectively. The clean heat transfer area is obtained from the design equation (2.8) using the clean overall heat transfer coefficient U_c, as:

$$A_c = \frac{q}{U_c \Delta T_m} \tag{7.1}$$

After combining the fouling factors, the design overall heat transfer coefficient U_d is obtained, allowing the evaluation of the design heat transfer area as:

$$A_d = \frac{q}{U_d \Delta T_m} \tag{7.2}$$

Commonly, the design heat transfer area A_d is rounded up to adhere to some standardization or requirement of the proposed design (e.g., available pipe length, to enforce an even number of exchangers in a bank or by adding safety factors). After that, the actual heat transfer area A for the exchanger is obtained, giving the actual (final) overall heat transfer coefficient U.

The relation among these areas is such that: $A > A_d > A_c$. Therefore, for a heat exchanger, the "design margin" can be defined as the relative difference between the actual (final) heat transfer area (A) and a reference heat transfer area required to perform properly the specified service (A_d or A_c). These quantities are named over-design and over-surface, respectively.

7.2.1 Over-design

The excess area between the actual (A) and design (A_d) areas is usually named over-design, evaluated in the form:

$$\text{Over-design}(\%) = 100 \left(\frac{A}{A_d} - 1 \right) = 100 \left(\frac{U_d}{U} - 1 \right) \tag{7.3}$$

7.2.2 Over-surface

The over-surface is a measure of the total excess area taking as reference the heat transfer area (A_c) required by a clean heat exchanger to meet the duty:

$$\text{Over-surface}(\%) = 100\left(\frac{A}{A_c} - 1\right) = 100\left(\frac{U_c}{U} - 1\right) \tag{7.4}$$

7.2.3 Fouling over-surface

The impact of the provision of fouling resistances upfront in the design can be quantified using the definition of "fouling over-surface," which is expressed in the following form:

$$\text{Fouling over-surface }(\%) = 100\left(\frac{A_d}{A_c} - 1\right) = 100\left(\frac{U_c}{U_d} - 1\right) \tag{7.5}$$

Fouling factors should be prescribed with care, since they can introduce an excessive over-design, not necessary to meet the actual process requirements. In fact, the application of high fouling factors can be the very cause of the deposit formation, as a result of the lower fluid velocities. Even though the indiscriminate use of fouling factors is not recommended, if a process stream is known to be prone to solid deposition, they must be applied by the designer. As a general (but not rigid) rule, in typical situations, the excess area introduced by the fouling resistances should outcome a fouling over-surface of around 30% or less.

7.2.4 Design margin ranges

There is no absolute rule for defining the margin built in a heat exchanger design. After all, this is a designer/engineer call, based on their specific knowledge about the process in question and the accuracy of available tools (methods, software, etc.) and physical property data. As a general reference, an over-design of 5–10% may be considered acceptable.

For designs involving dirty streams, the over-surface can be appreciably higher, since it includes the used fouling factors. In such cases, typical values may easily approach 40–50%.

7.3 Physical properties of chemical components

Physical properties of substances are a cornerstone of any chemical process design. They are reasonably susceptible to theoretical and/or experimental uncertainties and must be used with much care. The designer should be always suspicious about physical property data and cross-check it twice against multiple sources, if possible. On the other hand, physical properties of chemical compounds are not abundantly available, and having multiple data sources to compare with is usually a luxury. The situation is worse for mixtures, since the variation of compositions and concentrations creates unlimited possibilities.

Nowadays, the day-to-day routine of chemical process designers and analysts involve the evaluation of physical properties directly from process simulation computer programs. These simulators allow the estimation of physical properties for pure components and mixtures for a given pressure and temperature, by the application of some appropriate mixing rule. There are paid (e.g., Aspen [98], Hysys [99], Pro/II [100], Chemcad [101]) and free (e.g., COCO [102] and EMSO [103]) chemical process simulators, and a few of them have the invaluable bonus of being free and open source (e.g., DWSim [104] and ASCEND [105]). A comprehensive list of commercial simulators may be found in Refs. [106, 107].

A chemical process simulator that I personally recommend is DWSim [108]. Its open-source nature makes it an ideal tool for teaching purposes, since the students may see in bare eyes all the models and methods that go on under the hood when performing a chemical process simulation. Besides that, this great piece of software is continuously becoming more reliable and powerful for practical industrial applications.

Although the daily engineering practice comprises the use of property packages integrated in the process simulator software, we need "in paper" data to learn the subject matter presented herein. To offer support in the solution of the problems proposed in this text, Appendix C brings a complete set of the required physical properties for several chemical components. The provided plots of dynamic viscosity, specific heat, specific mass, and thermal conductivity are based on the same methods available in the DWSim simulator, which means that the student or chemical process professional interested in learning or validating by themselves the calculations underneath can do that by direct inspection of the simulator source code.

It is noteworthy that the physical properties given in Appendix C have the sole purpose of being instructional support material. Even though this data may be considered with enough accuracy for some practical applications, the reader is encouraged to use experimental data from manuals and handbooks available in the technical literature wherever possible.

7.4 Sizing

Sizing a heat exchanger comprises essentially the thermal and hydraulic specification of a new equipment that should be built to accomplish a given heat transfer service. This includes the selection of a heat exchanger type and configuration, and determination of its dimensions, flow arrangement, and, in some cases, the inlet and outlet temperatures or flow rate of a utility fluid.

For a heat exchanger that contacts two process streams, their operating temperatures and flow rates are dictated by the process, i.e., by the operations upstream and downstream the heat exchanger. If the equipment transfers heat from/to a process stream and a utility fluid, such as cooling water, steam, or thermal fluids, there is more freedom for the specification of the outlet conditions and flow rates of the utility, since the impact on the operation of the process is mostly limited to the equipment producing or recycling such fluids.

Once defined the flow arrangement and heat exchanger type/model, the sizing procedure aims to answer two questions:

1. What is the required heat transfer area (size) to match the service?
2. Does the exchanger with the designed size exceed the available pressure drop in the process line?

Notably, the most important question is: (1) the required heat transfer area. Question (2) is a verification step; however, its importance should not be underestimated, since the installation of a heat exchanger with exceeded pressure drop may cause the flow shutdown of the whole equipment train in a process line.

7.4.1 The design variables

The designer may specify the surface necessary to perform a heat transfer duty in several ways. Provided the type/model of the heat exchanger is selected, some design variable is tuned to yield the necessary heat transfer area.

Typically, the design variable is the dimension or the number of the equipment surface element, which varies according to the heat exchanger class or geometry. For example, in the case of planar heat exchangers, the surface elements are "plates"; therefore, with complete freedom, the designer can select the plate size and shape and calculate the number of these plates needed to attain the required duty. In some cases, the plate size and shape are restricted, for whatever reason, such as available stock, space limitations, existent supplier contracts, standardization/code, and recommended practice. Therefore, the designer cannot choose the plate characteristics, and the only remaining design variable is the quantity of plates to be assembled.

For tubular heat exchangers, e.g., double-pipe, multi-tube hairpins, or shell and tube, the surface elements are, as expected, tubes, and the design variable may be the diameter, length, or number of tubes. Sometimes, all of them are to be specified by the designer; however, it is commonplace to have the tube set selected by some heuristic rule and the length of the tubes to be the design variable used to define the heat exchanger area. If there are space restrictions limiting the length of the exchanger, the diameter or number of tubes becomes the design variable adjusted to fulfill the heat transfer area.

Usually, several double-pipe exchangers (hairpins) should be associated in banks (or batteries) to accomplish an ordinary energy duty. This fact introduces the number of hairpins as a design variable for tuning the heat transfer surface. Hence, for this type of heat exchanger, depending on the previously imposed design restrictions, the engineer should possibly deal with three design variables:
1. Inner and outer pipe diameters
2. Pipe length
3. Number of hairpins (associated to form a bank)

If all three variables are unrestricted, there are literally infinite designs to choose from, and some optimization method upon an objective function is the only way to find the best solution. For "manual" designs, some heuristics are necessary to select a pipe set, with specified diameters and length. After the hairpin pipe diameters and length are defined, the sizing task reduces to the determination of the number of hairpins connected in the bank.

7.4.2 Sizing outline

The problem of designing the double-pipe heat exchanger may be summarized in the following steps:
1   **Thermal design (sizing)**
   1.1   Collect process data parameters
      1.1.1   Fluid temperatures
      1.1.2   Physical properties
   1.2   Energy balance using mean bulk temperatures of the fluids
      1.2.1   Determine any unknown fluid temperature or flow rate
      1.2.2   Or just check the energy balance between fluids for correctness
   1.3   Select a heat exchanger type: double-pipe, multi-tube, or finned heat exchanger in the present case
   1.4   Select the design variable
      1.4.1   Pipe length, if diameter is specified
      1.4.2   Pipe diameter, if length is specified
      1.4.3   Number of hairpins, if the service should be performed by a set of standardized hairpin exchangers

1.5 Evaluate the actual mean temperature difference to be used in the heat exchanger design equation
1.6 Overall heat transfer coefficient
    1.6.1 Calculate the inner pipe heat transfer coefficient
    1.6.2 Calculate the annulus heat transfer coefficient
    1.6.3 Calculate wall temperature
    1.6.4 Iterate steps 1.6.1–1.6.3 until convergence of the wall temperature is achieved
    1.6.5 Calculate the clean overall heat transfer coefficient U_c
    1.6.6 Calculate the design (required) overall heat transfer coefficient U_d by introducing the fouling factors R_{di} and R_{do}
1.7 Heat transfer areas
    1.7.1 Calculate the design (required) heat transfer area A_d
    1.7.2 Specify the heat exchanger geometry to match the required heat transfer surface
        1.7.2.1 Number of hairpins
        1.7.2.2 Pipe diameter and/or number of internal tubes for multi-tube hairpins
    1.7.3 Calculate the actual (final) heat transfer area A
    1.7.4 Calculate the actual (final) overall heat transfer coefficient U for the exchanger
1.8 Calculate the over-surface and over-design
2 **Hydraulic design (verify if allowable pressure drop is not exceeded)**
2.1 Evaluate the inner pipe pressure drop
2.2 Evaluate the annulus pressure drop

Worked Example 7.1
Determine the required number of hairpins to heat a benzene stream of 0.7 kg/s from 20 to 55 °C using 2-butanol cooling from 60 to 40 °C. There are available double-pipe heat exchangers with 5 m length, built with NPS 32 × 50 mm, sch 40, stainless steel grade 316 pipes. For this service, the hot fluid should be allocated in the inner pipe.

Solution
1) Physical properties of fluids
  a) Mean temperatures of fluids
    The four process temperatures are specified; therefore, we can estimate the necessary fluid properties using the arithmetic mean of the inlet and outlet temperatures for each stream:

$$T_{1m} = \frac{T_{1,in}}{2} + \frac{T_{1,out}}{2} = \frac{293.15}{2} + \frac{328.15}{2} = 310.65\,K$$

$$T_{2m} = \frac{T_{2,\text{in}}}{2} + \frac{T_{2,\text{out}}}{2} = \frac{313.15}{2} + \frac{333.15}{2} = 323.15\,\text{K}$$

b) Fluid properties at these mean stream temperatures are evaluated using Appendices C.9 and C.10 for 2-butanol and benzene, respectively:

Fluid 1 (cold) → annulus	Fluid 2 (hot) → tube
Benzene	2-Butanol
310.65 K – 101,325 Pa	323.15 K – 101,325 Pa
$\mu = 0.000511$ s · Pa	$\mu = 0.00134$ s · Pa
$\mu = 0.511$ cP	$\mu = 1.34$ cP
$\rho = 860.0$ kg/m^3	$\rho = 865.0$ kg/m^3
$C_p = 2,370.0$ J/(kg · K)	$C_p = 1,530.0$ J/(kg · K)
$k = 0.13$ W/(m · K)	$k = 0.14$ W/(m · K)
Pr = 5.58	Pr = 24.5

c) The energy balance is used to determine the heat load and the 2-butanol flow rate:

$$Cp_2 W_2(T_{2,\text{out}} - T_{2,\text{in}}) + W_1 Cp_1(T_{1,\text{out}} - T_{1,\text{in}}) = 0$$

Solving for W_2, we have the 2-butanol flow rate:

$$W_2 = \frac{W_1 Cp_1(T_{1,\text{out}} - T_{1,\text{in}})}{Cp_2(T_{2,\text{in}} - T_{2,\text{out}})} = \frac{0.7 \cdot 1,530.0(328.15 - 293.15)}{2,370.0(-313.15 + 333.15)} = 0.791\,\text{kg/s}$$

Checking the enthalpy change for both fluids, we confirm they are about the same within round-off approximations. Therefore, the energy balance is correct:

$$q_1 = W_1 Cp_1(T_{1,\text{out}} - T_{1,\text{in}}) = 0.7 \cdot 1,530.0(-293.15 + 328.15) = 37,485.0\ \text{W}$$

$$q_2 = Cp_2 W_2(T_{2,\text{out}} - T_{2,\text{in}}) = 0.791 \cdot 2,370.0(313.15 - 333.15) = -37,493.0\ \text{W}$$

2) Logarithmic mean temperature difference

For counterflow arrangement, temperatures of fluids at the heat exchanger terminals are

$$T_{1,\text{in}} = 293.15\,\text{K} \quad \rightarrow \quad T_{1,\text{out}} = 328.15\,\text{K}$$
$$T_{2,\text{out}} = 313.15\,\text{K} \quad \leftarrow \quad T_{2,\text{in}} = 333.15\,\text{K}$$

And the logarithmic mean temperature difference is evaluated from:

$$\Delta T_{lm,c} = \frac{T_{1,in} - T_{2,out} + T_{2,in} - T_{1,out}}{\ln\left(\dfrac{T_{1,in} - T_{2,out}}{T_{1,out} - T_{2,in}}\right)} = \frac{293.15 - 313.15 - 328.15 + 333.15}{\ln\left(\dfrac{293.15 - 313.15}{328.15 - 333.15}\right)} = -10.82\,K$$

Notice that the order for taking the fluid temperature differences at the terminals was chosen purposefully to illustrate the fact that the negative sign of ΔT_{lm} is irrelevant in this context. It simply indicates that the terminal temperature differences were calculated from cold to hot fluid, in this order, and that the heat transfer rate should be used with negative sign when calculating the heat transfer area from the design equation $q = UA\,\Delta T_{lm}$.

3) Pipe dimensions
   The inner pipe is NPS 32 mm – sch 40 and the external pipe is specified as NPS 50 mm/sch 40; from Tab. 1 in Appendix A.1 the inner and outer diameters for the inner pipe are:

$$ID = D_i = 35.08\,mm$$

$$OD = D_o = 42.0\,mm$$

For the external pipe, only the inner diameter is necessary in the heat transfer calculations, which is:

$$ID = D_s = 52.48\,mm$$

4) Tube heat transfer coefficient
   Cross section of the inner pipe, mass flux, and Reynolds number are:

$$S_i = \frac{\pi D_i^2}{4} = \frac{\pi 0.03508^2}{4} = 0.00096652\,m^2$$

$$G_i = \frac{W_i}{S_i} = \frac{0.791}{0.00096652} = 818.4\,kg/(m^2 \cdot s)$$

$$Re_i = \frac{D_i G_i}{\mu_i} = \frac{0.03508}{0.00134}818.4 = 21,425.0 \rightarrow \text{fully turbulent}$$

With this Reynolds number in common flow conditions, the regime is turbulent. The Sieder–Tate equation (4.10) may be used to estimate the heat transfer coefficient. As the first approximation, let us disregard the viscosity correction by assuming $\phi_i = (\mu_i/\mu_w)^{0.14} = 1$ as follows:

$$Nu = 0.027 Pr^{0.33} Re^{0.8} \left(\frac{\mu_i}{\mu_w}\right)^{0.14}$$

$$h_i = \frac{0.027k}{D_i}\sqrt[3]{Pr}Re^{\frac{4}{5}} = \frac{0.027\sqrt[3]{24.5}}{0.03508}0.13 \cdot 21,425.0^{\frac{4}{5}} = 847.3\,W/(m^2 \cdot K)$$

Notice that the application of eq. (4.10) assumes fully developed flow. Although the whole pipe length is not known at this time, for the minimum number of only one hairpin in the bank, the corresponding number of diameters at the entrance surpasses by far the needed to achieve full development:

$$\#\text{Diameters} = \frac{L_{hp}}{D_i} = \frac{5.0}{0.03508} = 142.53$$

5) Annulus heat transfer coefficient

The annulus equivalent diameter used in the Reynolds number is taken here as equal to the hydraulic diameter:

$$D_e = D_s - D_o = 0.05248 - 0.042 = 0.01048 \text{ m}$$

$$S_o = \frac{\pi}{4}\left(D_s^2 - D_o^2\right) = \frac{\pi}{4}\left(0.05248^2 - 0.042^2\right) = 0.00077766 \text{ m}^2$$

$$G_o = \frac{W_o}{S_o} = \frac{0.7}{0.00077766} = 900.14 \text{ kg}/(\text{m}^2 \cdot \text{s})$$

$$Re_o = \frac{D_e G_o}{\mu_o} = \frac{0.01048}{0.000511}\,900.14 = 18,461.0 \rightarrow \text{fully turbulent}$$

For fully developed turbulent flow, and ignoring the viscosity correction as $\phi_o = (\mu_o/\mu_w)^{0.14} = 1$, the annulus film coefficient is estimated from the Sieder–Tate equation:

$$Nu = 0.027 Pr^{0.33} Re^{0.8} \left(\frac{\mu_o}{\mu_w}\right)^{0.14}$$

$$h_o = \frac{0.027k}{D_e}\sqrt[3]{Pr}Re^{\frac{4}{5}} = \frac{0.027\sqrt[3]{5.58}}{0.01048}\,18,461.0^{\frac{4}{5}} = 1,655.8 \text{ W}/(\text{m}^2 \cdot \text{K})$$

6) Variable properties correction factor

a) Wall temperature and correction factors

To further improve the accuracy of the heat transfer coefficient, the wall temperature is calculated using the first approximations of the heat transfer coefficients:

$$t_w = \frac{D_i h_i t_i + D_o h_o t_o}{D_i h_i + D_o h_o} = \frac{0.03508 \cdot 323.15 \cdot 847.3 + 0.042 \cdot 1,655.8 \cdot 310.65}{0.03508 \cdot 847.3 + 0.042 \cdot 1,655.8} = 314.39 \text{ K}$$

The fluid properties at the wall (from Appendices C.9 and C.10) are applied to correct the pipe and annulus film coefficients as follows:

Fluid 1 (cold)	Fluid 2 (hot)
Benzene	2-Butanol
314.39 K – 101, 325 Pa	314.39 K – 101, 325 Pa
$\mu = 0.000488$ s · Pa	$\mu = 0.00177$ s · Pa

$$\phi_i = \left(\frac{\mu_i}{\mu_w}\right)^{0.14} = \left(\frac{0.00134}{0.00177}\right)^{0.14} = 0.96179$$

$$\phi_o = \left(\frac{\mu_o}{\mu_w}\right)^{0.14} = \left(\frac{0.000511}{0.000488}\right)^{0.14} = 1.0065$$

The corrected heat transfer coefficients are

$$h_{i,\,corr} = h_i \phi_i = 0.96179 \cdot 847.3 = 814.92 \ W/(m^2 \cdot K)$$

$$h_{o,\,corr} = h_o \phi_o = 1.0065 \cdot 1,655.8 = 1,666.6 \ W/(m^2 \cdot K)$$

7) Overall heat transfer coefficient
The clean overall heat transfer coefficient, including the wall conductive resistance, the internal and external convective resistances is:

$$U_c = \frac{1}{\frac{D_o}{2k}\ln\left(\frac{D_o}{D_i}\right) + \frac{1}{h_o} + \frac{D_o}{D_i h_i}} = \frac{1}{\frac{0.042\ln\left(\frac{0.042}{0.03508}\right)}{2 \cdot 16.269} + \frac{1}{1,666.6} + \frac{0.042}{0.03508 \cdot 814.92}}$$

$$= 434.48 \ W/(m^2 \cdot K)$$

The provision for fouling for both fluids is given by the required fouling resistances R_{di} and R_{do}:

$$R_{di} = 0.0002 \ (m^2 \cdot K)/W$$

$$R_{do} = 0.0002 \ (m^2 \cdot K)/W$$

Therefore, the design overall heat transfer coefficient used to size the exchanger area is given by:

$$U_d = \frac{1}{R_{do} + \frac{1}{U_c} + \frac{D_o R_{di}}{D_i}} = \frac{1}{0.0002 + \frac{0.0002}{0.03508}0.042 + \frac{1}{434.48}} = 364.82 \ W/(m^2 \cdot K)$$

8) Number of hairpins
a) Heat transfer area
Solving the design equation for the required heat transfer area A_d for performing the service, using the calculated design overall heat transfer coefficient, we get:

$$q = A_d U_d \Delta T_{lm}$$

$$A_d = \frac{q}{U_d \Delta T_{lm}} = \frac{37,485.0}{10.82 \cdot 364.82} = 9.4962 \ m^2$$

b) Total pipe length and number of exchangers

The hairpin linear heat transfer area (based on the external surface of the inner pipe) is:

$$A_{\text{linear}} = \pi D_o = \pi 0.042 = 0.13195 \text{ m}^2/\text{m}$$

Therefore, given the total required area of 9.6685 m², the necessary pipe length and number of exchangers in the bank are:

$$L = \frac{A_d}{A_{\text{linear}}} = \frac{9.4962}{0.13195} = 71.968 \text{ m}$$

$$N_{\text{hp}} = \frac{L}{2L_{\text{hp}}} = \frac{71.968}{2 \cdot 5.0} = 7.1968 \rightarrow 8 \text{ hairpins}$$

Worked Example 7.2

Consider a service where 10,000 lb/h of benzene is heated from 60 to 120 °F in a bank of double-pipe heat exchangers using a stream of aniline cooled from 150 to 100 °F. There are available hairpins NPS 1.25 × 2 in – Sch 40, 16 ft long, manufactured in 316 stainless steel with thermal conductivity $k = 9.4$ BTU/(h ft °F) and absolute roughness $\varepsilon = 0.000895$ in. The annulus fluid return is of bonnet type. Due to pumping restrictions, the allowable pressure drop for each line is 20 psi. Propose a feasible design, specifying the number and configuration of an exchanger bank for performing this duty. A fouling factor of $R_d = 0.001$ (ft² h °F)/BTU should be provisioned for each stream.

Solution

This same problem was proposed in Ref. [28] as a solved example. Let us take that as a benchmark, approaching the same design, however, relying on a distinct source of physical properties and some alternative equations for heat transfer coefficients and friction factors. For easier comparison, if the reader is interested in doing so, the solution will be developed in US customary units.

Design 1

For the first configuration, we assume that benzene (fluid 1) is allocated in the inner tube and aniline (fluid 2) flowing inside the annulus.
1) Physical properties of fluids
   a) Mean temperatures of fluids
      The mean temperatures for each stream will be used for initial estimation of the physical properties:

$$T_{1m} = \frac{T_{1,\text{in}}}{2} + \frac{T_{1,\text{out}}}{2} = \frac{120.00}{2} + \frac{60.000}{2} = 90.000 \text{ °F}$$

$$T_{2m} = \frac{T_{2,\text{in}}}{2} + \frac{T_{2,\text{out}}}{2} = \frac{100.00}{2} + \frac{150.00}{2} = 125.00 \text{ °F}$$

b) Fluid properties at these mean stream temperatures are evaluated using Appendices C.10 and C.12 for benzene and aniline, respectively:

Fluid 1 (cold)	Fluid 2 (hot)
Benzene	Aniline
90.0°F – 14.7 psi	125.0°F – 14.7 psi
$\mu = 1.32$ lb/(ft·h)	$\mu = 4.31$ lb/(ft·h)
$\mu = 0.546$ cP	$\mu = 1.78$ cP
$\rho = 54.0$ lb/ft^3	$\rho = 64.3$ lb/ft^3
$Cp = 0.36$ BTU/(lb·°F)	$Cp = 0.414$ BTU/(lb·°F)
$k = 0.082$ BTU/(ft·h·°F)	$k = 0.0985$ BTU/(ft·h·°F)
Pr = 5.8	Pr = 18.1

c) The energy balance is used to determine the heat load and the flow rate of aniline:

$$Cp_2 W_2(T_{2,\text{out}} - T_{2,\text{in}}) + W_1 Cp_1(T_{1,\text{out}} - T_{1,\text{in}}) = 0$$

Isolating W_2 and evaluating, we have:

$$W_2 = \frac{W_1 Cp_1(T_{1,\text{out}} - T_{1,\text{in}})}{Cp_2(T_{2,\text{in}} - T_{2,\text{out}})} = \frac{0.36000 \cdot 10,000(120.00 - 60.000)}{0.41400(-100.00 + 150.00)} = 10,435 \text{ lb/h}$$

The heat load can be confirmed to be the same for both streams as:

$$q_1 = W_1 Cp_1(T_{1,\text{out}} - T_{1,\text{in}}) = 0.36000 \cdot 10,000(120.00 - 60.000)$$
$$= (2.1600e + 5) \text{ BTU/h}$$
$$q_2 = Cp_2 W_2(T_{2,\text{out}} - T_{2,\text{in}}) = 0.41400 \cdot 10,435(100.00 - 150.00)$$
$$= (-2.1600e + 5) \text{ BTU/h}$$

2) Logarithmic mean temperature difference
For counterflow arrangement, temperatures of fluids at the heat exchanger terminals are:

Terminal 1		Terminal 2
$T_{1,\text{in}} = 60.0$ °F	$\rightarrow$	$T_{1,\text{out}} = 120.0$ °F
$T_{2,\text{out}} = 100.0$ °F	$\leftarrow$	$T_{2,\text{in}} = 150.0$ °F
$\Delta T_1 = -40.000$ °F		$\Delta T_2 = -30.000$ °F

With the temperature differences at the terminals, the log mean temperature difference is:

$$\Delta T_1 = T_{1,\text{in}} - T_{2,\text{out}} = 60.000 - 100.00 = -40.000\,^\circ\text{F}$$

$$\Delta T_2 = T_{1,\text{out}} - T_{2,\text{in}} = 120.00 - 150.00 = -30.000\,^\circ\text{F}$$

$$\Delta T_{\text{lm,c}} = \frac{\Delta T_1 - \Delta T_2}{\ln\left(\dfrac{\Delta T_1}{\Delta T_2}\right)} = \frac{(-40.000) - (-30.000)}{\ln\left(\dfrac{-40.000}{-30.000}\right)} = -34.761\,^\circ\text{F}$$

Note that the negative sign in this result is irrelevant, since it is just a consequence of taking the temperature differences from the cold fluid (1) to the hot fluid (2). The negative sign can be dropped when appropriate. Also note that the heat transfer rates q_1 and q_2 calculated previously have opposite signs, since the hot fluid loses and the cold fluid gains enthalpy.

3) Pipe dimensions
   The detailed size specifications for both pipes are obtained from Appendix A.1 as:

NPS 1(1/4) in – schedule 40

ID = D_i = 1.38 in (inner diameter)
OD = D_o = 1.66 in (outer diameter)

NPS 2 in – schedule 40

ID = D_s = 2.067 in (inner diameter)
OD = 2.375 in (outer diameter)

---

4) Inner pipe heat transfer coefficient
   Cross section of the inner pipe, mass flux, and Reynolds number are:

$$S_i = \frac{\pi D_i^2}{4} = \frac{\pi 0.11500^2}{4} = 0.010387\,\text{ft}^2$$

$$G_i = \frac{W_i}{S_i} = \frac{10,000}{0.010387} = (9.6274e+5)\,\text{lb}/(\text{ft}^2\cdot\text{h})$$

$$\text{Re}_i = \frac{D_i G_i}{\mu_i} = \frac{(9.6274e+5)0.11500}{1.3200} = 83,875$$

Therefore, the regime is turbulent, and the Sieder–Tate equation (4.10) may be used to estimate the heat transfer coefficient. As the first approximation, let us disregard the viscosity correction by assuming $\phi_i = (\mu_i/\mu_w)^{0.14} = 1$ as follows:

$$\mathrm{Nu} = 0.027 \mathrm{Pr}^{0.33} \mathrm{Re}^{0.8} \left(\frac{\mu_i}{\mu_w} \right)^{0.14}$$

$$h_i = \frac{0.027k}{D_i} \sqrt[3]{\mathrm{Pr} \mathrm{Re}^{\frac{4}{5}}} = \frac{0.027 \sqrt[3]{5.8000}}{0.11500} 0.082000 \cdot 83,875^{\frac{4}{5}}$$

$$= 300.51 \ \mathrm{BTU}/\left(\mathrm{ft}^2 \cdot \mathrm{h} \cdot {}^\circ\mathrm{F}\right)$$

Notice that the application of eq. (4.10) assumes fully developed flow. Although the whole pipe length is not known at this time, for the minimum number of only one hairpin in the bank, the corresponding number of diameters at the entrance surpasses by far the needed to achieve the full development in turbulent flow, which is about 10–20 diameters. The number of diameters for just one hairpin leg is:

$$\#\mathrm{Diameters} = \frac{L_{hp}}{D_i} = \frac{16.000}{0.11500} = 139.13$$

5) Annulus heat transfer coefficient
   The annulus equivalent diameter used in the Reynolds number is taken here as equal to the hydraulic diameter:

$$D_e = D_s - D_o = 0.17225 - 0.13833 = 0.033920 \ \mathrm{ft}$$

$$S_o = \frac{\pi}{4}\left(D_s^2 - D_o^2\right) = \frac{\pi}{4}\left(0.17225^2 - 0.13833^2\right) = 0.0082741 \ \mathrm{ft}^2$$

$$G_o = \frac{W_o}{S_o} = \frac{10,435}{0.0082741} = (1.2612e + 6) \ \mathrm{lb}/\left(\mathrm{ft}^2 \cdot \mathrm{h}\right)$$

$$\mathrm{Re}_o = \frac{D_e G_o}{\mu_o} = \frac{(1.2612e + 6)0.033920}{4.3100} = 9,925.7$$

The annulus Reynolds number is still in the transition region, although quite near the turbulent region. For developing transition flow, the convective heat transfer may be estimated by the Hausen's equation (4.30). The developing flow term will be ignored, since for just one exchanger leg, the number of diameters is significant, given by:

$$\#\mathrm{Diameters} = \frac{L_{hp}}{D_e} = \frac{16.000}{0.033917} = 471.74$$

Also, ignoring the viscosity correction as $\phi_o = (\mu_o/\mu_w)^{0.14} = 1$, the annulus film coefficient is estimated as:

$$\mathrm{Nu} = \mathrm{Pr}^{0.33}\left(0.116\mathrm{Re}^{0.67} - 14.5\right)\left(\left(\frac{D_e}{L}\right)^{0.67} + 1\right)\left(\frac{\mu}{\mu_w}\right)^{0.14}$$

$$h_o = \frac{0.116\sqrt[3]{Pr}}{D_e}k\left(Re^{\frac{2}{3}} - 125.0\right) = \frac{0.116\sqrt[3]{18.100}}{0.033920}0.098500\left(9,925.7^{\frac{2}{3}} - 125.0\right)$$

$$= 297.93 \text{ BTU}/\left(\text{ft}^2 \cdot \text{h} \cdot {}^\circ\text{F}\right)$$

6) Variable properties correction factor
   a) Wall temperature and correction factors
      To improve the accuracy of the heat transfer coefficient further, the wall temperature is calculated using the first approximations of the heat transfer coefficients:

$$t_w = \frac{D_i h_i t_i + D_o h_o t_o}{D_i h_i + D_o h_o} = \frac{0.11500 \cdot 300.51 \cdot 90.000 + 0.13833 \cdot 125.00 \cdot 297.93}{0.11500 \cdot 300.51 + 0.13833 \cdot 297.93}$$

$$= 109.04 \,{}^\circ\text{F}$$

The fluid properties at the wall (from Appendices C.9 and C.10) are applied to correct the pipe and annulus film coefficients as follows:

Fluid 1 (cold)	Fluid 2 (hot)
Benzene	Aniline
109.0°F – 14.7 psi	109.0°F – 14.7 psi
$\mu = 1.16 \text{ lb}/(\text{ft} \cdot \text{h})$	$\mu = 5.31 \text{ lb}/(\text{ft} \cdot \text{h})$

$$\phi_i = \left(\frac{\mu_i}{\mu_{i,w}}\right)^{0.14} = \left(\frac{1.3200}{1.1600}\right)^{0.14} = 1.0183$$

$$\phi_o = \left(\frac{\mu_o}{\mu_{o,w}}\right)^{0.14} = \left(\frac{4.3100}{5.3100}\right)^{0.14} = 0.97121$$

Then, we have the corrected heat transfer coefficients as follows:

$$h_{i,\text{corr}} = h_i\phi_i = 1.0183 \cdot 300.51 = 306.01 \text{ BTU}/\left(\text{ft}^2 \cdot \text{h} \cdot {}^\circ\text{F}\right)$$

$$h_{o,\text{corr}} = h_o\phi_o = 0.97121 \cdot 297.93 = 289.35 \text{ BTU}/\left(\text{ft}^2 \cdot \text{h} \cdot {}^\circ\text{F}\right)$$

7) Overall heat transfer coefficient
   The clean overall heat transfer coefficient, including the wall conductive resistance, and the internal and external convective resistances, is:

$$U_c = \left(\frac{D_o}{2k}\ln\left(\frac{D_o}{D_i}\right) + \frac{1}{h_o} + \frac{D_o}{D_i h_i}\right)^{-1}$$

$$= \left(\frac{0.13833 \ln\left(\frac{0.13833}{0.11500}\right)}{2 \cdot 9.4000} + \frac{1}{289.35} + \frac{0.13833}{0.11500 \cdot 306.01} \right)^{-1} = 114.34 \; \text{BTU}/\left(\text{ft}^2 \cdot \text{h} \cdot {}^\circ\text{F}\right)$$

The fouling resistances required for both fluids are R_{di} and R_{do}:

$$R_{di} = 0.001 \; \left(\text{ft}^2 \cdot \text{h} \cdot {}^\circ\text{F}\right)/\text{BTU}$$

$$R_{do} = 0.001 \; \left(\text{ft}^2 \cdot \text{h} \cdot {}^\circ\text{F}\right)/\text{BTU}$$

Therefore, the design overall heat transfer coefficient used to size the exchanger area is given by:

$$U_d = \left(R_{do} + \frac{1}{U_c} + \frac{D_o R_{di}}{D_i} \right)^{-1} = \left(0.0010000 + \frac{1}{114.34} + \frac{0.0010000}{0.11500} 0.13833 \right)^{-1}$$

$$= 91.335 \; \text{BTU}/\left(\text{ft}^2 \cdot \text{h} \cdot {}^\circ\text{F}\right)$$

8) Number of hairpins
   a) Heat transfer area
      From the heat exchanger design equation, the clean heat transfer area is:

$$A_c = \frac{q}{U_c \Delta T_{lm}} = \frac{(-2.1600e+5)}{-34.761 \cdot 114.34} = 54.345 \; \text{ft}^2$$

And the design required heat transfer surface is

$$A_d = \frac{q}{U_d \Delta T_{lm}} = \frac{(-2.1600e+5)}{-34.761 \cdot 91.335} = 68.034 \; \text{ft}^2$$

   b) Total pipe length and number of exchangers
      The heat transfer area per unit length, based on the external surface of the inner pipe, is:

$$A_{\text{linear}} = \pi D_o = \pi 0.13833 = 0.43458 \; \left(\text{ft}^2/\text{ft}\right)$$

Therefore, the necessary pipe length and number of exchangers in the bank and the actual specified surface are:

$$L = \frac{A_d}{A_{\text{linear}}} = \frac{68.034}{0.43458} = 156.55 \; \text{ft}$$

$$N_{hp} = \frac{L}{2 L_{hp}} = \frac{156.55}{2 \cdot 16.000} = 4.8922 \rightarrow 5 \; \text{hairpins}$$

$$A = 2 A_{\text{linear}} L_{hp} N_{hp} = 2 \cdot 0.43458 \cdot 16.000 \cdot 5.0000 = 69.533 \; \text{ft}^2$$

The actual overall heat transfer and fouling factor for the exchanger are:

$$U = \frac{q}{A \Delta T_{lm}} = \frac{(-2.1600e + 5)}{-34.761 \cdot 69.533} = 89.366 \ \text{BTU}/(\text{ft}^2 \cdot \text{h} \cdot \text{°F})$$

$$R_d = \frac{U_c - U}{U_c U} = \frac{114.34 - 89.366}{114.34 \cdot 89.366} = 0.0024441 \ (\text{ft}^2 \cdot \text{h} \cdot \text{°F})/\text{BTU}$$

Taking the required heat transfer surface as reference, the over-design is only about 2%, calculated as follows:

$$\text{Over-design}(\%) = 100\left(\frac{A}{A_d} - 1\right) = 100\left(\frac{69.533}{68.034} - 1\right) = 2.2033\%$$

$$\text{Over-design}(\%) = 100\left(\frac{U_d}{U} - 1\right) = 100\left(\frac{91.335}{89.366} - 1\right) = 2.2033\%$$

Because of the fouling factors, the over-surface is significantly higher:

$$\text{Over-surface}(\%) = 100\left(\frac{A}{A_c} - 1\right) = 100\left(\frac{69.533}{54.345} - 1\right) = 27.947\%$$

$$\text{Over-surface}(\%) = 100\left(\frac{U_c}{U} - 1\right) = 100\left(\frac{114.34}{89.366} - 1\right) = 27.946\%$$

The fouling over-surface is evaluated as

$$\text{Fouling over-surface}(\%) = 100\left(\frac{A_d}{A_c} - 1\right) = 100\left(\frac{68.034}{54.345} - 1\right) = 25.189\%$$

$$\text{Fouling over-surface}(\%) = 100\left(\frac{U_c}{U_d} - 1\right) = 100\left(\frac{114.34}{91.335} - 1\right) = 25.187\%$$

9) Pressure drop
   a) Pressure drop in the tube
      (1) Friction factor

$$f_{Fi} = \frac{0.41}{\ln^2\left(0.23\left(\dfrac{\epsilon}{D_i}\right)^{\frac{10}{9}} + \dfrac{6.9}{Re_i}\right)} = \frac{0.41}{\ln^2\left(0.23\left(\dfrac{(7.4583e - 5)}{0.11500}\right)^{\frac{10}{9}} + \dfrac{6.9}{83,875}\right)}$$

$$= 0.0052745$$

$$\phi_{f,i} = \frac{1.02}{\left(\dfrac{\mu_i}{\mu_{i,w}}\right)^{0.14}} = \frac{1.02}{\left(\dfrac{1.3200}{1.1600}\right)^{0.14}} = 1.0017$$

$$f_{Fi,corr} = f_{Fi}\phi_{f,i} = 0.0052745 \cdot 1.0017 = 0.0052835$$

(2) Distributed pressure drop – tube

$$G_i = 962740.0 \ \text{lb}/(\text{ft}^2 \cdot \text{h})$$

$$\Delta P_{i,\text{dist}} = \frac{4G_i^2 L_{hp} N_{hp} f_{Fi}}{D_i \rho_i} = \frac{4(9.6274e+5)^2 \cdot 16.000 \cdot 5.0000 \cdot 0.0052835}{0.11500 \cdot 54.000}$$

$$= (2.5235e+11) \ \text{lb}/(\text{ft} \cdot \text{h}^2) = 4.2027 \ \text{psi}$$

(3) Localized pressure drop – tube

$$D_{i(\text{in})} = 1.38 \ \text{in}$$

$$K_i = 0.7 + \frac{2,000}{\text{Re}_i} + \frac{0.7}{D_{i(\text{in})}} = 0.7 + \frac{2,000}{83,875} + \frac{0.7}{1.3800} = 1.2311$$

$$\Delta P_{i,\text{loc}} = \frac{G_i^2 K_i}{2\rho_i}(2N_{hp}-1) = \frac{(9.6274e+5)^2 \cdot 1.2311}{2 \cdot 54.000}(2 \cdot 5.0000 - 1)$$

$$= (9.5089e+10) \ \text{lb}/(\text{ft} \cdot \text{h}^2) = 1.5836 \ \text{psi}$$

(4) Total pressure drop – tube

$$\Delta P_i = \Delta P_{i,\text{dist}} + \Delta P_{i,\text{loc}} = (2.5235e+11) + (9.5089e+10)$$

$$= (3.4744e+11) \ \text{lb}/(\text{ft} \cdot \text{h}^2)\frac{\text{lb}}{(\text{ft} \cdot \text{h}^2)} = 5.7864 \ \text{psi}$$

b) Pressure drop in the annulus

(1) Friction factor

$$f_{Fo} = \frac{0.41}{\ln^2\left(0.23\left(\dfrac{10}{D_e}\right)^{\frac{10}{9}} + \dfrac{6.9}{\text{Re}_o}\right)} = \frac{0.41}{\ln^2\left(0.23\left(\dfrac{(7.4583e-5)}{0.033920}\right)^{\frac{10}{9}} + \dfrac{6.9}{9,925.7}\right)}$$

$$= 0.0084696$$

$$\phi_{f,o} = \frac{1.02}{\left(\dfrac{\mu_o}{\mu_{o,w}}\right)^{0.14}} = \frac{1.02}{\left(\dfrac{4.3100}{5.3100}\right)^{0.14}} = 1.0502$$

$$f_{Fo,\text{corr}} = f_{Fo}\phi_{f,o} = 0.0084696 \cdot 1.0502 = 0.0088948$$

(2) Distributed pressure drop – annulus

$$\Delta P_{o,dist} = \frac{4G_o^2 L_{hp} N_{hp} f_{Fo}}{D_h \rho_o} = \frac{4(1.2612e+6)^2 \cdot 16.000 \cdot 5.0000 \cdot 0.0088948}{0.033920 \cdot 64.300}$$

$$= (2.0758e+12) \text{ lb}/(\text{ft} \cdot \text{h}^2) = 34.571 \text{ psi}$$

(3) Localized pressure drop – annulus

$$D_{h(in)} = 0.40704 \text{ in}$$

$$K_o = 1.75 + \frac{2,000}{Re_o} + \frac{1.75}{D_{h(in)}} = 1.75 + \frac{2,000}{9,925.7} + \frac{1.75}{0.40704} = 6.2508$$

$$\Delta P_{o,loc} = \frac{G_o^2 K_o N_{hp}}{2\rho_o} = \frac{(1.2612e+6)^2 \cdot 6.2508 \cdot 5.0000}{2 \cdot 64.300}$$

$$= (3.8657e+11) \text{ lb}/(\text{ft} \cdot \text{h}^2) \frac{lb}{(ft \cdot h^2)} = 6.4381 \text{ psi}$$

(4) Total pressure drop – annulus

$$\Delta P_o = \Delta P_{o,dist} + \Delta P_{o,loc} = (2.0758e+12) + (3.8657e+11)$$

$$= (2.4624e+12) \text{ lb}/(\text{ft} \cdot \text{h}^2) = 41.01 \text{ psi}$$

From the last result, we can see that the previous design is not practicable, once the aniline pressure drop in the annulus exceeds by far the allowable exchanger pressure drop. A secondary, but also important, issue is the barely turbulent Reynolds number (Re ~ 10,000) for the annulus fluid. Generally, it is not recommended to design an exchanger in the transition regime, because of the instabilities on the heat transfer coefficients and friction factor, which may cause sudden variations on the heat exchanger performance and vibration due to significant oscillations on the pressure drop.

Design 2

Our second attempt for sizing this exchanger should approach the previously identified problems. In order to reduce the hot fluid pressure drop, let us try a series–parallel bank by dividing the annulus fluid (aniline) in two separate streams with half flow rate. The drawback from halving the aniline stream is to reduce the Reynolds number further into the transition region, as we can calculate as follows:
-   Two parallel streams of aniline through the annulus:

$$G_o = \frac{W_o}{S_o} = \frac{(10,435/2)}{0.0082741} = (6.3058e+5) \text{ lb}/(\text{ft}^2 \cdot \text{h})$$

$$Re_o = \frac{D_e G_o}{\mu_o} = \frac{0.033920(6.3058e+5)}{4.3100} = 4,962.7 \rightarrow \text{transition}$$

We are going to refuse this alternative, to avoid the transition regime difficulties. Noticing the benzene is less viscous than the aniline and that the flow rates are similar, one possibility is to switch the fluid allocation by placing the benzene in the annulus divided into two parallel streams, and the aniline passing through the inner tube.
- Two parallel streams of benzene through the annulus:

$$G_o = \frac{W_o}{A_o} = \frac{(10,000.0/2)}{0.0082741} = (6.0430e+5) \text{ lb}/(\text{ft}^2 \cdot \text{h})$$

$$Re_o = \frac{D_e G_o}{\mu_o} = \frac{0.033920(6.0430e+5)}{1.3200} = 15,529 \rightarrow \text{ fully turbulent}$$

Therefore, the change of fluid allocation turns the annulus stream into fully turbulent flow. At this point, the analysis leads us to the following configuration:
i. Aniline allocated in the inner tube
ii. Benzene allocated in the annulus
iii. Divide the annulus (hot fluid) into two parallel streams

The division of the benzene stream (fluid 1) into two parallel streams through the annulus channel and the allocation of aniline (fluid 2) inside the internal pipe results in the design flow rates of

$$W_o = W_1 = 5,000.0 \text{ lb/h (annulus in parallel)}$$

$$W_i = W_2 = 10,435.0 \text{ lb/h (tube in series)}$$

1) Mean temperature difference for series–parallel arrangement

The mean temperature for a hairpin bank with the hot fluid passing in series and the cold fluid partitioned in $n_c = 2$ parallel streams is given from eqs. (6.5)–(6.7), as follows:

$$P_1 = \frac{T_2 - t_1}{T_1 - t_1} = \frac{100.00 - 60.000}{150.00 - 60.000} = 0.44444$$

$$R_1 = \frac{T_1 - T_2}{n_c(t_2 - t_1)} = \frac{-100.00 + 150.00}{2.0000(120.00 - 60.000)} = 0.41667$$

$$F_{sp} = \frac{P_1 + R_1(-P_1 + 1) - 1}{R_1 n_c \ln\left(\frac{1}{R_1}\left(R_1\left(\frac{1}{P_1}\right)^{\frac{1}{n_c}} - \left(\frac{1}{P_1}\right)^{\frac{1}{n_c}} + 1\right)\right)}$$

$$= \frac{0.41667(-0.44444 + 1) + 0.44444 - 1}{0.41667 \cdot 2.0000 \ln\left(\frac{1}{0.41667}\left(0.41667\left(\frac{1}{0.44444}\right)^{\frac{1}{2.0000}} - \left(\frac{1}{0.44444}\right)^{\frac{1}{2.0000}} + 1\right)\right)}$$

$$= 0.32300$$

Hence, with eq. (6.4):

$$\Delta T_m = F_{sp}(T_1 - t_1) = 0.32300(150.00 - 60.000) = 29.070\ ^\circ F$$

2) Inner pipe heat transfer coefficient

$$G_i = \frac{W_i}{S_i} = \frac{10,435}{0.010387} = (1.0046e+6)\ lb/(ft^2 \cdot h)$$

$$Re_i = \frac{D_i G_i}{\mu_i} = \frac{0.11500(1.0046e+6)}{4.3100} = 26,805$$

Using the turbulent Sieder and Tate equation (4.10) assuming initially $\phi_i = (\mu_i/\mu_w)^{0.14} = 1$:

$$Nu = 0.027 Pr^{0.33} Re^{0.8}\left(\frac{\mu}{\mu_w}\right)^{0.14}$$

$$h_i = \frac{0.027 k \sqrt[3]{Pr}}{D_i} Re^{\frac{4}{5}} = \frac{0.027 \cdot 0.098500 \cdot \sqrt[3]{18.100}}{0.11500} 26,805^{\frac{4}{5}}$$

$$= 211.79\ BTU/(ft^2 \cdot h \cdot {}^\circ F)$$

3) Annulus heat transfer coefficient

$$G_o = \frac{W_o}{S_o} = \frac{5,000.0}{0.0082741} = (6.0430e+5)\ lb/(ft^2 \cdot h)$$

$$Re_o = \frac{D_e G_o}{\mu_o} = \frac{0.033920(6.0430e+5)}{1.3200} = 15,529$$

From eq. (4.10) with $\phi_o = (\mu_o/\mu_w)^{0.14} = 1$:

$$h_o = \frac{0.027 k \sqrt[3]{Pr}}{D_e} Re^{\frac{4}{5}} = \frac{0.027 \cdot 0.082000 \cdot \sqrt[3]{5.8000}}{0.033920} 15,529^{\frac{4}{5}} = 264.31\ BTU/(ft^2 \cdot h \cdot {}^\circ F)$$

4) Variable properties correction factor
   a) Wall temperature and correction factors

$$t_w = \frac{D_i h_i t_i + D_o h_o t_o}{D_i h_i + D_o h_o} = \frac{0.11500 \cdot 125.00 \cdot 211.79 + 0.13833 \cdot 264.31 \cdot 90.000}{0.11500 \cdot 211.79 + 0.13833 \cdot 264.31}$$

$$= 103.99\ ^\circ F$$

Fluid 1 (cold)	Fluid 2 (hot)
Benzene	Aniline
$104.0\,°F - 14.7\,psi$	$104.0\,°F - 14.7\,psi$
$\mu = 1.2\ lb/(ft \cdot h)$	$\mu = 5.71\ lb/(ft \cdot h)$

The correction factor for nonisothermal flow and the new heat transfer coefficients are:

$$\phi_i = \left(\frac{\mu_i}{\mu_{i,w}}\right)^{0.14} = \left(\frac{4.3100}{5.7100}\right)^{0.14} = 0.96139$$

$$\phi_o = \left(\frac{\mu_o}{\mu_{o,w}}\right)^{0.14} = \left(\frac{1.3200}{1.2000}\right)^{0.14} = 1.0134$$

Using the viscosity correction factors, the heat transfer coefficients are:

$$h_{i,corr} = h_i\phi_i = 0.96139 \cdot 211.79 = 203.61\ BTU/\left(ft^2 \cdot h \cdot °F\right)$$

$$h_{o,corr} = h_o\phi_o = 1.0134 \cdot 264.31 = 267.85\ BTU/\left(ft^2 \cdot h \cdot °F\right)$$

5) Overall heat transfer coefficient

$$U_c = \left(\frac{D_o}{2k}\ln\left(\frac{D_o}{D_i}\right) + \frac{1}{h_o} + \frac{D_o}{D_i h_i}\right)^{-1} = \left(\frac{0.13833\ln\left(\frac{0.13833}{0.11500}\right)}{2 \cdot 9.4000} + \frac{1}{267.85} + \frac{0.13833}{0.11500 \cdot 203.61}\right)^{-1}$$
$$= 90.907\ BTU/\left(ft^2 \cdot h \cdot °F\right)$$

Considering the required fouling factors for each stream:

$$R_{di} = 0.001\ \left(ft^2 \cdot h \cdot °F\right)/BTU$$

$$R_{do} = 0.001\ \left(ft^2 \cdot h \cdot °F\right)/BTU$$

The design overall heat transfer coefficient is:

$$U_d = \left(R_{do} + \frac{1}{U_c} + \frac{D_o R_{di}}{D_i}\right)^{-1} = \left(0.0010000 + \frac{0.0010000}{0.11500}0.13833 + \frac{1}{90.907}\right)^{-1}$$
$$= 75.740\ BTU/\left(ft^2 \cdot h \cdot °F\right)$$

6) Number of hairpins
   a) Heat transfer areas

$$A_c = \frac{q}{U_c\Delta T_m} = \frac{(2.1600e + 5)}{29.070 \cdot 90.907} = 81.736\ ft^2$$

$$A_d = \frac{q}{U_d\Delta T_m} = \frac{(2.1600e + 5)}{29.070 \cdot 75.740} = 98.103\ ft^2$$

b) Total pipe length and number of exchangers

$$L = \frac{A_d}{A_{linear}} = \frac{98.103}{0.43458} = 225.74 \text{ ft}$$

$$N_{hp} = \frac{L}{2L_{hp}} = \frac{225.74}{2 \cdot 16.000} = 7.0544$$

Therefore, we must use two identical banks in parallel with four exchangers each. The total number of hairpins for the service is:

$$N_{hp,total} = 8.0$$

The total heat transfer area available from all exchangers is:

$$A = 2A_{linear}L_{hp}N_{hp} = 2 \cdot 0.43458 \cdot 16.000 \cdot 8.0000 = 111.25 \text{ ft}^2$$

The actual overall heat transfer coefficient and fouling factor are:

$$U = \frac{q}{A\Delta T_m} = \frac{(2.1600e+5)}{111.25 \cdot 29.070} = 66.790 \text{ BTU}/(\text{ft}^2 \cdot \text{h} \cdot °\text{F})$$

$$R_d = \frac{U_c - U}{U_c U} = \frac{90.907 - 66.788}{66.788 \cdot 90.907} = 0.0039725 \text{ (ft}^2 \cdot \text{h} \cdot °\text{F})/\text{BTU}$$

Since the required surface is 98.103 ft², the over-design is evaluated as:

$$\text{Over-design}(\%) = 100\left(\frac{A}{A_d} - 1\right) = 100\left(\frac{111.25}{98.103} - 1\right) = 13.401\%$$

$$\text{Over-design}(\%) = 100\left(\frac{U_d}{U} - 1\right) = 100\left(\frac{75.740}{66.788} - 1\right) = 13.404\%$$

And the over-surface is given by:

$$\text{Over-surface}(\%) = 100\left(\frac{A}{A_c} - 1\right) = 100\left(\frac{111.25}{81.736} - 1\right) = 36.109\%$$

$$\text{Over-surface}(\%) = 100\left(\frac{U_c}{U} - 1\right) = 100\left(\frac{90.907}{66.788} - 1\right) = 36.113\%$$

Over-surface for fouling:

$$\text{Fouling over-surface }(\%) = 100\left(\frac{A_d}{A_c} - 1\right) = 100\left(\frac{98.103}{81.736} - 1\right) = 20.024\%$$

$$\text{Fouling over-surface}(\%) = 100\left(\frac{U_c}{U_d} - 1\right) = \left(\frac{90.907}{75.740} - 1\right) = 20.025\%$$

7) Pressure drop
a) Pressure drop in the tube

i) Friction factor

Using the Haaland equation, the isothermal tube friction factor is:

$$f_{Fi} = \frac{0.41}{\ln^2\left(0.23\left(\frac{\epsilon}{D_i}\right)^{\frac{10}{9}} + \frac{6.9}{Re_i}\right)} = \frac{0.41}{\ln^2\left(0.23\left(\frac{(7.4583e-5)}{0.11500}\right)^{\frac{10}{9}} + \frac{6.9}{26,805}\right)} = 0.0063480$$

For turbulent flow (Re > 10,000), the corrected friction factor is evaluated as:

$$\phi_{f,i} = \frac{1.02}{\left(\frac{\mu_i}{\mu_{i,w}}\right)^{0.14}} = \frac{1.02}{\left(\frac{4.3100}{5.7100}\right)^{0.14}} = 1.0610$$

$$f_{Fi,corr} = f_{Fi}\phi_{f,i} = 0.0063480 \cdot 1.0610 = 0.0067352$$

ii) Distributed pressure drop – tube

$$\Delta P_{i,dist} = \frac{4G_i^2 L_{hp} N_{hp} f_{Fi}}{D_i \rho_i} = \frac{4(1.0046e+6)^2 \cdot 16.000 \cdot 8.0000 \cdot 0.0067352}{0.11500 \cdot 64.300}$$

$$= (4.7065e+11)\, lb/(ft \cdot h^2) = 7.8384\, psi$$

iii) Localized pressure drop – tube

(1) Pressure loss coefficient

$$D_{i(in)} = 1.38\ in$$

$$K_i = 0.7 + \frac{2,000}{Re_i} + \frac{0.7}{D_{i(in)}} = 0.7 + \frac{2,000}{26,805} + \frac{0.7}{1.3800} = 1.2819$$

(2) Localized pressure drop

$$\Delta P_{i,loc} = \frac{G_i^2 K_i}{2\rho_i}(2N_{hp} - 1) = \frac{(1.0046e+6)^2 \cdot 1.2819}{2 \cdot 64.300}(2 \cdot 8.0000 - 1)$$

$$= (1.5090e+11)\, lb/(ft \cdot h^2) = 2.5131\, psi$$

iv) Total pressure drop – tube

$$\Delta P_i = \Delta P_{i,dist} + \Delta P_{i,loc} = (1.5090e+11) + (4.7065e+11)$$

$$= (6.2155e+11)\, lb/(ft \cdot h^2) = 10.351\ psi$$

b) Pressure drop in the annulus

i) Friction factor

$$f_{Fo} = \frac{0.41}{\ln^2\left(0.23\left(\frac{\epsilon}{D_e}\right)^{\frac{10}{9}} + \frac{6.9}{Re_o}\right)} = \frac{0.41}{\ln^2\left(0.23\left(\frac{(7.4583e-5)}{0.033920}\right)^{\frac{10}{9}} + \frac{6.9}{15,529}\right)} = 0.0077709$$

$$\phi_{f,o} = \frac{1.02}{\left(\frac{\mu_o}{\mu_{o,w}}\right)^{0.14}} = \frac{1.02}{\left(\frac{1.3200}{1.2000}\right)^{0.14}} = 1.0065$$

$$f_{Fo,corr} = f_{Fo}\phi_{f,o} = 0.0077709 \cdot 1.0065 = 0.0078214$$

ii) Distributed pressure drop – annulus

$$\Delta P_{o,dist} = \frac{4G_o^2 L_{hp} N_{hp} f_{Fo}}{D_h \rho_o} = \frac{4(6.0430e+5)^2 \cdot 16.000 \cdot 4.0000 \cdot 0.0078214}{0.033920 \cdot 54.000}$$

$$= (3.9919e+11) \ \text{lb}/\left(\text{ft} \cdot \text{h}^2\right) = 6.6482 \ \text{psi}$$

iii) Localized pressure drop – annulus
  (1) Pressure loss coefficient

$$D_{h(in)} = 0.40704 \ \text{in}$$

$$K_o = 1.75 + \frac{2,000}{Re_o} + \frac{1.75}{D_{h(in)}} = 1.75 + \frac{2,000}{15,529} + \frac{1.75}{0.40704} = 6.1781$$

  (2) Localized pressure drop

$$\Delta P_{o,loc} = \frac{G_o^2 K_o N_{hp}}{2\rho_o} = \frac{(6.0430e+5)^2 \cdot 4.0000 \cdot 6.1780}{2 \cdot 54.000}$$

$$= (8.3558e+10) \ \text{lb}/\left(\text{ft} \cdot \text{h}^2\right) = 1.3916 \ \text{psi}$$

iv) Total pressure drop – annulus

$$\Delta P_o = \Delta P_{o,dist} + \Delta P_{o,loc} = (3.9919e+11) + (8.3558e+10)$$

$$= (4.8275e+11) \ \text{lb}/\left(\text{ft} \cdot \text{h}^2\right) = 8.0399 \ \text{psi}$$

Conclusion

The last configuration of eight hairpins with the annulus connected in two parallel banks of four exchangers, and the inner pipe connected in series, matches all the design requirements of heat transfer area and pressure drop, with the additional benefit that neither stream is operating in the transition flow regime. Also, the over-design of about 13% is a reasonable value.

7.5 Rating

The quantitative evaluation of how an existing heat exchanger equipment – or a proposed design – performs against a given heat transfer duty is usually named "analysis" or "rating." The heat exchanger specification should be available with sufficient detail to allow the evaluation of the several heat transfer parameters necessary to estimate the overall heat transfer coefficient for the equipment, such as the pertinent nondimensional numbers for each process stream, e.g., Prandtl, Reynolds, Nusselt, Graetz, and friction factors. Additionally, the pressure drop developed should be calculated, to check its accordance within the allowable pressure drop in the process line. Therefore, in summary, the designer rates a heat exchanger to answer the following questions:

1. Is the heat transfer surface (size) enough to transfer the required heat between the fluids?
2. Is the added pressure drop for both streams acceptable for the line in which the heat exchanger will be installed?

To answer the first question, it is necessary to determine how the analyzed heat exchanger would perform under the specified process conditions. From this calculation, we get the required heat transfer surface to meet the proposed service. The heat exchanger (or bank) is considered "thermally satisfactory" if the available heat transfer area is greater than the required one. In other words, the equipment must have some excess area, quantified as over-design, as calculated from eq. (7.3), and repeated below for convenience:

$$\text{Over-design}(\%) = 100\left(\frac{A}{A_d} - 1\right) = 100\left(\frac{U_d}{U} - 1\right) \qquad (7.6)$$

For the purpose of rating, A is the effective heat transfer surface of the tested exchanger, and A_d is the required heat transfer area, including the provisioned fouling resistances. The same result can be obtained using the fouled (U_d) and actual (U) overall heat transfer coefficients.

Along with being thermally satisfactory, the heat exchanger must not block the process line, nor impair the product flow rate when installed; therefore, we have to answer the second question, to make sure the developed pressure drop does not exceed the available pressure drop in the line.

7.5.1 Rating outline

1 Thermal design
    1.1   Collect process data parameters
         1.1.1   Fluid temperatures
         1.1.2   Physical properties
    1.2   Energy balance using mean bulk temperatures of the fluids
         1.2.1   Determine any unknown fluid temperature or flow rate
         1.2.2   Or just check the energy balance between fluids for correctness
    1.3   Select a heat exchanger type: double-pipe heat exchanger in the present case
    1.4   Select the design variable
         1.4.1   Pipe length, if diameter is specified
         1.4.2   Pipe diameter, if length is specified
         1.4.3   Number of hairpins, if the service should be performed by a set of standardized hairpin exchangers
    1.5   Evaluate the actual mean temperature difference to be used in the heat exchanger design equation
    1.6   Overall heat transfer coefficient
         1.6.1   Calculate the inner pipe heat transfer coefficient
         1.6.2   Calculate the annulus heat transfer coefficient
         1.6.3   Calculate wall temperature
         1.6.4   Iterate steps 1.6.1–1.6.3 until convergence of the wall temperature is achieved
         1.6.5   Calculate the clean overall heat transfer coefficient U_c
         1.6.6   Calculate the fouled overall heat transfer coefficient U_d by introducing the fouling factors R_{di} and R_{do}
    1.7   Heat transfer areas
         1.7.1   Calculate the required clean heat transfer area A_c, using U_c
         1.7.2   Calculate the required fouled heat transfer area A_d, using U_d
         1.7.3   Calculate the available heat transfer area A, from the equipment geometric specifications
         1.7.4   Calculate the required overall heat transfer coefficient U, using the available area A, the heat duty q, and the appropriate mean temperature difference (e.g., ΔT_{lm})
    1.8   Calculate the over-design excess area, from eq. (7.6)
2 Hydraulic design (verify if the allowable pressure drop is not exceeded)
    2.1   Evaluate the inner pipe pressure drop
    2.2   Evaluate the annulus pressure drop

7.6 The data sheet

Once the design procedure is done, all the data necessary to undertake the detailed mechanical design and building the heat exchanger must be provided to the manufacturer. Accordingly, fundamental information such as exchanger type, piping specs, operating conditions (flow rates, temperatures, stream pressure, pressure drop, fouling factors, etc.), performance requirements, and any special requests are packed in a formulary generally called "data sheet," "specification sheet," or "spec-sheet."

The data sheet is filled in progressively in steps. In some cases, if the heat exchanger type is proprietary or patented, the process engineer just specify the fluids and operating conditions, and let the contractor do the thermal and mechanical designs. A common situation is to have the engineering staff of a chemical plant performing the thermal and hydraulic calculations, based on their better knowledge about the process. Then, the data sheet with the derived design information filled in goes to the heat exchanger vendor for accomplishing the complete mechanical specification.

There is no rigid format for the fields included in a heat exchanger data sheet, and, in fact, many manufacturers set their own standard; however, typically they consist in some modification of the format defined by the TEMA standard [109]. The illustration of a typical TEMA data sheet is presented in Fig. 7.1. Though this data sheet is originally targeted to shell-and-tube heat exchangers, it is suitable for double-pipe, multi-tube, and finned heat exchangers as well. The lines in the data sheet are numbered for precise and easy reference of the information therein.

Therefore, the data sheet is a quite important document, and the designer must insure its correctness. Commonly, it is the sole document forwarded to the equipment supplier. Although the detailed calculation log of the design procedure may comprise dozens of pages, the exchanger construction will be based on the few pages in the data sheet!

Unless stated otherwise, the manufacturer's guarantee will cover the thermal performance and the mechanical integrity when the heat exchanger operates strictly in the conditions given in the data sheet.

The information comprising the data sheet is usually grouped in the sections described bellow.

Identification and control

This section (Fig. 7.2) records customer and management information about the design task, e.g., the process service, the chemical plant and its location, and equipment id. An overview of the equipment is included, such as the total heat transfer area (surface per unit, line 6), number of connected exchangers (shells per unit, line 6), and flow arrangement (connected in, line 5).

	HEAT EXCHANGER SPECIFICATION SHEET				
1	CUSTOMER	CUSTOMER NO.	SPEC. NO	REVISION NO.	TAG
2	LOCATION		PAGE	OF	
3	PLANT	SECTION	PROJECT NO.	EQUIP. NO	
4	SERVICE				
5	Size	TEMA Type	Connected in (series/parallel)		
6	Surface per Unit		Shells per Unit	Surface per Shell	m²
7		PERFORMANCE OF ONE UNIT			
8			Shell Side	Tube Side	
9	Fluid Name				
10	Flow Total	kg/h			
11	Vapor	kg/h (in/out)			
12	Liquid	kg/h (in/out)			
13	Steam	kg/h (in/out)			
14	Water	kg/h (in/out)			
15	Noncondensable	kg/h (in/out)			
16	Temperature (In/Out)	°C (in/out)			
17	Density	kg/m³			
18	Viscosity	cP			
19	Molecular Weight, vapor				
20	Specific Heat	J/(kg °C)			
21	Thermal Conductivity	W/(m °C)			
22	Latent Heat	J/kg			
23	Inlet Pressure	kPa(a) (inlet)			
24	Velocity	m/s			
25	Press Drop Allow/Calc	kPa(g)			
26	Fouling Factor	(m² °C)/W			
27	Heat Exchanged	W	MTD (corrected) °C		
28	Heat Transfer Coefficient	W/(m² °C)			
29	Service Coeff.	W/(m² °C)	Dirty	Clean	
30		CONSTRUCTION DATA FOR ONE SHELL			
31		Shell Side	Tube Side	Sketch	
32	Design/Test Press	kPa(g)			
33	Design Temperature	°C			
34	No. Passes per Shell				
35	Corrosion Allowance	mm			
36	Connections Size & Rating	In			
37		Out			
38		Intermediate			
39	Tubes No.	OD, mm	Gauge	Length, m.	Pitch layout, deg.
40	Type		Material	Pitch ratio	
41	Shell	OD, mm	ID, mm	Material	
42	Channel or Bonnet	OD, mm	Thick	Channel Cover	
43	Tubesheet Type				
44	Floating Heat Cover		Impingement Protection		
45	Baffles Cross (number)	% Cut (d)		Spacing C/C, mm	
46	Baffles Long	Seal Type No			
47	Supports Tube	U-Bend		Type	
48	Bypass Seal Arrangement		Tube-Tubesheet Joint		
49	Expansion Joint No		Type		
50	ρV²-Inlet Nozzle	Bundle Entrance		Bundle Exit	
51	Gaskets - Shell Side		Tube Side		
52	Floating Heat Cover		Supports		
53	Code Requirements		TEMA Class		
54	Weight per shell kg	Filled w/water		Bundle	
55	Notes				
56					
57					
58	REVISION NO.				
59	QUALITY LEVEL				
60	PREPARED BY				
61	CHECKED BY				
62	APPROVED BY				
63	PROCESS SPEC. VER.				
64	DATE				

Fig. 7.1: Typical heat exchanger data sheet (adapted from Ref. [109]).

#					
1	CUSTOMER	CUSTOMER NO.	SPEC. NO.	REVISION NO.	TAG
2	LOCATION		PAGE	OF	
3	PLANT	SECTION	PROJECT NO.	EQUIP. NO.	
4	SERVICE				
5	Size	TEMA Type		Connected in (series/parallel)	
6	Surface per Unit	Shells per Unit		Surface per Shell	m^2

Fig. 7.2: Identification and control section of a typical data sheet.

Streams and physical properties

The flow rates, inlet/outlet temperatures, and physical properties of the fluids are listed here (see Fig. 7.3). The fluid allocation is also specified. The physical properties are evaluated at the average bulk temperatures (design temperature, line 33). Notice that intermediary temperatures (e.g., wall temperatures) used in the calculations do not appear in the data sheet.

Thermal and hydraulic performance

A summary of the thermal and hydraulic calculations is compiled in this section (Fig. 7.4). Although the heat transfer coefficient for each stream is not part of the standard TEMA data sheet, it is useful information during the design development and revision steps.

Basic construction information (Fig. 7.5) for the inner tube – or tube bundle, in case of multi-tube exchangers – is presented in lines 31–54. This is not the mechanical design, which is considerably more detailed in the specification of each assembled component in the heat exchanger, such as gaskets, flanges, valves, bolts, and welding.

Annotations

Any important assumption and special requirement must be recorded in the annotation section (Fig. 7.6). The area for annotation can grow considerably to several pages in a practical design; therefore, the notes need to be numbered to facilitate the identification. It is usual to fill some fields in the data sheet with a reference to a given note.

Revision control

The specification and acquisition of heat exchangers typically involve a considerable cost. For safety, the design process is subjected to overseeing and cooperation of several engineers. This section (Fig. 7.7) serves the purpose of tracking the revisions during the design phase.

Worked Example 7.3
A residual hot water stream of 19,842 lb/h must be cooled down from 158 to 140°F before treatment. The energy should be recovered to a product stream of benzoic acid being heated from 95 to 122°F. There is available a mounted bank of 12 carbon steel double-pipe heat exchangers of length 16 ft, built with NPS 2 in × 3-1/2 in – Sch 40 pipes. The hairpins have the tube and annulus both connected in series,

8			Shell Side	Tube Side
9	Fluid Name			
10	Flow Total	kg/h		
11	Vapor	kg/h	(in/out)	
12	Liquid	kg/h	(in/out)	
13	Steam	kg/h	(in/out)	
14	Water	kg/h	(in/out)	
15	Noncondensable	kg/h	(in/out)	
16	Temperature (In/Out)	°C	(in/out)	
17	Density	kg/m³		
18	Viscosity	cP		
19	Molecular Weight, vapor			
20	Specific Heat	J/(kg °C)		
21	Thermal Conductivity	W/(m °C)		
22	Latent Heat	J/kg		

Fig. 7.3: Streams and physical properties section of a typical data sheet.

23	Inlet Pressure	kPa(a)	(inlet)		
24	Velocity	m/s			
25	Press Drop Allow/Calc	kPa(g)			
26	Fouling Factor	(m² °C)/W			
27	Heat Exchanged	W			MTD (corrected) °C
28	Heat Transfer Coefficient	W/(m² °C)			
29	Service Coeff.	W/(m² °C)	Dirty		Clean

Fig. 7.4: Thermal and hydraulic performance section of a typical data sheet.

#		Shell Side	Tube Side	Sketch
31				
32	Design/Test Press	kPa(g)		
33	Design Temperature	°C		
34	No. Passes per Shell			
35	Corrosion Allowance	mm		
36	Connections Size &	In		
37	Rating	Out		
38		Intermediate		
39	Tubes No.	OD, mm	Gauge	Length, m.	Pitch layout, deg.
40	Type		Material		Pitch ratio
41	Shell	OD, mm	ID, mm	Material	
42	Channel or Bonnet	OD, mm	Thick	Channel Cover	
43	Tubesheet Type				
44	Floating Heat Cover			Impingement Protection	
45	Baffles Cross (number)		% Cut (d)		Spacing C/C, mm
46	Baffles Long		Seal Type No		
47	Supports Tube		U-Bend		Type
48	Bypass Seal Arrangement			Tube-Tubesheet Joint	
49	Expansion Joint No.			Type	
50	ρV²-Inlet Nozzle		Bundle Entrance	Bundle Exit	
51	Gaskets - Shell Side			Tube Side	
52	Floating Heat Cover			Supports	
53	Code Requirements			TEMA Class	
54	Weight per shell	kg	Filled w/water	Bundle	

Fig. 7.5: Construction information section of a typical data sheet.

55	Notes
56	
57	

Fig. 7.6: General notes section of a typical data sheet.

58	REVISION NO.							
59	QUALITY LEVEL							
60	PREPARED BY							
61	CHECKED BY							
62	APPROVED BY							
63	PROCESS SPEC. VER.							
64	DATE							

Fig. 7.7: Document revision control section of a typical data sheet.

and the annulus return is of bonnet type. Fouling factors of 0.002 and 0.001 ft^2 h ° F/BTU are required for water and benzoic acid, respectively. Consider that water is allocated in the tube and an allowable pressure drop for the tube and annulus is 20 psi. Determine if the available equipment can perform this service.

Solution
1) Fluid physical properties
   a) Mean temperature of the fluids

Since the benzoic acid flow rate is not specified, we need to determine its value from the energy balance using the average bulk temperatures of the fluids. The process temperatures are given as follows:

$$\text{Water:} \qquad T_{1,\,in} = 158\,°F \qquad \rightarrow \qquad T_{1,\,out} = 140\,°F$$

$$\text{Benzoic acid:} \quad T_{2,\,out} = 122\,°F \qquad \leftarrow \qquad T_{2,\,in} = 95\,°F$$

The average bulk temperatures and the heat capacities for both fluids are

$$T_{1m} = \frac{T_{1,\,in}}{2} + \frac{T_{1,\,out}}{2} = \frac{140.00}{2} + \frac{158.00}{2} = 149.00\,°F$$

$$T_{2m} = \frac{T_{2,\,in}}{2} + \frac{T_{2,\,out}}{2} = \frac{122.00}{2} + \frac{95.000}{2} = 108.50\,°F$$

Fluid 1 (hot) – tube	Fluid 2 (cold) – annulus
Water	Benzoic acid
609.0 °R = 149.0 °F 14.7 psi	568.0 °R = 108.0 °F 14.7 psi
$Cp = 0.983$ BTU/(lb · °F)	$Cp = 0.324$ BTU/(lb · °F)

   b) Benzoic acid flow rate

From the enthalpy balance, the mass flow rate of the benzoic acid can be evaluated as follows:

$$Cp_2 W_2 (T_{2,\,out} - T_{2,\,in}) + W_1 Cp_1 (T_{1,\,out} - T_{1,\,in}) = 0$$

$$W_2 = \frac{W_1 Cp_1 (T_{1,\,out} - T_{1,\,in})}{Cp_2 (T_{2,\,in} - T_{2,\,out})} = \frac{0.98300 \cdot 19842.(140.00 - 158.00)}{0.32400(-122.00 + 95.000)} = 40,133 \ \text{lb/h}$$

Let us check the heat load for both fluids using the four process temperatures and the respective flow rates:

$$q_1 = W_1 Cp_1(-T_{1,\text{in}} + T_{1,\text{out}}) = 0.98300 \cdot 19842.(140.00 - 158.00) = (-3.5108e + 5) \ \text{BTU/h}$$

$$q_2 = Cp_2 W_2(-T_{2,\text{in}} + T_{2,\text{out}}) = 0.32400 \cdot 40,133(122.00 - 95.000) = (3.5108e + 5) \ \text{BTU/h}$$

Within the accuracy of five significant figures used in the calculations, the energy exchange is balanced.

Using Appendices C.12 and C.13, the physical properties at the bulk temperatures are estimated as:

Fluid 1 (hot)	Fluid 2 (cold)
Water	Benzoic acid
609.0 °R = 149.0 °F 14.7 psi	568.0 °R = 108.0 °F 14.7 psi
$\mu = 1.04$ lb/(ft · h)	$\mu = 34.6$ lb/(ft · h)
$\mu = 0.43$ cP	$\mu = 14.3$ cP
$\rho = 64.0$ lb/ft^3	$\rho = 73.7$ lb/ft^3
$Cp = 0.983$ BTU/(lb · °F)	$Cp = 0.324$ BTU/(lb · °F)
$k = 0.38$ BTU/(ft · h · °F)	$k = 0.0949$ BTU/(ft · h · °F)
Pr = 2.68	Pr = 118.0

2) Logarithmic mean temperature difference

For counterflow arrangement, the fluid temperatures at the heat exchanger terminals are:

Terminal 1	Terminal 2
$T_{1,\text{in}} = 158$ °F $\rightarrow$	$T_{1,\text{out}} = 140$ °F
$T_{2,\text{out}} = 122$ °F $\leftarrow$	$T_{2,\text{in}} = 95$ ° F
$\Delta T_1 = 36.000$ °F	$\Delta T_2 = 45.000$ °F

With the temperature differences at the terminals, the log mean temperature differences are:

$$\Delta T_1 = T_{1,\text{in}} - T_{2,\text{out}} = -122.00 + 158.00 = 36.000 \ ^\circ F$$

$$\Delta T_2 = T_{1,\text{out}} - T_{2,\text{in}} = 140.00 - 95.000 = 45.000 \ ^\circ F$$

$$\Delta T_{\text{lm, c}} = \frac{\Delta T_1 - \Delta T_2}{\ln\left(\frac{\Delta T_1}{\Delta T_2}\right)} = \frac{36.000 - 45.000}{\ln\left(\frac{36.000}{45.000}\right)} = 40.333 \ ^\circ F$$

Notice that, considering its bulk viscosity of 14.3 cP, the benzoic acid is a relatively viscous stream, which indicates a possible increase in the overall heat transfer coefficient if a parallel flow arrangement is employed. In such case, as a later analysis refinement, it is worth to verify the performance of this exchanger in parallel flow.

3) Pipe dimensions
The detailed size specifications for both pipes are obtained from Appendix A.1 as:

Inner tube: NPS 2 in – Sch STD/40/40S

ID = D_i = 2.067 in (inner diameter)
OD = D_o = 2.375 in (outer diameter)

Outer tube: NPS 3 – 1/2 in – Sch STD/40/40S

ID = D_s = 3.548 in (inner diameter)
OD = 4.0 in (outer diameter)

Carbon steel

Absolute roughness: ϵ = 0.001378 in = 0.00011483 ft

Heat conductivity: k = 29.467 BTU/(ft · h · °F)

---

4) Inner pipe heat transfer coefficient
The cross section of the inner pipe, mass flux, and Reynolds number are:

$$S_i = \frac{\pi D_i^2}{4} = \frac{\pi 0.17225^2}{4} = 0.023303 \text{ ft}^2$$

$$G_i = \frac{W_i}{S_i} = \frac{19,842}{0.023303} = (8.5148e+5) \text{ lb}/(\text{ft}^2 \cdot \text{h})$$

$$Re_i = \frac{D_i G_i}{\mu_i} = \frac{(8.5148e+5)0.17225}{1.0400} = (1.4103e+5)$$

Therefore, the flow regime is turbulent. Considering the carbon steel absolute roughness, the isothermal friction factor coefficient can be estimated with the Haaland's equation as:

$$f_{Fi} = \frac{0.41}{\ln^2\left(0.23\left(\frac{\epsilon}{D_i}\right)^{\frac{10}{9}} + \frac{6.9}{Re_i}\right)} = \frac{0.41}{\ln^2\left(0.23\left(\frac{0.00011483}{0.17225}\right)^{\frac{10}{9}} + \frac{6.9}{(1.4103e+5)}\right)} = 0.0050019$$

The convective heat transfer coefficient can be evaluated from the Petukhov–Popov equation (4.11):

$$K_1 = 13.6 f_{Fi} + 1 = 13.6 \cdot 0.0050019 + 1 = 1.0680$$

$$K_2 = 11.7 + \frac{1.8}{\sqrt[3]{Pr_i}} = 11.7 + \frac{1.8}{\sqrt[3]{2.6800}} = 12.996$$

$$Nu_i = \frac{Pr_i Re_i f_{Fi}}{2K_1 + \sqrt{2} K_2 \sqrt{f_{Fi}} \left(Pr_i^{\frac{2}{3}} - 1 \right)} = \frac{(1.4103e+5)0.0050019 \cdot 2.6800}{\sqrt{2}\sqrt{0.0050019} \cdot 12.996 \left(2.6800^{\frac{2}{3}} - 1 \right) + 2 \cdot 1.0680}$$

$$= 565.33$$

$$h_i = \frac{k Nu_i}{D_i} = \frac{0.38000}{0.17225} 565.33 = 1,247.2 \ \text{BTU}/\left(\text{ft}^2 \cdot \text{h} \cdot {}^\circ\text{F}\right)$$

5) Annulus heat transfer coefficient

$$D_e = D_s - D_o = 0.29567 - 0.19792 = 0.097750 \ \text{ft}$$

$$S_o = \frac{\pi}{4}\left(-D_o^2 + D_s^2\right) = \frac{\pi}{4}\left(-0.19792^2 + 0.29567^2\right) = 0.037894 \ \text{ft}^2$$

$$G_o = \frac{W_o}{S_o} = \frac{40,133}{0.037894} = (1.0591e+6) \ \text{lb}/\left(\text{ft}^2 \cdot \text{h}\right)$$

$$Re_o = \frac{D_e G_o}{\mu_o} = \frac{(1.0591e+6)0.097750}{34.600} = 2,992.1$$

$$Nu_o = Pr_o^{0.33}\left(0.116 Re_o^{0.67} - 14.5\right) = 118.00^{0.33}\left(0.116 \cdot 2,992.1^{0.67} - 14.5\right) = 47.366$$

$$h_o = \frac{k Nu_o}{D_e} = \frac{0.094900}{0.097750} 47.366 = 45.985 \ \text{BTU}/\left(\text{ft}^2 \cdot \text{h} \cdot {}^\circ\text{F}\right)$$

6) Variable physical property factor
   a) Wall temperature

Disregarding the wall conductive and fouling resistances, the wall temperature may be estimated as:

$$t_w = \frac{D_i h_i t_i + D_o h_o t_o}{D_i h_i + D_o h_o} = \frac{0.17225 \cdot 1,247.2 \cdot 149.00 + 0.19792 \cdot 108.50 \cdot 45.985}{0.17225 \cdot 1,247.2 + 0.19792 \cdot 45.985} = 147.35 \ {}^\circ\text{F}$$

Fluid 1 (hot)	Fluid 2 (cold)
Water	Benzoic acid
607.0 °R = 147.0 °F 14.7 psi	607.0 °R = 147.0 °F 14.7 psi
$\mu = 1.05$ lb/(ft·h)	$\mu = 15.9$ lb/(ft·h)
$\mu = 0.434$ cP	$\mu = 6.57$ cP
$\rho = 64.0$ lb/ft³	$\rho = 72.4$ lb/ft³
$Cp = 0.983$ BTU/(lb·°F)	$Cp = 0.337$ BTU/(lb·°F)
$k = 0.38$ BTU/(ft·h·°F)	$k = 0.0922$ BTU/(ft·h·°F)
Pr = 2.72	Pr = 58.1

Due to the strong change in the benzoic acid viscosity, it is recommended the use of the Petukhov or Gnielinski factors to account for the effect of the varying physical properties on the heat transfer coefficients. Then, from eq. (4.25), we have:

$$\phi_i = \left(\frac{Pr_i}{Pr_{i,w}}\right)^{0.11} = \left(\frac{2.6800}{2.7200}\right)^{0.11} = 0.99837$$

$$\phi_o = \left(\frac{Pr_o}{Pr_{o,w}}\right)^{0.11} = \left(\frac{118.00}{58.100}\right)^{0.11} = 1.0811$$

And the corrected heat transfer coefficients[20] are:

$$h_{i,corr} = h_i\phi_i = 0.99837 \cdot 1,247.2 = 1,245.2 \text{ BTU}/\left(\text{ft}^2 \cdot \text{h} \cdot °\text{F}\right)$$

$$h_{o,corr} = h_o\phi_o = 1.0811 \cdot 45.985 = 49.714 \text{ BTU}/\left(\text{ft}^2 \cdot \text{h} \cdot °\text{F}\right)$$

7) Overall heat transfer coefficient
Using the corrected convective coefficients, the clean overall heat transfer coefficient is given by:

$$U_c = \cfrac{1}{\cfrac{D_o}{2k}\ln\left(\cfrac{D_o}{D_i}\right) + \cfrac{1}{h_o} + \cfrac{D_o}{D_ih_i}} = \cfrac{1}{\cfrac{0.19792\ln\left(\frac{0.19792}{0.17225}\right)}{2\cdot 29.467} + \cfrac{1}{49.714} + \cfrac{0.19792}{0.17225\cdot 1,245.2}}$$

$$= 46.502 \text{ BTU}/\left(\text{ft}^2 \cdot \text{h} \cdot °\text{F}\right)$$

---

20 Even in this case, where the cold stream is highly viscous, the divergence from the Sieder and Tate correction factor, based on the viscosity on the wall, is not quite significant: $\phi_i = (\mu_i/\mu_{i,w})^{0.14} = (1.0400/1.0500)^{0.14} = 0.99866$ and $\phi_o = (\mu_o/\mu_{o,w})^{0.14} = (34.600/15.900)^{0.14} = 1.1150$.

The required fouling factors for both streams are:

$$R_{di} = 0.002 \ (ft^2 \cdot h \cdot °F)/BTU$$

$$R_{do} = 0.001 \ (ft^2 \cdot h \cdot °F)/BTU$$

Therefore, the fouled overall heat transfer coefficient is:

$$U_d = \frac{1}{R_{do} + \frac{1}{U_c} + \frac{D_o R_{di}}{D_i}} = \frac{1}{0.0010000 + \frac{0.0020000}{0.17225} 0.19792 + \frac{1}{46.502}} = 40.319 \ BTU/(ft^2 \cdot h \cdot °F)$$

8) Heat transfer areas
   a) Required heat transfer areas

The required heat transfer surface for performing the service is:

$$A_c = \frac{q}{U_c \Delta T_{lm}} = \frac{(3.5108e + 5)}{40.333 \cdot 46.502} = 187.19 ft^2$$

Including the fouling resistances, the required surface increases to:

$$A_d = \frac{q}{U_d \Delta T_{lm}} = \frac{(3.5108e + 5)}{40.319 \cdot 40.333} = 215.89 \ ft^2$$

9) Heat exchanger performance
   a) Available heat transfer area

The bank is composed of $N_{hp} = 12$ hairpins associated in series–series; hence, the actual heat transfer surface is given by:

$$A = 2A_{linear}L_{hp}N_{hp} = 2 \cdot 0.62178 \cdot 12.000 \cdot 16.000 = 238.76 \ ft^2$$

For the required heat duty and mean temperature difference, the actual overall heat transfer coefficient and fouling factor are:

$$U = \frac{q}{A \Delta T_{lm}} = \frac{(3.5108e + 5)}{238.76 \cdot 40.333} = 36.457 \ BTU/(ft^2 \cdot h \cdot °F)$$

$$R_d = \frac{U_c - U}{U_c U} = \frac{-36.457 + 46.502}{36.457 \cdot 46.502} = 0.0059251 \ (ft^2 \cdot h \cdot °F)/BTU$$

As we see, the actual fouling factor provided by the whole area of the bank surpasses the value specified for the service. Additionally, using the required surface in the fouled condition, the over-design margin is calculated as:

$$\text{Over-design}(\%) = \frac{100A}{A_d} - 100 = -100 + \frac{100}{215.89}238.76 = 10.593\%$$

These results state that the available hairpin bank is enough to perform the requested heat duty, with some operational margin. Subsequently, it is necessary to confirm that the available pressure drop for both streams is not exceeded.

10) Pressure drop
    a) Inner tube pressure drop

The previously calculated Reynolds number points to turbulent flow:

$$Re_i = 141,030.0 \rightarrow \text{fully turbulent}$$

Then we can evaluate the friction factor from the Haaland's equation:

$$f_{Fi} = \frac{0.41}{\ln^2\left(0.23\left(\frac{\varepsilon}{D_i}\right)^{\frac{10}{9}} + \frac{6.9}{Re_i}\right)} = \frac{0.41}{\ln^2\left(0.23\left(\frac{0.00011483}{0.17225}\right)^{\frac{10}{9}} + \frac{6.9}{(1.4103e + 5)}\right)}$$

$$= 0.0050019$$

To include the effect of variable physical properties, the correction factor and adjusted friction factor are:

$$\phi_{f,i} = \frac{1.02}{\left(\frac{\mu_i}{\mu_{i,w}}\right)^{0.14}} = \frac{1.02}{\left(\frac{1.0400}{1.0500}\right)^{0.14}} = 1.0214$$

$$f_{F,corr} = f_{F0}\phi_{f,i} = 0.0050019 \cdot 1.0214 = 0.0051089$$

i) Distributed pressure drop

$$\Delta P_{i,dist} = \frac{4L_{hp}N_{hp}f_F}{D_i\rho_i}G_i^2 = \frac{4(8.5148e + 5)^2 \cdot 0.0051089 \cdot 12.000}{0.17225 \cdot 64.000}16.000$$

$$= (2.5805e + 11) \text{ lb}/(\text{ft} \cdot \text{h}^2) = 4.2976 \text{ psi}$$

ii) Localized pressure drop
With the inner tube diameter expressed in inches, the friction loss coefficient for the internal pipe and the localized pressure drop are calculated as follows:

$$D_{i(in)} = 2.067 \text{ in}$$

$$K_i = 0.7 + \frac{2,000}{Re_i} + \frac{0.7}{D_{i(in)}} = 0.7 + \frac{0.7}{2.0670} + \frac{2,000}{(1.4103e + 5)} = 1.0528$$

$$\Delta P_{i,\,loc} = \frac{G_i^2 K_i}{2\rho_i}\left(2N_{hp} - 1\right) = \frac{(8.5148e+5)^2 \cdot 1.0528}{2 \cdot 64.000}\left(2 \cdot 12.000 - 1\right)$$

$$= (1.3716e + 11)\ lb/\left(ft \cdot h^2\right) = 2.2843\ psi$$

iii) Total pressure drop

$$\Delta P_i = \Delta P_{i,\,dist} + \Delta P_{i,\,loc} = (1.3716e + 11) + (2.5805e + 11)$$

$$= (3.9521e + 11)\ lb/\left(ft \cdot h^2\right) = 6.582\ psi$$

b) Annulus pressure drop

The annulus fluid operates in the transition range, as indicated by the previously calculated Reynolds number:

$$Re_o = 2,992.1 \rightarrow transition$$

The friction factor accounting for variable physical properties is evaluated as follows:

$$f_{Fo} = 2.3 \cdot 10^{-8} Re_o^{1.5} + 0.0054 = 2.3 \cdot 10^{-8} \cdot 2,992.1^{1.5} + 0.0054 = 0.0091644$$

$$\phi_{f,o} = \frac{1.02}{\left(\frac{\mu_o}{\mu_{o,w}}\right)^{0.14}} = \frac{1.02}{\left(\frac{34.600}{15.900}\right)^{0.14}} = 0.91480$$

$$f_{F,corr} = f_{Fo}\phi_{f,o} = 0.0091644 \cdot 0.91480 = 0.0083836$$

i) Distributed pressure drop

And the distributed pressure drop comes from

$$\Delta P_{o,\,dist} = \frac{4L_{hp}N_{hp}f_F}{D_h\rho_o}G_o^2 = \frac{4(1.0591e + 6)^2 \cdot 0.0083836 \cdot 12.000}{0.097750 \cdot 73.700}16.000$$

$$= (1.0025e + 12)\ lb/\left(ft \cdot h^2\right) = 16.696\ psi$$

ii) Localized pressure drop

Taking the hydraulic diameter as the annulus equivalent diameter, the localized pressure drop for an annulus with a straight pipe return can be evaluated as:

$$D_{h(in)} = 1.173\ in$$

$$K_o = 2.8 + \frac{2,000}{Re_o} + \frac{2.8}{D_{h(in)}} = 2.8 + \frac{2,000}{2,992.1} + \frac{2.8}{1.1730} = 5.8555$$

$$\Delta P_{o,loc} = \frac{G_o^2 K_o N_{hp}}{2\rho_o} = \frac{(1.0591e + 6)^2 \cdot 12.000 \cdot 5.8555}{2 \cdot 73.700}$$

$$= (5.3471e + 11)\ lb/\left(ft \cdot h^2\right) = 8.9052\ psi$$

iii) Total pressure drop

$$\Delta P_o = \Delta P_{o,dist} + \Delta P_{o,loc} = (1.0025e + 12) + (5.3471e + 11)$$

$$= (1.5372e + 12) \ lb/(ft \cdot h^2) = 25.601 \ psi$$

Conclusion

The battery of heat exchangers has enough surface to undergo the requested service, however, for the given process conditions, the developed pressure drop in the annulus cannot be accepted, implying that the available equipment is not suitable, at least in the current configuration.

Worked Example 7.4

In an attempt to fulfill the service specified in Worked Example 7.3, a series–parallel configuration with the annulus fluid divided into two parallel streams was proposed to overcome the problem of exceeding benzoic acid pressure drop. Rate this design and determine if it can accomplish the heat duty adequately.

Solution

We can anticipate that splitting the benzoic acid into two halved parallel streams should solve the annulus pressure drop excess, since a decrease to about 1/8 of the original pressure loss is expected. Nevertheless, considering that the division of the stream generally impairs the heat exchanger performance, we need to test if the available heat transfer surface is still enough to achieve the demanded heat duty.

Much of the calculation done in the former problem is valid; hence, let us just adapt the necessary steps.

1) Benzoic acid flow rate

The cold fluid will pass through the annulus in two parallel streams, therefore, the actual flow rate used in the calculations is:

$$W_2 = \frac{W_{2,total}}{n} = \frac{40,133}{2.0000} = 20,066 \ lb/h$$

2) Mean temperature difference

For the hot fluid in series and the cold fluid split in n_c parallel streams, the effective mean temperature difference is given by eq. (6.5), hence, evaluating the terms:

$$P_1 = \frac{T_2 - t_1}{T_1 - t_1} = \frac{140.00 - 95.000}{158.00 - 95.000} = 0.71429$$

$$R_1 = \frac{T_1 - T_2}{n_c(-t_1 + t_2)} = \frac{-140.00 + 158.00}{2.0000(122.00 - 95.000)} = 0.33333$$

The series–parallel temperature difference factor F_{sp} is

$$F_{sp} = \frac{P_1 + R_1(-P_1 + 1) - 1}{R_1 n_c \ln\left(\frac{1}{R_1}\left(R_1\left(\frac{1}{P_1}\right)^{\frac{1}{n_c}} - \left(\frac{1}{P_1}\right)^{\frac{1}{n_c}} + 1\right)\right)}$$

$$= \frac{0.33333(-0.71429 + 1) + 0.71429 - 1}{0.33333 \cdot 2.0000 \ln\left(\frac{1}{0.33333}\left(0.33333\left(\frac{1}{0.71429}\right)^{\frac{1}{2.0000}} - \left(\frac{1}{0.71429}\right)^{\frac{1}{2.0000}} + 1\right)\right)} = 0.62604$$

Finally, the effective temperature difference in this configuration is:

$$\Delta T_m = F_{sp}(T_1 - t_1) = 0.62604(158.00 - 95.000) = 39.441 \text{ °F}$$

3) Tube heat transfer coefficient
The isothermal heat transfer coefficient for the inner fluid remains unaltered, then:

$$h_i = 1,247.2 \text{ BTU}/(\text{ft}^2 \cdot \text{h} \cdot \text{°F})$$

4) Annulus heat transfer coefficient
The Reynolds number for the halved benzoic acid flow rate is:

$$G_o = \frac{W_o}{S_o} = \frac{20,066}{0.037894} = (5.2953e + 5) \text{ lb}/(\text{ft}^2 \cdot \text{h})$$

$$Re_o = \frac{D_e G_o}{\mu_o} = \frac{(5.2953e + 5)0.097750}{34.600} = 1,496.0 \rightarrow \text{laminar}$$

Using the Sieder–Tate equation for the laminar regime, the Nusselt number and convective coefficient are:

$$Nu_o = 1.86\left(\frac{D_e Pr_o}{L}Re_o\right)^{0.33} = 1.86\left(\frac{0.097750}{16.000}118.00 \cdot 1,496.0\right)^{0.33} = 19.096$$

$$h_o = \frac{k Nu_o}{D_e} = \frac{0.094900}{0.097750}19.096 = 18.539 \text{ BTU}/(\text{ft}^2 \cdot \text{h} \cdot \text{°F})$$

5) Variable properties correction factor
   a) Wall temperature

$$t_w = \frac{D_i h_i t_i + D_o h_o t_o}{D_i h_i + D_o h_o} = \frac{0.17225 \cdot 1,247.2 \cdot 149.00 + 0.19792 \cdot 108.50 \cdot 18.539}{0.17225 \cdot 1,247.2 + 0.19792 \cdot 18.539} = 148.32 \text{ °F}$$

The difference between this wall temperature and the one obtained with the previous arrangement is neglectable; therefore, we are going to use the isothermal heat transfer coefficients to calculate the overall heat transfer coefficients.

6) Overall heat transfer coefficient

The clean overall heat transfer coefficient is given by:

$$U_c = \frac{1}{\frac{D_o}{2k}\ln\left(\frac{D_o}{D_i}\right) + \frac{1}{h_o} + \frac{D_o}{D_i h_i}} = \frac{1}{\frac{0.19792\ln\left(\frac{0.19792}{0.17225}\right)}{2 \cdot 29.467} + \frac{1}{18.539} + \frac{0.19792}{0.17225 \cdot 1247.2}}$$

$$= 18.074 \text{ BTU}/\left(\text{ft}^2 \cdot \text{h} \cdot {}^\circ\text{F}\right)$$

With the fouling factors:

$$R_{di} = 0.002 \ \left(\text{ft}^2 \cdot \text{h} \cdot {}^\circ\text{F}\right)/\text{BTU}$$

$$R_{do} = 0.001 \ \left(\text{ft}^2 \cdot \text{h} \cdot {}^\circ\text{F}\right)/\text{BTU}$$

The fouled overall heat transfer coefficient can be evaluated as:

$$U_d = \frac{1}{R_{do} + \frac{1}{U_c} + \frac{D_o R_{di}}{D_i}} = \frac{1}{0.0010000 + \frac{0.0020000}{0.17225}0.19792 + \frac{1}{18.074}}$$

$$= 17.057 \text{ BTU}/\left(\text{ft}^2 \cdot \text{h} \cdot {}^\circ\text{F}\right)$$

From the effective mean temperature difference and overall heat transfer coefficients, the clean and fouled heat transfer surfaces are:

$$A_c = \frac{q}{U_c \Delta T_m} = \frac{(1.1392e + 17)}{(1.0556e + 13)21.912} = 492.51 \text{ ft}^2$$

$$A_d = \frac{q}{U_d \Delta T_m} = \frac{(3.5108e + 5)}{17.057 \cdot 39.441} = 521.86 \text{ ft}^2$$

The total available area on the exchanger battery is:

$$A = 2A_{linear}L_{hp}N_{hp} = 2 \cdot 0.62178 \cdot 12.000 \cdot 16.000 = 238.76 \text{ ft}^2$$

Therefore, the excess area given by the over-design margin is:

$$\text{Over} - \text{design}(\%) = \frac{100A}{A_d} - 100 = \frac{100}{521.86}238.76 - 100 = -54.248\%$$

Conclusion

The proposed series–parallel association cannot match the required heat duty. As indicated by the negative design margin of about –54%, the splitting of the cold fluid into two parallel streams lowered severely the overall heat transfer coefficient to the point that the whole heat transfer surface of the bank is not enough to achieve the specified heat transfer rate.

Problems

(7.1) A stream of 1.4 kg/s of hot water must be cooled from 73 to 35 °C before discharge, using cooling water at 20 °C. The exit temperature of the cooling water must not exceed 45 °C. There is available 4.5 m long stainless steel double-pipe hairpins (bonnet type), with tube diameters DN 50 × 80 mm/40S. The fouling factors are 0.00021 and 0.00035 m²K/W for the hot water and cooling water, respectively. A pressure loss of 70 kPa is allowed for both streams. Determine the (a) number of exchangers with their tube and annulus connected in series required to perform this service and answer, (b) tube and annulus heat transfer coefficient, (c) clean overall heat transfer coefficient, (d) fouling overall heat transfer coefficient, (e) tube and annulus pressure drop, and (f) resulting over-surface of this project.

Answer:
(a) $N_{hp} = 9$ exchangers
(b) $h_i = 4{,}312.4$ W/(m²·K); $h_o = 4{,}775.0$ W/(m²·K)
(c) $U_c = 1{,}212.3$ W/(m²·K)
(d) $U_d = 695.88$ W/(m²·K)
(e) $\Delta P_i = 23.387$ kPa; $\Delta P_o = 42.065$ kPa
(f) Over-surface = 77.6%

(7.2) Examine an alternative design by changing the fluid allocation in the service presented in Problem 7.1. Determine the (a) number of exchangers with their tube and annulus connected in series required to perform this service and answer, (b) tube and annulus heat transfer coefficient, (c) clean overall heat transfer coefficient, (d) fouling overall heat transfer coefficient, (e) tube and annulus pressure drop, and (f) resulting over-surface of this project. (g) Do you think this new design is feasible? Why?

Answer:
(a) $N_{hp} = 9$ exchangers
(b) $h_i = 3{,}564.7$ W/(m²·K); $h_o = 5{,}708.5$ W/(m²·K)
(c) $U_c = 1181.3$ W/(m²·K)
(d) $U_d = 695.5$ W/(m²·K)
(e) $\Delta P_i = 10.187$ kPa; $\Delta P_o = 96.533$ kPa
(f) Over-surface = 73.1%
(g) No. Although the design margin is about the same of the previous configuration, the permitted pressure drop is exceeded in the annulus

(7.3) Design a bank of double-pipe exchangers to recover heat from a 360 kg/h flue gas stream, using water at an initial temperature of 20 °C. The needed water exit temperature is 90 °C. The flue gas is allocated in the inner pipe and must be cooled from 350 to 175 °C. A few 2 m long DN 40 × 65 mm/40S stainless steel double-pipe (bonnet-type return) hairpins are available. A fouling factor of 0.00018 m² K/W is provided for the water stream. It was verified that to avoid excessive noise, the velocity of the gas stream should not exceed 20 m/s. Assume that the flue gas has the same physical properties of the air and determine the:

(a) Number of gas parallel streams in the tube
(b) Required number of heat exchangers in the bank
(c) Final overall heat transfer coefficient
(d) Final fouling factor
(e) Percentage of heat transfer over-surface
(f) Pressure drop in the tube and annulus

Answer:
(a) Six parallel streams
(b) Six heat exchangers
(c) $U = 25.645 \ \text{W/(m}^2 \cdot \text{K)}$
(d) $R_d = 0.010301 \ \text{(m}^2 \cdot \text{K)/W}$
(e) Over-surface = 35.904%
(f) $\Delta P_i = 0.47740 \ \text{kPa}$; $\Delta P_o = 0.13603 \ \text{kPa}$

(7.4) During winter, a 468 kg/h flow rate of ambient air at −10 °C must be heated to 20 °C before feeding the ventilation system of a chemical plant facility. To reduce energy costs, it is used a 137 kg/h stream of exhausted combustion gases (mostly CO_2) at 250 °C. There is in standby a few 3 m long DN 40 × 65 mm/40S carbon steel double-pipe hairpins. The annulus has a straight-pipe-type return. Because of noise generation restrictions, the maximum velocity for both streams is set to 50 m/s. The fouling factors are 0.00176 and 0.00035 m²·K/W for the air and combustion gases, respectively. The allowable pressure drop is 80 kPa for both streams. Design a battery of exchangers to accomplish this service. (a) Choose the fluid allocation (justify). (b) Define the configuration of the heat exchangers battery (justify). Determine the (c) heat transfer coefficients for the tube and annulus, (d) final overall heat transfer coefficient, (e) percentage of heat transfer over-surface, and (f) pressure drop in the tube and annulus.

Answer:
(a) Combustion gases in the tube and air in the annulus (justify!)
(b) Tube in series and annulus divided into two streams (justify!)
(c) $h_i = 81.976$ W/(m^2·K), $h_o = 155.08$ W/(m^2·K)
(d) $U = 11.295$ W/(m^2·K)
(e) Over-surface = 323.04%
(f) $\Delta P_i = 3.7927$ kPa; $\Delta P_o = 20.469$ kPa

(7.5) Assume that the exchanger required in Problem 7.4 will be installed in a housed remote place, in a way that no restrictions on noise production holds anymore. Redesign the heat exchanger bank to accomplish this service and determine the (a) heat transfer coefficients for the tube and annulus, (b) final overall heat transfer coefficient, (c) percentage of heat transfer over-surface, and (d) pressure drop in the tube and annulus.

Answer:
(a) $h_i = 81.976$ W/(m^2·K), $h_o = 265.17$ W/(m^2·K)
(b) $U = 22.462$ W/(m^2·K)
(c) Over-surface = 143.92%
(d) $\Delta P_i = 1.6812$ kPa; $\Delta P_o = 79.311$ kPa

8 Multi-tube heat exchangers

A method to attenuate the disadvantage of low compactness of double-pipe exchangers is to use an internal bundle of tubes instead of a single tube-inside-tube design. Such type of heat exchanger is commonly referred to as "multi-tube" or "multi-pipe" exchanger, and Fig. 8.1 is representative of its structure. The installation of multiple internal tubes may increase the heat transfer area per unit length of the exchanger a few times, in comparison with a regular double-pipe exchanger.

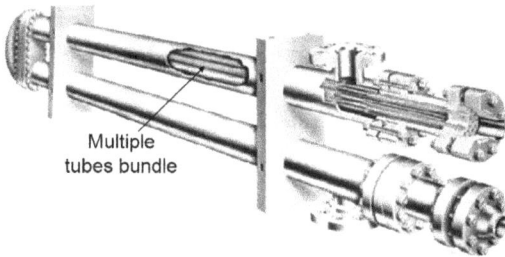

Multiple
tubes bundle

Fig. 8.1: Multi-tube hairpin heat exchanger (adapted from Ref. [110]).

When the process requires a temperature cross between the cold and hot fluids, usually the multi-tube exchanger is the most effective option for transferring greater heat loads. Standard shell pipes with diameters up to 900 mm (36 in), and surface areas as large as 930 m^2 (10,000 ft^2) per exchanger section [110], are found in the market, but the most common range is about 80–400 mm (3–16 in) [28].

8.1 Tube count

In a multi-tube heat exchanger, the quantity of tubes comprising the tube bundle can differ significantly, ranging from only two to as many as a few hundred, depending on the outer tube (shell) size. Furthermore, the number of tubes installed inside a shell with a given internal diameter may also vary from a manufacturer to another because of particularities in the mechanical construction. Typically, manufacturers offer standardized tables for tube counting, which must be consulted for selecting up front a valid number of tubes when designing a multi-tube heat exchanger. With the sole purpose of convenient reference, Tab. 4 in Appendix A.4 shows a tube counting table, compiled from a few heat exchanger suppliers.

https://doi.org/10.1515/9783110585872-008

8.2 Design method

Within practical accuracy, the method outlined in Chapter 7 for designing and rating regular double-pipe heat exchangers may be adapted using the hydraulic diameter as an equivalent diameter for the channel delimited by the tube bundle and the external tube shell.

8.2.1 Cross-flow area

Given the mass flow rates for the internal and external fluids in a multi-tube exchanger, the cross-sectional area used in the calculation of the mass fluxes must be evaluated accordingly.

For the internal fluid, the total cross section S_i is the summation of the individual cross sections over all N_t tubes in the bundle, with the internal diameter D_i:

$$S_i = \frac{\pi N_t}{4} D_i^2 \tag{8.1}$$

The cross-flow area S_o for the external channel is delimited by the external diameter of the internal tubes D_o and the internal diameter of the outer tube D_s:

$$S_o = \frac{\pi}{4} \left(D_s^2 - D_o^2 N_t \right) \tag{8.2}$$

8.2.2 Shell equivalent diameter

From the definition of hydraulic diameter, we have:

$$D_h = 4 \times \frac{\text{flow cross section}}{\text{wetted perimeter}}$$

Defining as D_o the outer diameter of a single tube in the bundle containing N_t tubes, and the inner diameter of the outer tube (shell) as D_s, the flow cross section and wetted perimeter are given by:

$$\text{flow cross section} = \frac{\pi}{4} \left(D_s^2 - D_o^2 N_t \right)$$

$$\text{wetted perimeter} = \pi (D_s + D_o N_t)$$

Substituting in the hydraulic diameter expression, the equivalent diameter (D_e) to be used on the multi-tube exchanger calculations is

$$D_e = D_h = \frac{D_s^2 - D_o^2 N_t}{D_s + D_o N_t} \tag{8.3}$$

Worked Example 8.1

Consider a service where 1.8 kg/s of acrylic acid is cooled from 70 to 45 °C before storage. Nearby the acrylic acid storage tank, there is a line of 1 kg/s of isopropanol with 32 °C that can be used as the cooling fluid. There are available multi-tube hairpins 5 m long built-in carbon steel with a bundle of nine tubes DN 10 mm-sch 40 inside a shell DN 80 mm-sch 40. The carbon steel absolute roughness is estimated as $\varepsilon = 0.035$ mm and its thermal conductivity is $k = 51$ W/mK. Design a bank of multi-tube heat exchangers to perform this service assuming a fouling factor of $R_d = 0.0001$ m²K/W and the allowable pressure drop of 90 kPa for each stream. The shell fluid return is of "straight pipe" type.

Solution

1) Physical properties of fluids
    a) Mean temperatures of fluids
        Within the problem specification, we do not have the outlet temperature of the cooling fluid, isopropanol; therefore, the bulk mean temperature cannot be determined. Let us use the energy balance to calculate the exit temperature of isopropanol.
        Assuming a countercurrent flow, the given operating temperatures are

$$\text{Acrylic acid:} \quad T_{1,in} = 343.15 \text{ K} \quad \rightarrow \quad T_{1,out} = 318.15 \text{ K}$$
$$\text{Isopropanol:} \quad T_{2,out} = ? \quad \quad \leftarrow \quad T_{2,in} = 305.15 \text{ K}$$

As a first approximation, we are not going to guess the exit temperature of the cold fluid, instead, let us evaluate the specific heats using the entrance temperature of the cold fluid and the mean bulk temperature of the hot fluid, given by:

$$T_{1m} = \frac{T_{1,in}}{2} + \frac{T_{1,out}}{2} = \frac{318.15}{2} + \frac{343.15}{2} = 330.65 \text{ K}$$

The specific heat for both fluids can be obtained from Appendices C.11 and C.12:

Fluid 1 (hot)	Fluid 2 (cold)
Acrylic acid	Isopropanol
330.65 K – 101, 325 Pa	305.15 K – 101, 325 Pa
$Cp = 1,870.0$ J/(kg·K)	$Cp = 2,520.0$ J/(kg·K)

Assuming no heat lost for the environment, from the energy balance between both streams, we have:

$$Cp_2 W_2 (T_{2,out} - T_{2,in}) + W_1 Cp_1 (T_{1,out} - T_{1,in}) = 0$$

Solving for the isopropanol exit temperature, we get:

$$T_{2,out} = \frac{1}{Cp_2 W_2} (Cp_2 T_{2,in} W_2 + T_{1,in} W_1 Cp_1 - T_{1,out} W_1 Cp_1)$$

$$= \frac{1}{1.0000 \cdot 2,520.0} (1.0000 \cdot 2,520.0 \cdot 305.15 - 1.8000 \cdot 1,870.0 \cdot 318.15$$

$$+ 1.8000 \cdot 1,870.0 \cdot 343.15) = 338.54 \text{ K}$$

In addition, the mean bulk temperature of the cold fluid is now evaluated as:

$$T_{2m} = \frac{T_{2,in}}{2} + \frac{T_{2,out}}{2} = \frac{305.15}{2} + \frac{338.54}{2} = 321.85 \text{ K}$$

Finally, let us check the energy balance, using the calculated outlet temperature:

$$q_1 = W_1 Cp_1 (T_{1,out} - T_{1,in}) = 1.8000 \cdot 1,870.0 (318.15 - 343.15) = -84,150 \text{ W}$$

$$q_2 = W_2 Cp_2 (T_{2,out} - T_{2,in}) = 1.0000 \cdot 2,520.0 (-305.15 + 338.54) = 84,143 \text{ W}$$

As we see, the enthalpy changes of both streams are in reasonable agreement within the accuracy of five significant figures used in the calculations.

Using the obtained mean bulk temperatures, from Appendices C.12 and C.11, the physical properties are estimated as:

Fluid 1 (hot)	Fluid 2 (cold)
Acrylic acid	Isopropanol
330.65 K – 101, 325 Pa	321.85 K – 101, 325 Pa
$\mu = 0.000651 \text{ s} \cdot \text{Pa}$	$\mu = 0.00108 \text{ s} \cdot \text{Pa}$
$\mu = 0.651 \text{ cP}$	$\mu = 1.08 \text{ cP}$
$\rho = 1,090.0 \text{ kg/m}^3$	$\rho = 771.0 \text{ kg/m}^3$
$Cp = 1,870.0 \text{ J/(kg} \cdot \text{K)}$	$Cp = 2,610.0 \text{ J/(kg} \cdot \text{K)}$
$k = 0.15 \text{ W/(m} \cdot \text{K)}$	$k = 0.133 \text{ W/(m} \cdot \text{K)}$
$Pr = 8.15$	$Pr = 21.3$

2) Logarithmic mean temperature difference

For counterflow arrangement, the temperatures of fluids at the heat exchanger terminals are:

Terminal 1	Terminal 2
$T_{1,in} = 343.15$ K $\rightarrow$	$T_{1,out} = 318.15$ K
$T_{c,out} = 338.54$ K $\leftarrow$	$T_{2,in} = 305.15$ K
$\Delta T_1 = 4.6100$ K	$\Delta T_2 = 13.000$ K

With the temperature differences at the terminals, the log mean temperature difference is:

$$\Delta T_1 = T_{1,in} - T_{2,out} = -338.54 + 343.15 = 4.6100\text{K}$$

$$\Delta T_2 = T_{1,out} - T_{2,in} = -305.15 + 318.15 = 13.000\text{K}$$

$$\Delta T_{lm,c} = \frac{\Delta T_1 - \Delta T_2}{\ln\left(\frac{\Delta_1}{\Delta T_2}\right)} = \frac{4.6100 - 13.000}{\ln\left(\frac{4.6100}{13.000}\right)} = 8.0928\text{K}$$

3) Pipe dimensions

The detailed size specifications for both pipes are obtained from Appendix A.1 as follows:

DN 10 mm – schedule 40

$D_i = $ ID $= 12.48$ mm (inner diameter)
$D_o = $ OD $= 17.1$ mm (outer diameter)

DN 80 mm – schedule 40

$D_s = $ ID $= 77.92$ mm (inner diameter)
OD $= 88.9$ mm (outer diameter)

4) Heat transfer coefficient of inner pipes

The internal bundle is composed of $N_t = 9$ tubes; therefore, the total flow cross section, mass flux, and Reynolds number for the internal fluid are:

$$S_i = \frac{\pi N_t}{4} D_i^2 = \frac{\pi 9.0000}{4} 0.012480^2 = 0.0011009 \text{ m}^2$$

$$G_i = \frac{W_i}{S_i} = \frac{1.0000}{0.0011009} = 908.35 \text{ kg}/(\text{m}^2 \cdot \text{s})$$

$$\text{Re}_i = \frac{D_i G_i}{\mu_i} = \frac{0.012480}{0.0010800} 908.35 = 10,496 \text{ (fully turbulent flow)}$$

For this Reynolds number, the flow regime is considered fully turbulent, and then we may apply the Petukhov–Popov equation (4.11) to estimate the heat transfer coefficient. Since the pipe's absolute roughness is known, for increased accuracy, we can use a friction equation that considers the effect of the surface roughness on the flow, such as the Haaland's equation:

$$f_F = \frac{0.41}{\ln^2\left(0.23\left(\frac{\varepsilon}{D_i}\right)^{\frac{10}{9}} + \frac{6.9}{Re_i}\right)} = \frac{0.41}{\ln^2\left(0.23\left(\frac{(3.5000e-5)}{0.012480}\right)^{\frac{10}{9}} + \frac{6.9}{10,496}\right)} = 0.0085752$$

The Petukhov–Popov's parameters are evaluated as follows:

$$K_1 = 13.6 f_F + 1 = 13.6 \cdot 0.0085752 + 1 = 1.1166$$

$$K_2 = 11.7 + \frac{1.8}{\sqrt[3]{Pr}} = 11.7 + \frac{1.8}{\sqrt[3]{21.300}} = 12.349$$

Solving the Petukhov–Popov equation for h_i, and applying the previously calculated parameters, we have:

$$h_i = \frac{Pr Re_i f_F k}{D_i\left(2K_1 + \sqrt{2}K_2 Pr^{\frac{2}{3}}\sqrt{f_F} - \sqrt{2K_2}\sqrt{f_F}\right)}$$

$$= \frac{0.0085752 \cdot 0.13300 \cdot 10,496.21.300}{0.012480\left(\sqrt{2}\sqrt{0.0085752} \cdot 12.349 \cdot 21.300^{\frac{2}{3}} - \sqrt{2}\sqrt{0.0085752} \cdot 12.349 + 2 \cdot 1.1166\right)}$$

$$= 1,566.5 \ W/(m^2 \cdot K)$$

This heat transfer coefficient assumes fully developed flow, a condition that is physically valid in the present case because for a single hairpin leg the number of tube diameters is already high:

$$\# \text{Diameters} = \frac{L_{hp}}{D_i} = \frac{5.0000}{0.012480} = 400.64 \gg 10 \text{ diameters (fully developed flow)}$$

Considering the relatively low viscosity of both fluids, the viscosity correction factor for variable physical properties will be ignored, i.e., $\phi_i = (\mu_i/\mu_w)^{0.14} = 1$.

5) Heat transfer coefficient of the shell
The annulus equivalent diameter used in the Reynolds number is taken here as equal to the hydraulic diameter considering nine tubes in the internal bundle:

$$D_e = D_h = \frac{D_s^2 - D_o^2 N_t}{D_s + D_o N_t} = \frac{0.077920^2 - 0.017100^2 \cdot 9.0000}{0.017100 \cdot 9.0000 + 0.077920} = 0.014838 \ m$$

$$S_o = \frac{\pi}{4}\left(D_s^2 - D_o^2 N_t\right) = \frac{\pi}{4}\left(0.077920^2 - 0.017100^2 \cdot 9.0000\right) = 0.0027016 \ m^2$$

$$G_o = \frac{W_o}{S_o} = \frac{1.8000}{0.0027016} = 666.27 \text{ kg}/(\text{m}^2 \cdot \text{s})$$

$$Re_o = \frac{D_e G_o}{\mu_o} = \frac{0.014838}{0.00065100} 666.27 = 15,186 \text{ (turbulent flow)}$$

This Reynolds number indicates turbulent flow, and the number of equivalent diameters for the shell in a single exchanger is:

$$\#\text{Diameters} = \frac{L_{hp}}{D_e} = \frac{5.0000}{0.014838} = 336.97 \gg 10 \text{ diameters (fully developed flow)}$$

Therefore, for further accuracy, we can use the Petukhov–Popov equation combined with Haaland's friction factor as follows:

$$f_F = \frac{0.41}{\ln^2\left(0.23\left(\frac{\varepsilon}{D_e}\right)^{\frac{10}{9}} + \frac{6.9}{Re_o}\right)} = \frac{0.41}{\ln^2\left(0.23\left(\frac{3.5000e-5}{0.014838}\right)^{\frac{10}{9}} + \frac{6.9}{15,186}\right)} = 0.0078639$$

$$K_1 = 13.6 f_F + 1 = 13.6 \cdot 0.0078639 + 1 = 1.1069$$

$$K_2 = 11.7 + \frac{1.8}{\sqrt[3]{Pr}} = 11.7 + \frac{1.8}{\sqrt[3]{8.1500}} = 12.594$$

Solving for h_o, and substituting the calculated parameters, we have:

$$h_o = \frac{Pr Re_o f_F k}{D_e\left(2K_1 + \sqrt{2}K_2 Pr^{\frac{2}{3}}\sqrt{f_F} - \sqrt{2}K_2\sqrt{f_F}\right)}$$

$$= \frac{0.0078639 \cdot 0.15000 \cdot 15,186 \cdot 8.1500}{0.014838\left(\sqrt{2}\sqrt{0.0078639} \cdot 12.594 \cdot 8.1500^{\frac{2}{3}} - \sqrt{2}\sqrt{0.0078639} \cdot 12.594 + 2 \cdot 1.0954\right)}$$

$$= 1,404.0 \text{ W}/(\text{m}^2 \cdot \text{K})$$

The factor for correction of variable physical properties will be ignored also for the shell fluid, due to the low viscosity, then $\phi_o = (\mu_o/\mu_w)^{0.14} = 1$.

6) Overall heat transfer coefficient
The overall heat transfer coefficient, including the wall conductive resistance and the internal and external convective resistances, is:

$$U_c = \frac{1}{\frac{D_o}{2k}\ln\left(\frac{D_o}{D_i}\right) + \frac{1}{h_o} + \frac{D_o}{D_i h_i}} = \frac{1}{\frac{0.017100\,\ln\left(\frac{0.017100}{0.012480}\right)}{2 \cdot 51.000} + \frac{1}{1,404.0} + \frac{0.017100}{0.012480 \cdot 1,566.5}}$$

$$= 609.85 \text{ W}/(\text{m}^2 \cdot \text{K})$$

The fouling resistances required for both fluids are R_{di} and R_{do}:

$$R_{di} = 0.0001 \ (m^2 \cdot K)/W$$

$$R_{do} = 0.0001 \ (m^2 \cdot K)/W$$

Therefore, the final design overall heat transfer coefficient used to size the exchanger area is given by:

$$U_d = \frac{1}{R_{do} + \frac{1}{U_c} + \frac{D_o R_{di}}{D_i}} = \frac{1}{0.00010000 + \frac{0.00010000}{0.012480} 0.017100 + \frac{1}{609.85}} = 532.83 \ W/(m^2 \cdot K)$$

7) Number of hairpins

a) Heat transfer area

From the heat exchanger design equation, the required clean heat transfer surface is:

$$A_c = \frac{q}{U_c \Delta T_{lm}} = \frac{84,143}{609.85 \cdot 8.0928} = 17.049 \ m^2$$

And the fouled heat transfer area is calculated as:

$$A_d = \frac{q}{U_d \Delta T_{lm}} = \frac{84,143}{532.83 \cdot 8.0928} = 19.513 \ m^2$$

b) Total pipe length and number of exchangers

The heat transfer area per unit length, based on the external surface of the internal bundle, is:

$$A_{linear} = \pi D_o N_t = \pi 0.017100 \cdot 9.0000 = 0.48349 \ (m^2/m)$$

Therefore, the necessary pipe length L, number of exchangers N_{hp} in the bank, and the final specified surface are:

$$L = \frac{A_d}{A_{linear}} = \frac{19.513}{0.48349} = 40.359 \ m$$

$$N_{hp} = \frac{L}{2L_{hp}} = \frac{40.359}{2 \cdot 5.0000} = 4.0359 \rightarrow 5 \ \text{hairpins}$$

The final design heat transfer area, overall heat transfer, and fouling factor for the exchanger are:

$$A = 2A_{linear}L_{hp}N_{hp} = 2 \cdot 0.48349 \cdot 5.0000^2 = 24.174 m^2$$

$$U = \frac{q}{A \Delta T_{lm}} = \frac{84,143}{24.174 \cdot 8.0928} = 430.10 \ W/(m^2 \cdot K)$$

$$R_d = \frac{U_c - U}{U_c U} = \frac{609.85 - 430.09}{430.09 \cdot 609.85} = 0.00068535 \ (\text{m}^2 \cdot \text{K})/\text{W}$$

8) Pressure drop
   a) Pressure drop in the internal tubes
   The friction factor evaluated previously from Haaland's equation is:

$$f_{\text{Fi}} = 0.0085752$$

(1) Distributed pressure drop – tube
For a bank with five heat exchangers, the distributed pressure drop along the internal tube bundle is:

$$\Delta P_{i,\text{dist}} = \frac{4 L_{hp} N_{hp} f_{\text{Fi}}}{D_i \rho_i} G_i^2 = \frac{4 \cdot 0.0085752 \cdot 5.0000^2}{0.012480 \cdot 771.00} 908.35^2$$

$$= 73,533 \ \text{kg}/(\text{m} \cdot \text{s}^2) = 73.533 \text{kPa}$$

(2) Localized pressure drop – tube

$$D_{i(mm)} = 12.48 \text{mm}$$

$$K_i = 0.7 + \frac{2,000}{\text{Re}_i} + \frac{17.78}{D_{i(mm)}} = 0.7 + \frac{17.78}{12.480} + \frac{2,000}{10,496} = 2.3152$$

$$\Delta P_{i,\text{loc}} = \frac{G_i^2 K_i}{2 \rho_i} (2 N_{hp} - 1) = \frac{2.3152 \cdot 908.35^2}{2 \cdot 771.00} (2 \cdot 5.0000 - 1)$$

$$= 11,149 \ \text{kg}/(\text{m} \cdot \text{s}^2) = 11.149 \text{kPa}$$

(3) Total pressure drop – tube

$$\Delta P_i = \Delta P_{i,\text{dist}} + \Delta P_{i,\text{loc}} = 11,149 + 73,533 = 84,682 \ \text{kg}/(\text{m} \cdot \text{s}^2) = 84.682 \text{kPa}$$

b) Pressure drop in the shell
   (1) Friction factor – shell

The shell friction factor was calculated previously as follows:

$$f_{\text{Fo}} = 0.0078639$$

(2) Distributed pressure drop – shell

$$\Delta P_{o,\text{dist}} = \frac{4 L_{hp} N_{hp} f_{\text{Fo}}}{D_h \rho_o} G_o^2 = \frac{4 \cdot 0.0078639 \cdot 5.0000^2}{0.014838 \cdot 1,090.0} 666.27^2$$

$$= 21,584 \ \text{kg}/(\text{m} \cdot \text{s}^2) = 21.584 \text{kPa}$$

(3)   Localized pressure drop – shell

Considering a straight pipe return connecting the shells of the two legs of the exchanger, the localized pressure drop is evaluated from eqs. (5.40) and (5.32) as follows:

$$D_{h(mm)} = 14.838mm$$

$$K_o = 2.8 + \frac{2,000}{Re_o} + \frac{71.12}{D_{h(mm)}} = 2.8 + \frac{2,000}{15,186} + \frac{71.12}{14.838} = 7.7248$$

$$\Delta P_{o,loc} = \frac{G_o^2 K_o N_{hp}}{2\rho_o} = \frac{5.0000 \cdot 666.27^2 \cdot 7.7248}{2 \cdot 1,090.0} = 7,865.0 \text{ kg}/(m \cdot s^2) = 7.865 \text{ kPa}$$

(4)   Total pressure drop – shell

$$\Delta P_o = \Delta P_{o,dist} + \Delta P_{o,loc} = 21,584 + 7,865.0 = 29,449 \text{ kg}/(m \cdot s^2) = 29.449 kPa$$

Conclusion

The proposed service can be performed by a battery of five multi-tube hairpin exchangers with both tube and shell associated in series. The pressure loss for both fluids is below the allowable value of 90 kPa.

Problems

(8.1) A stream of isopropanol with a flow rate of 19,842.0 lb/h is cooled in the annulus (shell) of a multi-tube heat exchanger from 122 to 77 °F, using treated cooling water at 73.4 °F and flow rate of 23,810.0 lb/h. The available exchangers are carbon steel hairpins 16.4 ft long with 31 internal tubes of OD = 0.75 in and ID = 0.584. The external pipe is NPS 6 in schedule 40S, also made of carbon steel. The annulus return is a rod through straight pipe. Evaluate the (a) annulus flow area, (b) equivalent diameter, (c) internal and (d) external heat transfer coefficients, and (e) overall heat transfer coefficient. (f) How many exchangers are enough to meet this service?

Answer:
(a) $S_o = 0.10559 \text{ ft}^2$
(b) $D_e = 0.055033 \text{ ft}$
(c) $h_i = 570.43 \text{ BTU}/(ft^2 h \text{ °F})$
(d) $h_o = 40.055 \text{ BTU}/(ft^2 h \text{ °F})$
(e) $U_c = 36.339 \text{ BTU}/(ft^2 h \text{ °F})$
(f) Seven hairpins connected with tube and annulus in series

(8.2) The configuration used in Problem 8.1 produced a tube heat transfer coefficient one order of magnitude greater than the heat transfer coefficient of the annulus. Such substantial difference may indicate that an inverted fluid allocation can possibly improve the design, requiring less heat transfer area to perform the same heat load. Verify the feasibility of this alternative design allocating isopropanol in the tubes and the water in the annulus. Calculate (a) the internal and (b) external heat transfer coefficients, (c) the overall heat transfer coefficient, (d) the new number of exchangers necessary to perform the service, the pressure loss (e) in the tube, and (f) in the annulus.

Answer:
(a) $h_i = 89.126$ BTU/(ft^{2}h °F)
(b) $h_o = 285.65$ BTU/(ft^{2}h °F)
(c) $U_c = 54.901$ BTU/(ft^{2}h °F)
(d) Five hairpins connected with tube and annulus in series
(e) $\Delta P_i = 3.1721$ psi
(f) $\Delta P_o = 0.94645$ psi

(8.3) Cooling water is used to cool a benzene stream in a bank of multi-tube hairpin exchangers, where the water is the internal fluid. The average temperatures are 99.5 and 86.1 °F for benzene and water, respectively. The heat transfer coefficients are $h_i = 1,027.5$ BTU/(ft^2 h °F) and $h_o = 289.3$ BTU/(ft^2 h °F). The inner pipes are of carbon steel ($k = 26.001$ BTU/(ft h°F)), ID = 0.505 in, and OD = 0.625 in. The fouling factors for the fluids are $R_{di} = 0.0019874$ ft^2 h °F/BTU and $R_{do} = 0.0010221$ ft^2 h °F/BTU. Determine the clean overall heat transfer coefficient (a) disregarding the tube wall resistance and (b) including the pipe wall conductive resistance. Estimate the (c) internal and (d) external wall temperatures.

Answer:
(a) $U_c = 214.54$ BTU/(ft^{2}h °F)
(b) $U_c = 205.13$ BTU/(ft^{2}h °F)
(c) $t_{wi} = 89.4$ °F
(d) $t_{wo} = 90.0$ °F

(8.4) A battery of three multi-tube carbon steel hairpins is used to cool down 547.2 kg/h of benzene from 60.0 to 23.0 °C using 792 kg/h of cooling water at 20.0 °C, allocated inside the tubes. The heat exchangers are 5 m long, with outer pipe DN 100 schedule 40 and 19 inner tubes ID = 12.83 mm/OD = 15.88 mm. The annulus fluid return is a straight pipe, and the fouling factor is 0.00018 and 0.00035 m^{2}K/W for the benzene and cooling water, respectively. A pressure drop of 70 kPa is allowed in both streams. (a) Do you think the pipe set of these exchangers is adequate for this service? Determine (b) the heat load, (c) inner and outer convective heat transfer coefficients, (d) tubes and annulus pressure loss, and (e) the over-surface.

Answer:
(a) No (justify!)
(b) $q = 8{,}721.2$ W
(c) $h_i = 238.82$ W/(m$^2 \cdot$ K), $h_o = 45.462$ W/(m$^2 \cdot$ K)
(d) $\Delta P_i = 511.43$ Pa, $\Delta P_o = 115.3$ Pa
(e) Over-surface = 41.3%

(8.5) Reassessing the service presented in Problem 8.4, an engineer suggested the division of the cooling water in two parallel streams allocated inside the tubes as a presumably more cost-effective design. (a) Does the new proposed design make sense, in your judgment? Determine the (b) effective mean temperature difference, (c) heat transfer coefficient of the tube and annulus, (d) clean overall heat transfer coefficient, and (e) resulting over-surface. (f) How many hairpins are necessary to match the desired heat load using this configuration?

Answer:
(a) No (justify!)
(b) 10.77 °C
(c) $h_i = 189.53$ W/(m$^2 \cdot$ K), $h_o = 45.462$ W/(m$^2 \cdot$ K)
(d) $U_c = 35.008$ W/(m$^2 \cdot$ K)
(e) Over-surface = 63.9%
(f) $N_{hp} = 4$ exchangers

9 Finned tube heat exchangers

When specifying an exchanger with a relatively low heat-transfer coefficient for one or both fluids, the extension of the heat-transfer surface using fins may be an effective solution. If only one fluid is a gas, a very viscous liquid, or develops a low Reynolds number (small flow rate), possibly it will control the heat transfer process, and ultimately determine the size of the equipment. The low convective heat-transfer coefficient can be balanced by the additional area provided by the fins installed over the base pipe surface. In these cases, finned double-pipe heat exchangers often surpass the performance of shell-and-tube exchangers, being more cost-effective [44]. An in depth presentation of general finned surfaces and methods of heat transfer rate calculation is found in Ref. [30].

Fins can be attached to the pipe external surface by extrusion, soldering, or welding. In common low-pressure applications, the typical thickness of extruded or soldered fins up to 12.7 mm high is 0.5 mm, while for higher fins the ordinary thickness is 0.8 mm. Welded fins are 0.89 mm thick for heights up to 25.4 mm [48].

For services with very viscous or condensing fluids, the use of longitudinal fins is recommended, as exemplified in Figs. 9.1 and 9.2 [18]. The reason is that longitudinal fins, aligned with the fluid flow, produce significantly less pressure drop. For a condensing vapor, besides the pressure drop decrease, longitudinal fins contribute for less local accumulation of condensate, reduced corrosion, and more streamlined and uniform condensate outflow of the exchanger.

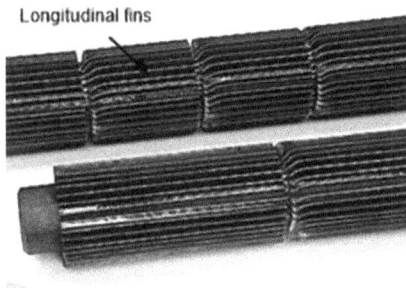

Fig. 9.1: Heat transfer tubes with longitudinal fins. [Adapted from Ref. [110]].

As a general guide, the recommendations on the use of fins are [95]:
1. One of the fluids is a gas or a viscous liquid with very low heat-transfer coefficient.
2. Condensation of organic vapors, using liquid, e.g., cooling water.
3. When one fluid operates at high pressure, it is allocated in a thicker inner pipe to reduce costs. The placement of fins on the external surface of the inner pipe may significantly decrease the total pipe length required for the service, reducing the exchanger cost even further.

https://doi.org/10.1515/9783110585872-009

4. One fluid is very corrosive. The corrosive fluid can be allocated inside the inner pipe made of more resistant and expensive material. External fins installed on the inner pipe can decrease the heat exchanger size and, consequently, the overall cost of the unit.

9.1 Fins count

The number and height of external longitudinal fins in a double-pipe or multi-tube exchanger depends on the number and outer diameter of the internal tubes, and on the inner diameter of the shell. In general, for a specified shell, the increase in the number of internal tubes implies a decrease in the fin height.

Even though the fins height and count can be determined by simple geometry, not all geometric possibilities are practically feasible, because of construction restrictions. Commonly, heat exchanger fabricators arrange standard tables with fin specifications according to given internal tube and shell diameters. It is recommended that the designer refers to these tables when selecting the tubing and fins dimensions of the new unit.

There is the commercial availability of finned double-pipe exchangers having 16 to 64 external longitudinal fins, with fin heights varying from 6.35 mm to 25.4 mm (0.25 in to 1 in). Detailed design data for several tube and shell sizes can be found in Tab. 5 of Appendix A.4.

Standard multi-tube exchangers may have 7 to 55 internal tubes, equipped with 16 or 20 fins, with height ranging from 5.33 mm to 11.13 mm. Tab. 6 from Appendix A.4 shows finned multi-tube exchanger specifications compiled from selected manufacturers. The cross-section illustration of a usual multi-tube heat exchanger with 16 longitudinal fins per tube is shown in Fig. 9.3.

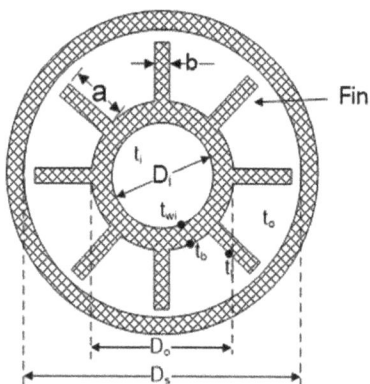

Fig. 9.2: Cross section of a double-pipe heat exchanger extended with longitudinal fins on the external surface of the internal tube.

Notice that any mechanically achievable configuration of fins and tube count can possibly be customized by a manufacturer, however sticking to standardized specifications, such as those found in Tabs. 5 and 6, often leads to the most cost-effective designs when purchasing time comes.

9.2 Design method

The design of finned tubular double- or multi-tube exchangers with longitudinal fins placed outside the internal tubes is very similar to the one used for "regular" unfinned heat exchangers. Since the longitudinal fins are installed along the pipe length, the fluid dynamics is comparable to the closed flow inside bare pipes and the concept of equivalent diameter (D_e) may be used with acceptable accuracy for most practical applications.

9.2.1 Cross-flow area

In general, let us consider a heat exchanger with a bundle of N_t tubes with inner diameter D_i, outer diameter D_o, and N_f fins over each one, placed in a shell of internal diameter D_s. The internal and external cross-flow section for the whole bundle are evaluated as:

$$S_i = \frac{\pi N_t}{4} D_i^2 \tag{9.1}$$

$$S_o = \frac{\pi}{4} \left(D_s^2 - D_o^2 N_t \right) - N_f N_t ab \tag{9.2}$$

where a and b are the fins height and width, respectively (Fig. 9.2).

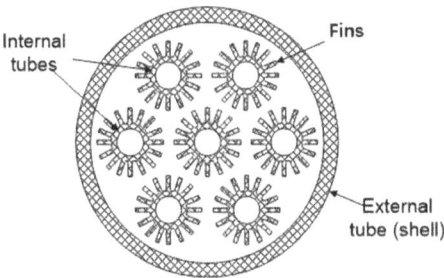

Fig. 9.3: Cross section of a multi-tube heat exchanger extended with longitudinal fins on the external surface of the internal tubes.

9.2.2 Shell equivalent diameter

The equivalent diameter is based on the definition of hydraulic diameter ($D_e = D_h = 4S_{flow}/P_{wetted}$), where the flow cross section (S_{flow}) comes from eq. (9.2) and the total wetted perimeter touched by the outer fluid for N_t tubes with N_f fins is:

$$P_{wetted} = \pi D_s + N_t(\pi D_o + 2N_f a) \tag{9.3}$$

Therefore, the equivalent diameter can be evaluated in the form:

$$D_e = D_h = \frac{\pi D_s^2 - N_t(\pi D_o^2 + 4N_f ab)}{\pi D_s + N_t(\pi D_o + 2N_f a)} \tag{9.4}$$

9.3 Fin efficiency

By far, the most found fins in extended tubular exchangers have a uniform rectangular cross section, normally designated as rectangular fins. For this type, the efficiency for a fin with width b and height a (Fig. 9.2) is given by:

$$\eta_f = \frac{1}{am}\tanh(am) \tag{9.5}$$

where

$$m = \sqrt{\frac{2h_o'}{bk_f}} \tag{9.6}$$

Note that the fin efficiency is dependent upon the effective heat transfer coefficient (h_o') acting on the fin surface, then, if a fouling factor R_{do} is provided for the external fluid, it must be combined with the outer convective heat transfer h_o as:

$$h_o' = \frac{1}{R_{do} + \frac{1}{h_o}} \tag{9.7}$$

9.4 Weighted fin efficiency

The weighted fin efficiency is a quite useful concept, because it allows the relation of the heat load from an extended (finned) surface with the full area exchanging the energy. From the physical reasoning that a hypothetical fin surface A_f with unitary efficiency (η_f) has the same temperature of the base surface t_b, we may state that the base (not finned) area A_b exchanges energy with unitary efficiency. Therefore, a kind of average efficiency weighted by the areas A_b and A_f may be defined in the form:

$$\eta_o = \frac{A_b \times 1 + A_f \times \eta_f}{A_{\text{total}}} = \frac{A_b + A_f \eta_f}{A_b + A_f} \tag{9.8}$$

The free unfinned external area and the total area of the fins for a bundle of N_t internal tubes with outer diameter D_o, length L, and N_f longitudinal fins installed are, respectively:

$$A_b = LN_t(\pi D_o - N_f b) \tag{9.9}$$

$$A_f = LN_f N_t(2a + b) \tag{9.10}$$

With eqs. (9.9) and (9.10) into (9.8), after simplification, we get the weighted fin efficiency as:

$$\eta_o = \frac{\pi D_o + N_f(\eta_f(2a + b) - b)}{\pi D_o + 2N_f a} \tag{9.11}$$

9.5 Overall heat transfer coefficient

With the installation of fins over the outside surface of the inner tubes, the effective heat transfer area becomes the sum of the remaining external surface not covered by fins (A_b) and the added surface from the fins (A_f). As usual, the overall heat transfer coefficient is referenced to the total external surface ($A_{\text{total}} = A_b + A_f$) participating in the heat exchange.

For a set of N_t inner tubes with N_f external fins each, we can devise the following heat transfer areas:

– Internal area of the inner tubes

$$A_i = \pi D_i L N_t \tag{9.12}$$

– External area of the inner tubes (without the fins)

$$A_o = \pi D_o L N_t \tag{9.13}$$

– Unfinned base external area (not covered by fins)

$$A_b = LN_t(\pi D_o - N_f b) \tag{9.14}$$

– Total area of the fins, including the tips

$$A_f = LN_f N_t(2a + b) \tag{9.15}$$

– Total external area: remaining base area plus the surface added by the fins

$$A_{\text{total}} = A_b + A_f = LN_t(\pi D_o + 2N_f a) \tag{9.16}$$

The following resistances can represent the thermal circuit for a finned inner tube:
1. Internal convective resistance (R_i)
2. Internal fouling resistance (R_{fi})
3. Wall conductive resistance (R_k)
4. Combined external resistance (R_o), comprising:
   a. Fin conductive resistance
   b. External fouling resistance
   c. External convective resistance

The internal convective and fouling resistances are:

$$R_i = \frac{1}{A_i h_i} \tag{9.17}$$

$$R_{fi} = \frac{R_{di}}{A_i} \tag{9.18}$$

Solving eq. (9.12) for the tube length L, the conductive resistance of the inner tube wall may be written as:

$$R_k = \frac{\ln\left(\frac{D_o}{D_i}\right)}{2\pi k L} = \frac{D_i N_t \ln\left(\frac{D_o}{D_i}\right)}{2 A_i k} \tag{9.19}$$

Due to the presence of the outer fouling, the effective heat transfer coefficient (h_o') perceived by the fin is a combination of the fouling and convective external resistances as calculated by eq. (9.7). Thus, using the definitions of fin efficiency (η_f) and weighted fin efficiency (η_o), the heat transfer rate through the outer surface of the finned tubes is given by:

$$q = h_o'(A_b + A_f \eta_f)(t_b - t_o) = h_o' \eta_o (A_b + A_f)(t_b - t_o) \tag{9.20}$$

Therefore, the combined resistance from eq. (9.7) into (9.20) is:

$$R_o = \frac{1}{h_o' \eta_o (A_b + A_f)} = \frac{R_{do} + \frac{1}{h_o}}{\eta_o (A_b + A_f)} = \frac{R_{do}}{\eta_o (A_b + A_f)} + \frac{1}{h_o \eta_o (A_b + A_f)} \tag{9.21}$$

The equivalent thermal resistance between the inner and outer fluids, referred to the total external area, is given as:

$$\frac{1}{U_d (A_b + A_f)} = R_i + R_{fi} + R_k + R_o \tag{9.22}$$

Substituting the resistances from eqs. (9.17)–(9.19) and (9.21) into eq. (9.22), we have:

$$\frac{1}{U_d (A_b + A_f)} = \frac{1}{A_i h_i} + \frac{R_{di}}{A_i} + \frac{D_i N_t \ln\left(\frac{D_o}{D_i}\right)}{2 A_i k} + \frac{R_{do}}{\eta_o (A_b + A_f)} + \frac{1}{h_o \eta_o (A_b + A_f)} \tag{9.23}$$

Using the areas given by eq. (9.12) to (9.15), and doing some algebraic manipulation, the design overall heat transfer coefficient for the heat exchanger is found in the form:

$$U_d = \left(\left(D_o + \frac{2aN_f}{\pi}\right)\left(\frac{1}{2k}\ln\left(\frac{D_o}{D_i}\right) + \frac{1}{D_i}\left(R_{di} + \frac{1}{h_i}\right)\right) + \frac{1}{\eta_o}\left(R_{do} + \frac{1}{h_o}\right)\right)^{-1}$$ (9.24)

The overall heat transfer coefficient for a clean exchanger can be easily obtained by setting $R_{di} = R_{do} = 0$ in eq. (9.24):

$$U_c = \left(\left(D_o + \frac{2aN_f}{\pi}\right)\left(\frac{1}{2k}\ln\left(\frac{D_o}{D_i}\right) + \frac{1}{D_i h_i}\right) + \frac{1}{h_o\eta_o}\right)^{-1}$$ (9.25)

Worked example 9.1

Determine the overall heat transfer coefficient for a double-pipe heat exchanger with an inner pipe having $D_o = 25.4$ mm and a thickness of $\tau = 2.77$ mm. The tube accommodates 20 fins with height $a = 11.1$ mm and thickness $b = 0.50$ mm. The tube-side and annulus heat-transfer coefficients are $h_i = 250$ W/m² K and $h_o = 125$ W/m² K, respectively. The thermal conductivity of the tube material is 46 W/m K, and fouling resistances of $R_{di} = R_{do}$ 0.0001 m² K/W should be provided for each stream [30].

Solution

1) Effective heat-transfer coefficient over the fin

The effective heat-transfer coefficient is needed for an evaluation of the fin efficiency, then, with the required external fouling factor R_{do} and convective coefficient h_o, from eq. (9.7):

$$h'_o = \frac{1}{R_{do} + \frac{1}{h_o}} = \frac{1}{0.000100 + \frac{1}{125}} = 123\frac{W}{(m^2 \cdot K)}$$

2) Regular and weighted fin efficiencies

From eqs. (9.5) and (9.6), the fin efficiency is:

$$m = \left(\frac{2h'_o}{bk_f}\right)^{0.5} = 2^{0.5}\left(\frac{2\cdot123}{0.000500\cdot46}\right)^{0.5} = 103\frac{1}{m}$$

$$\eta_f = \frac{1}{am}\tanh(am) = \frac{\tanh(0.0111\cdot103)}{0.0111\cdot103} = 0.713$$

And the weighted fin efficiency is given by eq. (9.11):

$$\eta_o = \frac{\pi D_o - N_f b + N_f \eta_f (2a + b)}{\pi D_o + 2N_f a}$$

$$= \frac{-0.000500 \cdot 20.0 + \pi 0.0254 + 0.713 \cdot 20.0(0.000500 + 2 \cdot 0.0111)}{2 \cdot 0.0111 \cdot 20.0 + \pi 0.0254} = 0.751$$

3) Overall heat transfer coefficient (U_d)
The inner diameter of the tube is:

$$D_i = D_o - 2\tau = 0.0254 - 2 \cdot 0.00277 = 0.0199m$$

Finally, using eq. (9.24), the design overall heat-transfer coefficient is:

$$U_d = \left(\left(D_o + \frac{2N_f}{\pi} a \right) \left(\frac{\ln\left(\frac{D_o}{D_i}\right)}{2k} + \frac{1}{D_i} \left(R_{di} + \frac{1}{h_i} \right) \right) + \frac{1}{\eta_o} \left(R_{do} + \frac{1}{h_o} \right) \right)^{-1}$$

$$= \left(\left(\frac{2}{\pi} 0.0111 \cdot 20.0 + 0.0254 \right) \left(\frac{\ln\left(\frac{0.0254}{0.0199}\right)}{2 \cdot 46.0} + \frac{1}{0.0199} \left(0.000100 + \frac{1}{250.} \right) \right) \right.$$

$$\left. + \frac{1}{0.751} \left(0.000100 + \frac{1}{125.} \right) \right)^{-1} = 21.9 \frac{W}{(m^2 \cdot K)}$$

9.6 Finned wall temperatures

As for regular (unfinned) heat exchangers, the wall temperatures are very important for a more accurate estimation of convective heat transfer coefficients for heat exchangers with extended surfaces. Accordingly, it is necessary to estimate the temperature of the tube internal surface (t_{wi}) and some average temperature over both the fins and remaining base surface (t_{wo}), which should be applied in the calculation of the viscosity correction factors ϕ_i and ϕ_o appearing in heat-transfer correlations appropriate for variable physical properties.

Considering the internal pipe equipped with longitudinal fins, as seen in Fig. 9.2, under steady state, we can write the following heat-transfer rates between the internal and external fluids, through the finned wall of the inner tube:

$$q = A_i h_i (t_{wi} - t_i) \tag{9.26}$$

$$q = \frac{2A_o k(t_b - t_{wi})}{D_o N_t \ln\left(\frac{D_o}{D_i}\right)} \tag{9.27}$$

$$q = A_b h_o (t_o - t_b) + A_f h_o (t_o - t_f) \tag{9.28}$$

Equating the heat rates from eqs. (9.26) to (9.27) and eqs. (9.27) to (9.28) gives two new equations with three unknown temperatures t_{wi}, t_b, and t_f. Using the definition of fin efficiency to close the system of equations,[21] we end up with the following:

$$A_i h_i (t_{wi} - t_i) = \frac{2 A_o k (t_b - t_{wi})}{D_o N_t \ln\left(\frac{D_o}{D_i}\right)} \tag{9.29}$$

$$\frac{2 A_o k (t_b - t_{wi})}{D_o N_t \ln\left(\frac{D_o}{D_i}\right)} = A_b h_o (t_o - t_b) + A_f h_o (t_o - t_f) \tag{9.30}$$

$$\eta_f = \frac{t_o - t_f}{t_o - t_b} \tag{9.31}$$

Equations (9.29) to (9.31) can be solved analytically for temperatures t_{wi}, t_b, and t_f, then, taking the heat transfer areas from eqs. (9.12) to (9.15) and substituting in the previous temperature solutions, we may obtain equations that are independent from the heat exchanger length, in the form:

$$t_{wi} = \frac{(K_4 + K_6) t_i + K_5 t_o}{K_4 + K_5 + K_6} \tag{9.32}$$

$$t_b = \frac{K_6 t_i + (K_4 + K_5) t_o}{K_4 + K_5 + K_6} \tag{9.33}$$

$$t_f = \frac{K_6 \eta_f t_i + (K_4 + K_5 + K_6 (1 - \eta_f)) t_o}{K_4 + K_5 + K_6} \tag{9.34}$$

Noticing that the outer fluid is exposed to the surface over the fins (A_f) and to some remaining base area left after the fins installation (A_b), the outer viscosity correction factor should be evaluated on a weighted average temperature taking into account both surface temperatures t_b and t_f, respectively:

$$t_{wo} = \frac{A_b t_b + A_f t_f}{A_b + A_f} \tag{9.35}$$

Substituting the previous results from eqs. (9.33) and (9.34) into eq. (9.35) and rearranging, t_{wo} is given as:

$$t_{wo} = \frac{K_1 t_b + K_2 t_f}{K_1 + K_2} \tag{9.36}$$

---

[21] It is worth remarking that we could use eqs. (9.26) and (9.28) to build the third equation to close the system, however such selection of equations leads to a particular solution for the temperatures t_{wi}, t_b, and t_f, which is limited to the case of unitary fin efficiency.

The eq. (9.36) can be transformed in the following equivalent expressions, allowing the estimation of the average outer wall temperature (t_{wo}) as:

$$t_{wo} = \frac{(K_1 + K_2\eta_f)t_b + K_2(1 - \eta_f)t_o}{K_1 + K_2} \tag{9.37}$$

$$t_{wo} = \frac{K_6(K_1 + K_2\eta_f)t_i + ((K_1 + K_2)(K_4 + K_5) + K_2K_6(1 - \eta_f))t_o}{(K_1 + K_2)(K_4 + K_5 + K_6)} \tag{9.38}$$

The K's parameters work likewise weights in eqs. (9.32)–(9.38), and are evaluated as:

$$K_1 = \pi D_o - N_f b \tag{9.39}$$

$$K_2 = N_f(2a + b) \tag{9.40}$$

$$K_3 = h_o(K_1 + K_2\eta_f) \tag{9.41}$$

$$K_4 = D_i h_i \ln\left(\frac{D_o}{D_i}\right) K_3 \tag{9.42}$$

$$K_5 = 2kK_3 \tag{9.43}$$

$$K_6 = 2\pi D_i h_i k \tag{9.44}$$

Worked example 9.2
An 18,000 lb/h stream of petroleum distillate oil is to be cooled from 250 °F to 150 °F with water ranging from 80 °F to 120 °F. A fouling factor of 0.002 h ft² °F/Btu is provisioned for each stream. The exchanger is built in carbon steel with 25 ft long 1.5 × 3 in – sch 40 hairpins. The inner tube has 24 carbon steel fins 0.5 in high and 0.035 in thick. The heat-transfer coefficients, assuming constant physical properties, for the tube and annulus were evaluated as $h_i = 1,436$ Btu/h ft² °F and $h_o = 70.5$ Btu/h ft² °F, respectively. The thermal conductivity of the carbon steel is 29.5 BTU/(ft h °F). Determine the wall temperatures t_{wi} and t_{wo} to be used in the calculation of the viscosity correction factors for considering the effect of variable physical properties on the heat transfer coefficients h_i and h_o [28].

Solution
1) Mean fluid temperatures
The average temperatures for the tube and annulus fluids between the inlet and outlet are:

$$t_i = \frac{t_{i,in}}{2} + \frac{t_{i,out}}{2} = \frac{150.00}{2} + \frac{250.00}{2} = 200.00\,°F$$

$$t_o = \frac{t_{o,in}}{2} + \frac{t_{o,out}}{2} = \frac{120.00}{2} + \frac{80.000}{2} = 100.00\,°F$$

2) Fin efficiency

The effective annulus heat transfer coefficient with the outer fouling resistance from eq. (9.7) is:

$$h'_o = \frac{1}{R_{do} + \frac{1}{h_o}} = \frac{1}{0.0020000 + \frac{1}{70.500}} = 61.788 \frac{BTU}{(ft^2 \cdot h \cdot °F)}$$

Using eq. (9.5) and (9.6), we have the fin efficiency η_f:

$$m = \left(\frac{2h'_o}{bk_f}\right)^{0.5} = \left(\frac{2 \cdot 61.788}{0.0029167 \cdot 29.500}\right)^{0.5} = 37.897 \frac{1}{ft}$$

$$\eta_f = \frac{1}{am} \tanh(am) = \frac{\tanh(0.041667 \cdot 37.897)}{0.041667 \cdot 37.897} = 0.58165$$

3) Wall temperatures

The K's parameters from eqs. (9.39)–(9.44) are:

$$K_1 = \pi D_o - N_f b = -0.0029167 \cdot 24.000 + \pi 0.15833 = 0.42741 ft$$

$$K_2 = N_f(2a + b) = 24.000(0.0029167 + 2 \cdot 0.041667) = 2.0700 ft$$

$$K_3 = h_o(K_1 + K_2\eta_f) = 70.500(0.42741 + 0.58165 \cdot 2.0700) = 115.02 \frac{BTU}{(ft \cdot h \cdot °F)}$$

$$K_4 = D_i K_3 h_i \ln\left(\frac{D_o}{D_i}\right) = 0.13417 \cdot 115.02 \cdot 1436.0 \ln\left(\frac{0.15833}{0.13417}\right) = 3,669.2 \frac{BTU^2}{(ft^2 \cdot h^2 \cdot °F^2)}$$

$$K_5 = 2K_3 k = 2 \cdot 115.02 \cdot 29.500 = 6,786.2 \frac{BTU^2}{(ft^2 \cdot h^2 \cdot °F^2)}$$

$$K_6 = 2\pi D_i h_i k = 2\pi 0.13417 \cdot 1,436.0 \cdot 29.500 = 35,712 \frac{BTU^2}{(ft^2 \cdot h^2 \cdot °F^2)}$$

The tube wall temperature is evaluated from eq. (9.32) as:

$$t_{wi} = \frac{K_5 t_o + (K_4 + K_6)t_i}{K_4 + K_5 + K_6} = \frac{100.00 \cdot 6786.2 + (35712. + 3669.2)200.00}{35712. + 3669.2 + 6786.2} = 185.30 \, °F$$

And the external base surface temperature t_b comes from eq. (9.33):

$$t_b = \frac{K_6 t_i + (K_4 + K_5)t_o}{K_4 + K_5 + K_6} = \frac{(3,669.2 + 6,786.2)100.00 + 200.00 \cdot 35,712}{35,712. + 3,669.2 + 6,786.2} = 177.35 \, °F$$

Using the calculated external base surface temperature t_b in eq. (9.37), we get the outer wall temperature t_{wo} as:

$$t_{wo} = \frac{(K_1 + K_2\eta_f)t_b + (K_2 - K_2\eta_f)t_o}{K_1 + K_2}$$

$$= \frac{100.00(-0.58165 \cdot 2.0700 + 2.0700) + 177.35(0.42741 + 0.58165 \cdot 2.0700)}{0.42741 + 2.0700}$$

$$= 150.53 \ ^\circ F$$

Equation (9.34) allows the evaluation of the effective (average) temperature over the fins (t_f) as:

$$t_f = \frac{K_6\eta_f t_i + (K_4 + K_5 - K_6\eta_f + K_6)t_o}{K_4 + K_5 + K_6}$$

$$= \frac{0.58165 \cdot 200.00 \cdot 35,712 + (-0.58165 \cdot 35,712 + 35,712 + 3,669.2 + 6,786.2)100.00}{35,712 + 3,669.2 + 6,786.2}$$

$$= 144.99 \ ^\circ F$$

Using the previous t_b and t_f values, we could alternatively get t_{wo} from eq. (9.36):

$$t_{wo} = \frac{K_1 t_b + K_2 t_f}{K_1 + K_2} = \frac{0.42741 \cdot 177.35 + 144.99 \cdot 2.0700}{0.42741 + 2.0700} = 150.53 \ ^\circ F$$

Or, directly from the process fluid temperatures t_i and t_o through eq. (9.38):

$$t_{wo} = \frac{K_6(K_1 + K_2\eta_f)t_i + (K_2K_6(1 - \eta_f) + (K_1 + K_2)(K_4 + K_5))t_o}{(K_1 + K_2)(K_4 + K_5 + K_6)}$$

$$= \frac{\begin{array}{c}(2.0700 \cdot 35,712(1 - 0.58165) + (0.42741 + 2.0700)(3,669.2 + 6786.2))100.00 \\ + 35,712(0.42741 + 0.58165 \cdot 2.0700) \cdot 200.00\end{array}}{(0.42741 + 2.0700)(35,712. + 3,669.2 + 6,786.2)}$$

$$= 150.53 \ ^\circ F$$

The preceding calculations for t_{wo} confirm the interesting result that the eqs. (9.36), (9.37), and (9.38) are equivalent. Therefore, selecting one of them for the determination of t_{wo} is mostly a matter of choice, but according to the calculation sequence of t_b and t_f, the appropriate choice may imply in the simplest arithmetic.

Worked example 9.3

In a refinery, 18,000.0 lb/h of a lubricant oil with physical properties similar to those of the n-docosane must be cooled from 250 °F to 150 °F, using raw water ranging from 85 °F to 120 °F. Design a heat exchanger bank composed of finned tube carbon steel 25 ft long hairpins with tubing 3 in × 1 ½ in, both schedule 40. The annulus fluid return is through a straight pipe. The inner tube accommodates 24 external longitudinal fins with height $a = 0.5$ in and thickness $b = 0.035$ in. The fouling resistances of $R_{di} = R_{do} = 0.002$ ft$^2 \cdot$ h $\cdot$ °F/BTU should be provided for each stream and a pressure loss of 15 psi is allowed.

Solution

Considering the given requirements, the first attempted configuration is a counter flow heat exchanger with both tube and annulus channels connected in series.

1) Fluids physical properties
   a) Mean bulk temperatures of the fluids

$$T_{1m} = \frac{T_{1,in}}{2} + \frac{T_{1,out}}{2} = \frac{150.00}{2} + \frac{250.00}{2} = 200.00 \; °F$$

$$T_{2m} = \frac{T_{2,in}}{2} + \frac{T_{2,out}}{2} = \frac{120.00}{2} + \frac{85.000}{2} = 102.50 \; °F$$

   b) Fluid properties at these mean stream temperatures are evaluated using Appendix C.13:

Fluid 1 (hot)	Fluid 2 (cold)
N-docosane	Water
660.0°R = 200.0 °F – 14.7psi	562.0°R = 102.0 °F – 14.7psi
$\mu = 4.66$ lb/(ft $\cdot$ h)	$\mu = 1.61$ lb/(ft $\cdot$ h)
$\mu = 1.93$ cP	$\mu = 0.666$ cP
$\rho = 69.0$ lb/ft^3	$\rho = 65.7$ lb/ft^3
Cp = 0.524 BTU/(lb $\cdot$ °F)	Cp = 0.974BTU/(lb $\cdot$ °F)
$k = 0.082$ BTU/(ft $\cdot$ h $\cdot$ °F)	$k = 0.364$BTU/(ft $\cdot$ h $\cdot$ °F)
Pr = 29.8	Pr = 4.32

2) Energy balance
   a) Water flow rate
   Assuming an adiabatic heat exchanger, from the energy balance between the fluids, the water flow rate is:

$$W_2 = \frac{W_1 Cp_1(T_{1,out} - T_{1,in})}{Cp_2(T_{2,in} - T_{2,out})} = \frac{0.52400 \cdot 18,000(150.00 - 250.00)}{0.97400(85.000 - 120.00)} = 27,668 \frac{lb}{h}$$

   b) Heat load
   Using the four process temperatures, we may confirm if the enthalpy change of both fluids matches:

$$q_1 = W_1 Cp_1(T_{1,\,\text{out}} - T_{1,\,\text{in}}) = 0.52400 \cdot 18,000(150.00 - 250.00) = (-9.4320e+5)\frac{\text{BTU}}{\text{h}}$$

$$q_2 = Cp_2 W_2(T_{2,\,\text{out}} - T_{2,\,\text{in}}) = 0.97400 \cdot 27,668(120.00 - 85.000) = (9.4320e+5)\frac{\text{BTU}}{\text{h}}$$

Therefore, the energy balance is correct within thecalculation accuracy.

3) Logarithmic mean temperature difference

For counter flow arrangement, the fluid temperatures at the heat exchanger termi-nals are:

Terminal 1		Terminal 2
$T_{1,\,\text{in}} = 250$ °F	$\rightarrow$	$T_{1,\,\text{out}} = 150$ °F
$T_{2,\,\text{out}} = 120$ °F	$\leftarrow$	$T_{2,\,\text{in}} = 85$ °F
$\Delta T_1 = 130.000$ °F		$\Delta T_2 = 65.000$ °F

Temperature differences at the terminals and the log mean temperature difference are:

$$\Delta T_1 = T_{1,\,\text{in}} - T_{2,\,\text{out}} = 250.00 - 120.00 = 130.00\,°F$$

$$\Delta T_2 = T_{1,\,\text{out}} - T_{2,\,\text{in}} = 150.00 - 85.000 = 65.000\,°F$$

$$\Delta T_{lm,\,c} = \frac{\Delta T_1 - \Delta T_2}{\ln\left(\frac{\Delta T_1}{\Delta T_2}\right)} = \frac{130.00 - 65.000}{\ln\left(\frac{130.00}{65.000}\right)} = 93.775\,°F$$

4) Pipe dimensions

The detailed size specifications for both pipes and fins are obtained from the Ap-pendix A.1 as:

Inner tube: NPS 1 ½ in – Sch STD/40/40S

ID = D_i = 1.61 in (inner diameter)
OD = D_o = 1.9 in (outer diameter)

Outer tube: NPS 3 in – Sch STD/40/40S

ID = D_s = 3.068 in (inner diameter)
OD = 3.5 in (outer diameter)

External fins

Height: $a = 0.5$ in
Thickness: $b = 0.145$ in
Count: $N_f = 24$

Carbon steel

Absolute roughness: $\epsilon = 0.001378$ in $= 0.00011483$ ft

---
Heat conductivity: $k = 26.001$ BTU/(ft $\cdot$ h $\cdot$ °F)

---

5) Inner pipe heat transfer coefficient

The cross section of the inner pipe, mass flux, and Reynolds number are:

$$S_i = \frac{\pi D_i^2}{4} = \frac{\pi 0.13417^2}{4} = 0.014138 \text{ft}^2$$

$$G_i = \frac{W_i}{S_i} = \frac{27,668}{0.014138} = (1.9570e+6)\frac{\text{lb}}{(\text{ft}^2 \cdot \text{h})}$$

$$\text{Re}_i = \frac{D_i G_i}{\mu_i} = \frac{(1.9570e+6)0.13417}{1.6100} = (1.6309e+5) \rightarrow \textit{fully turbulent}$$

Therefore, the flow regime is fully turbulent and the Petukhov–Popov equation (eq. (4.11)) is applicable.

To consider the effect of the carbon steel roughness on the heat transfer coefficient, we estimate the isothermal friction factor coefficient with the Haaland's equation as:

$$f_{Fi} = \frac{0.41}{\ln^2\left(0.23\left(\frac{\epsilon}{D_i}\right)^{\frac{10}{9}} + \frac{6.9}{\text{Re}_i}\right)} = \frac{0.41}{\ln^2\left(0.23\left(\frac{0.00011483}{0.13417}\right)^{\frac{10}{9}} + \frac{6.9}{(1.6309e+5)}\right)} = 0.0051392$$

Then, the convective heat transfer coefficient is:

$$K_1 = 13.6f_{Fi} + 1 = 13.6 \cdot 0.0051392 + 1 = 1.0699$$

$$K_2 = 11.7 + \frac{1.8}{\sqrt[3]{\text{Pr}_i}} = 11.7 + \frac{1.8}{\sqrt[3]{4.3200}} = 12.805$$

$$\text{Nu}_i = \frac{\text{Pr}_i \text{Re}_i f_{Fi}}{2K_1 + \sqrt{2}K_2\sqrt{f_{Fi}}\left(\text{Pr}_i^{\frac{2}{3}} - 1\right)}$$

$$= \frac{(1.6309e+5)0.0051392 \cdot 4.3200}{\sqrt{2}\sqrt{0.0051392} \cdot 12.805\left(4.3200^{\frac{2}{3}} - 1\right) + 2 \cdot 1.0699}$$

$$= 844.98$$

$$h_i = \frac{kNu_i}{D_i} = \frac{0.36400}{0.13417}844.98 = 2,292.4 \frac{\text{BTU}}{(\text{ft}^2 \cdot \text{h} \cdot \text{°F})}$$

6) Annulus heat transfer coefficient
   a) Equivalent diameter
   With a single internal tube ($N_t = 1$), the equivalent (hydraulic) diameter is evaluated as:

$$D_e = \frac{\pi D_s^2 - N_t\left(\pi D_o^2 + 4N_f ab\right)}{\pi D_s + N_t\left(\pi D_o + 2N_f a\right)}$$

$$= \frac{\pi 0.25567^2 - 1.0000\left(4 \cdot 0.0029167 \cdot 0.041667 \cdot 24.000 + \pi 0.15833^2\right)}{\pi 0.25567 + 1.0000\left(2 \cdot 0.041667 \cdot 24.000 + \pi 0.15833\right)}$$

$$= 0.034822 \text{ft}$$

b) Reynolds number

$$S_o = -N_f N_t ab - \frac{\pi}{4}\left(D_o^2 N_t - D_s^2\right)$$

$$= -0.0029167 \cdot 0.041667 \cdot 1.0000 \cdot 24.000 - \frac{\pi}{4}\left(0.15833^2 \cdot 1.0000 - 0.25567^2\right)$$

$$= 0.028734 \text{ft}^2$$

$$G_o = \frac{W_o}{S_o} = \frac{18,000}{0.028734} = (6.2644e + 5)\frac{\text{lb}}{\left(\text{ft}^2 \cdot \text{h}\right)}$$

$$\text{Re}_o = \frac{D_e G_o}{\mu_o} = \frac{(6.2644e + 5)0.034822}{4.6600} = 4,681.1 \rightarrow \text{transition}$$

Assuming initially $\phi_o = 1$ and ignoring the development length, the convective heat transfer coefficient estimated with the Hausen equation (eq. (4.30)) is:

$$\text{Nu}_o = \text{Pr}_o^{0.33}\left(0.116\text{Re}_o^{0.67} - 14.5\right) = 29.800^{0.33}\left(0.116 \cdot 4,681.1^{0.67} - 14.5\right) = 55.992$$

$$h_o = \frac{k\text{Nu}_o}{D_e} = \frac{0.082000}{0.034822}55.992 = 131.85 \frac{\text{BTU}}{\left(\text{ft}^2 \cdot \text{h} \cdot {}^\circ\text{F}\right)}$$

7) Variable physical property factor
   a) Fin efficiency
   The determination of the external wall temperature requires the fin efficiency, which is given for rectangular fins by:

$$h_o' = \frac{1}{R_{do} + \frac{1}{h_o}} = \frac{1}{0.0020000 + \frac{1}{131.85}} = 104.34 \frac{\text{BTU}}{\left(\text{ft}^2 \cdot \text{h} \cdot {}^\circ\text{F}\right)}$$

$$m = \left(\frac{2h_o{}'}{bk_f}\right)^{0.5} = \left(\frac{2 \cdot 104.34}{0.0029167 \cdot 26.001}\right)^{0.5} = 52.457\frac{1}{\text{ft}}$$

$$\eta_f = \frac{1}{am}\tanh(am) = \frac{\tanh(0.041667 \cdot 52.457)}{0.041667 \cdot 52.457} = 0.44610$$

b) Wall temperature

Calculating the K's parameters from eqs. (9.39) to (9.44), we have:

$$K_1 = \pi D_o - N_f b = -0.0029167 \cdot 24.000 + \pi 0.15833 = 0.42741\text{ft}$$

$$K_2 = N_f(2a + b) = 24.000(0.0029167 + 2 \cdot 0.041667) = 2.0700\text{ft}$$

$$K_3 = h_o(K_1 + K_2\eta_f) = 131.85(0.42741 + 0.44610 \cdot 2.0700) = 178.11\frac{\text{BTU}}{(\text{ft} \cdot \text{h} \cdot {}^\circ\text{F})}$$

$$K_4 = D_i K_3 h_i \ln\left(\frac{D_o}{D_i}\right) = 0.13417 \cdot 178.11 \cdot 2292.4 \ln\left(\frac{0.15833}{0.13417}\right) = 9,070.4\frac{\text{BTU}^2}{(\text{ft}^2 \cdot \text{h}^2 \cdot {}^\circ\text{F}^2)}$$

$$K_5 = 2K_3 k = 2 \cdot 178.11 \cdot 26.001 = 9,262.1\frac{\text{BTU}^2}{(\text{ft}^2 \cdot \text{h}^2 \cdot {}^\circ\text{F}^2)}$$

$$K_6 = 2\pi D_i h_i k = 2\pi 0.13417 \cdot 2,292.4 \cdot 26.001 = 50,248\frac{\text{BTU}^2}{(\text{ft}^2 \cdot \text{h}^2 \cdot {}^\circ\text{F}^2)}$$

The inner wall temperature and outer the temperature of the base external surface comes from eqs. (9.32) and (9.33):

$$t_{wi} = \frac{K_5 t_o + t_i(K_4 + K_6)}{K_4 + K_5 + K_6} = \frac{102.50(50,248 + 9,070.4) + 200.00 \cdot 9,262.1}{50,248 + 9,070.4 + 9,262.1} = 115.67\,{}^\circ\text{F}$$

$$t_b = \frac{K_6 t_i + t_o(K_4 + K_5)}{K_4 + K_5 + K_6} = \frac{102.50 \cdot 50,248 + 200.00(9,070.4 + 9,262.1)}{50,248 + 9,070.4 + 9,262.1} = 128.56\,{}^\circ\text{F}$$

The eq. (9.37) can evaluate the average outer wall temperature as:

$$
\begin{aligned}
t_{wo} &= \frac{K_2 t_o(1 - \eta_f) + t_b(K_1 + K_2\eta_f)}{K_1 + K_2} \\[6pt]
&= \frac{128.56(0.42741 + 0.44610 \cdot 2.0700) + 2.0700 \cdot 200.00(-0.44610 + 1)}{0.42741 + 2.0700} \\[6pt]
&= 161.36\,{}^\circ\text{F}
\end{aligned}
$$

c) Wall temperature correction factors
From Appendix C, the fluid properties at the wall are:

Fluid 1 (hot)	Fluid 2 (cold)
N-docosane	**Water**
$621.0°R = 161.0\ °F - 14.7\text{psi}$	$575.0°R = 116.0\ °F - 14.7\text{psi}$
$\mu = 6.57\ \text{lb}/(\text{ft}\cdot\text{h})$	$\mu = 1.4\ \text{lb}/(\text{ft}\cdot\text{h})$
$\mu = 2.72\ \text{cP}$	$\mu = 0.579\ \text{cP}$
$\rho = 70.2\ \text{lb}/\text{ft}^3$	$\rho = 65.3\ \text{lb}/\text{ft}^3$
$Cp = 0.502\ \text{BTU}/(\text{lb}\cdot°F)$	$Cp = 0.976\ \text{BTU}/(\text{lb}\cdot°F)$
$k = 0.0844\ \text{BTU}/(\text{ft}\cdot\text{h}\cdot°F)$	$k = 0.369\ \text{BTU}/(\text{ft}\cdot\text{h}\cdot°F)$
$Pr = 39.1$	$Pr = 3.7$

Equation (4.25) gives the Prandtl-based correction factor as:

$$\phi_i = \left(\frac{Pr_i}{Pr_{i,w}}\right)^{0.11} = \left(\frac{4.3200}{3.7000}\right)^{0.11} = 1.0172$$

$$\phi_o = \left(\frac{Pr_o}{Pr_{o,w}}\right)^{0.11} = \left(\frac{29.800}{39.100}\right)^{0.11} = 0.97056$$

Then, the corrected heat transfer coefficients are:

$$h_{i,corr} = h_i\phi_i = 1.0172 \cdot 2292.4 = 2{,}331.8\ \frac{\text{BTU}}{\left(\text{ft}^2\cdot\text{h}\cdot°F\right)}$$

$$h_{o,corr} = h_o\phi_o = 0.97056 \cdot 131.85 = 127.97\ \frac{\text{BTU}}{\left(\text{ft}^2\cdot\text{h}\cdot°F\right)}$$

8) Overall heat transfer coefficient
Using the corrected convective coefficients, the clean overall heat transfer coefficient is given by:

$$U_c = \left(\left(D_o + \frac{2N_f}{\pi}a\right)\left(\frac{\ln\left(\frac{D_o}{D_i}\right)}{2k} + \frac{1}{D_ih_i}\right) + \frac{1}{h_o\eta_o}\right)^{-1}$$

$$= \left(\left(\frac{\ln\left(\frac{0.15833}{0.13417}\right)}{2\cdot 26.001} + \frac{1}{0.13417\cdot 2{,}331.8}\right)\left(\frac{2}{\pi}0.041667\cdot 24.000 + 0.15833\right)\right.$$

$$\left. + \frac{1}{0.54089\cdot 127.97}\right)^{-1} = 51.231\ \frac{\text{BTU}}{\left(\text{ft}^2\cdot\text{h}\cdot°F\right)}$$

The required fouling factors for both streams are:

$$R_{di} = 0.002 \frac{\text{ft}^2 \cdot \text{h} \cdot {}^\circ\text{F}}{\text{BTU}}$$

$$R_{do} = 0.002 \frac{\text{ft}^2 \cdot \text{h} \cdot {}^\circ\text{F}}{\text{BTU}}$$

Therefore, the fouled overall heat transfer coefficient is:

$$U_d = \left(\left(D_o + \frac{2N_f}{\pi} a \right) \left(\frac{\ln\left(\frac{D_o}{D_i}\right)}{2k} + \frac{1}{D_i} \left(R_{di} + \frac{1}{h_i} \right) \right) + \frac{1}{\eta_o} \left(R_{do} + \frac{1}{h_o} \right) \right)^{-1}$$

$$= \left(\left(\frac{\ln\left(\frac{0.15833}{0.13417}\right)}{2 \cdot 26.001} + \frac{1}{0.13417} \left(0.0020000 + \frac{1}{2,331.8} \right) \right) \left(\frac{2}{\pi} 0.041667 \cdot 24.000 + 0.15833 \right) \right.$$

$$\left. + \frac{1}{0.54089} \left(0.0020000 + \frac{1}{127.97} \right) \right)^{-1} = 28.517 \frac{\text{BTU}}{\left(\text{ft}^2 \cdot \text{h} \cdot {}^\circ\text{F} \right)}$$

9) Number of hairpins
    a) Heat transfer area
        The required heat transfer surface for performing the service is:

$$A_c = \frac{q}{U_c \Delta T_{lm}} = \frac{(9.4320e + 5)}{51.231 \cdot 93.775} = 196.33 \text{ft}^2$$

Including the fouling resistances, the design required surface is:

$$A_d = \frac{q}{U_d \Delta T_{lm}} = \frac{(9.4320e + 5)}{28.517 \cdot 93.775} = 352.71 \text{ft}^2$$

    b) Total pipe length and number of exchangers
        The heat transfer area per unit length, based on the external surface of the inner pipe, is:

$$A_{linear} = N_t(\pi D_o + 2N_f a) = 1.0000(2 \cdot 0.041667 \cdot 24.000 + \pi 0.15833) = 2.4974 \frac{\text{ft}^2}{\text{ft}}$$

Therefore, the necessary pipe length and number of exchangers in the bank and the actual specified surface are:

$$L = \frac{A_d}{A_{linear}} = \frac{352.71}{2.4974} = 141.23 \text{ft}$$

$$N_{hp} = \frac{L}{2L_{hp}} = \frac{141.23}{2 \cdot 25.000} = 2.8246 \rightarrow 3 hairpins$$

$$A = 2A_{\text{linear}}L_{\text{hp}}N_{\text{hp}} = 2 \cdot 2.4974 \cdot 25.000 \cdot 3.0000 = 374.61 \text{ft}^2$$

The actual overall heat transfer and fouling factor for the exchanger are:

$$U = \frac{q}{A\Delta T_{lm}} = \frac{(9.4320e+5)}{374.61 \cdot 93.775} = 26.850 \frac{\text{BTU}}{(\text{ft}^2 \cdot \text{h} \cdot {}^\circ\text{F})}$$

$$R_d = \frac{U_c - U}{U_c U} = \frac{-26.850 + 51.231}{26.850 \cdot 51.231} = 0.017725 \frac{\text{ft}^2 \cdot \text{h} \cdot {}^\circ\text{F}}{\text{BTU}}$$

Taking the required heat transfer as reference, the over-design is only about 6%, as calculated below:

$$Over-design(\%) = \frac{100A}{A_d} - 100 = -100 + \frac{100}{352.71} 374.61 = 6.2091\%$$

Because of the fouling factors, the over-surface is considerably higher:

$$Over-surface(\%) = \frac{100A}{A_c} - 100 = -100 + \frac{100}{196.33} 374.61 = 90.806\%$$

The fouling over-surface is evaluated as:

$$Fouling over-surface(\%) = -100 + \frac{100A_d}{A_c} = -100 + \frac{100}{196.33} 352.71 = 79.652\%$$

10) Pressure drop
    a)   Pressure drop in the tube

$$Re_i = 163,090.0$$

$$f_{Fi} = \frac{0.41}{\ln^2\left(0.23\left(\frac{\epsilon}{D_i}\right)^{\frac{10}{9}} + \frac{6.9}{Re_i}\right)} = \frac{0.41}{\ln^2\left(0.23\left(\frac{0.00011483}{0.13417}\right)^{\frac{10}{9}} + \frac{6.9}{(1.6309e+5)}\right)} = 0.0051392$$

$$\phi_{f,i} = \frac{1.02}{\left(\frac{\mu_i}{\mu_{i,w}}\right)^{0.14}} = \frac{1.02}{\left(\frac{1.6100}{1.4000}\right)^{0.14}} = 1.0002$$

$$f_{F,\text{corr}} = f_{F0}\phi_{f,i} = 0.0051392 \cdot 1.0002 = 0.0051402$$

    (1)   Distributed pressure drop – tube

$$\Delta P_{i,\text{dist}} = \frac{4L_{\text{hp}}N_{\text{hp}}f_F}{D_i\rho_i} G_i^2 = \frac{4(1.9570e+6)^2 \cdot 0.0051402 \cdot 25.000}{0.13417 \cdot 65.700} 3.0000$$

$$= (6.6998e+11)\frac{\text{lb}}{(\text{ft} \cdot \text{h}^2)} = 11.158 \text{ psi}$$

(2) Localized pressure drop – tube

$$D_{i(in)} = 1.61\,in$$

$$K_i = 0.7 + \frac{2,000}{Re_i} + \frac{0.7}{D_{i(in)}} = 0.7 + \frac{0.7}{1.6100} + \frac{2,000}{(1.6309e+5)} = 1.1470$$

$$\Delta P_{i,loc} = \frac{G_i^2 K_i}{2\rho_i}(2N_{hp}-1) = \frac{(1.9570e+6)^2 \cdot 1.1470}{2 \cdot 65.700}(2 \cdot 3.0000 - 1)$$

$$= (1.6716e+11)\frac{lb}{(ft \cdot h^2)} = 2.7839\,psi$$

$$\Delta P_i = \Delta P_{i,dist} + \Delta P_{i,loc} = (1.6716e+11)+(6.6998e+11)$$

$$= (8.3714e+11)\frac{lb}{(ft \cdot h^2)} = 13.942\,psi$$

b) Pressure drop in the annulus
   i) Friction factor

$$f_{Fo} = \frac{0.41}{\ln^2\left(0.23\left(\frac{\epsilon}{D_e}\right)^{\frac{10}{9}} + \frac{6.9}{Re_o}\right)} = \frac{0.41}{\ln^2\left(0.23\left(\frac{0.00011483}{0.034822}\right)^{\frac{10}{9}} + \frac{6.9}{4,681.1}\right)} = 0.010400$$

$$\phi_{f,o} = \frac{1.02}{\left(\frac{\mu}{\mu_{o,w}}\right)^{0.14}} = \frac{1.02}{\left(\frac{4.6600}{6.5700}\right)^{0.14}} = 1.0703$$

$$f_{F,corr} = f_{Fo}\phi_{f,o} = 0.010400 \cdot 1.0703 = 0.011131$$

   ii) Distributed pressure drop – annulus

$$\Delta P_{o,dist} = \frac{4L_{hp}N_{hp}f_{Fo}}{D_h\rho_o}G_o^2 = \frac{4(6.2644e+5)^2 \cdot 0.011131 \cdot 25.000}{0.034822 \cdot 69.000}3.0000$$

$$= (5.4540e+11)\frac{lb}{(ft \cdot h^2)} = 9.0833\,psi$$

   iii) Localized pressure drop – annulus
For exchangers with straight pipe annulus return, we have:

$$D_{h(in)} = 0.41786in$$

$$K_o = 2.8 + \frac{2,000}{Re_o} + \frac{2.8}{D_{h(in)}} = 2.8 + \frac{2,000}{4,681.1} + \frac{2.8}{0.41786} = 9.9281$$

$$\Delta P_{o,\text{loc}} = \frac{G_o^2 K_o N_{hp}}{2\rho_o} = \frac{(6.2644e+5)^2 \cdot 3.0000 \cdot 9.9278}{2 \cdot 69.000}$$

$$= (8.4694e+10)\frac{\text{lb}}{(\text{ft}\cdot\text{h}^2)} = 1.4105 \text{ psi}$$

iv) Total pressure drop

$$\Delta P_o = \Delta P_{o,\text{dist}} + \Delta P_{o,\text{loc}} = (5.4540e+11) + (8.4694e+10)$$

$$= (6.3009e+11)\frac{\text{lb}}{(\text{ft}\cdot\text{h}^2)} = 10.494 \text{ psi}$$

Conclusion

The requested heat load can be held by three finned heat exchangers with their tube and annulus (shell) channels connected in series. Assuming that the required fouling factors were selected rigorously, the final over-design of about 6% can be considered appropriate. The pressure drop for both fluids is under the allowable design value of 15 psi.

Problems

(9.1) A stainless steel ($k = 12$ W/(m·K)) finned double pipe exchanger is used to transfer heat from a hot fluid to a cold fluid, allocated in the annulus and tube, respectively. The average bulk temperatures are 38 °C and 34 °C. The inner and outer convective heat transfer coefficients are $h_i = 3018.8$ W/(m²K) and $h_o = 159.85$ W/(m²K). The exchanger was provided with fouling factors of $R_{di} = 0.00035$ m²K/W and $R_{do} = 0.00018$ m²K/W. The exchanger specifications are:

Outer pipe	DN 65 mm sch 40S
Inner pipe	OD = 42.16 mm
	ID = 35.05 mm
No. of tubes	1
No. of fins	24
Fin height	7.95 mm
Fin thickness	0.889 mm

Determine the (a) fin efficiency, (b) weighted fin efficiency, (c) inner and (d) outer wall temperatures, (e) clean, and (f) design overall heat transfer coefficients.

Answer:
(a) $\eta_f = 0.64535$
(b) $\eta_o = 0.72201$
(c) $t_{wi} = 34.54\ °C$
(d) $t_{wo} = 35.82\ °C$
(e) $U_c = 87.183\ W/(m^2 \cdot K)$
(f) $U_d = 74.887\ W/(m^2 \cdot K)$

(9.2) A finned multi-tube heat exchanger has the following specifications:

Outer pipe	DN 200 mm sch 40S
Inner pipe	OD = 22.23 mm
	ID = 18.01 mm
No. of tubes	19
No. of fins	16
Fin height	7.95 mm
Fin thickness	0.889 mm

Determine the (a) internal and (b) external flow areas, (c) annulus equivalent diameter and (d) the heat transfer area per unit length.

Answer:
(a) $S_i = 0.0048403\ m^2$
(b) $S_o = 0.02276\ m^2$
(c) $D_e = 0.013393\ m$
(d) $A_{linear} = 6.1605\ (m^2/m)$

(9.3) In a plasticizer production plant, benzoic acid is heated in the tubes of a finned multi-tube heat exchanger from 23 °C to 40 °C, using 7,560 kg/h of acetic acid ranging from 50 °C to 25 °C. The available exchangers are 5 m long carbon steel hairpins of bonnet type and with the same specifications from Problem 9.2. Design a bank of exchangers in series to perform this service, and then confirm its viability for an allowable pressure drop of 90 kPa. Determine: (a) the number of necessary exchangers, (b) the internal and (c) external heat transfer coefficients, the (d) clean and (e) fouled overall heat transfer coefficients, (f) the tube, and (g) annulus pressure drop. (h) Is the developed design feasible?

Answer:
(a) N_{hp} = 19 exchangers
(b) h_i = 133.94 W/(m²·K)
(c) h_o = 75.117 W/(m²·K)
(d) U_c = 17.487 W/(m²·K)
(e) U_d = 17.487 W/(m²·K)
(f) ΔP_i = 380.72 kPa
(g) ΔP_o = 4.0054 kPa
(h) No, the tube pressure loss is in excess

(9.4) Noticing that the bank designed in the Problem 9.3 exceeded significantly the allowable pressure loss in the tube (benzoic acid), attempt, as a possible solution, a series–parallel arrangement with the tube divided in two parallel streams. Calculate: (a) the number of required exchangers, (b) the clean overall heat transfer coefficient, the pressure loss (c) in the tubes, and (d) annulus. (e) Is this series–parallel arrangement really a practical solution?

Answer:
(a) N_{hp} = 82 exchangers
(b) U_c = 14.639 W/(m²·K)
(c) ΔP_i = 400.02 kPa
(d) ΔP_o = 17.287 kPa
(e) No, the number of exchangers is very large, and the tube pressure drop is even greater

(9.5) A battery of four finned multi-tube stainless steel heat exchangers is available to heat 23,578 lb/h of toluene from 73.4 °F to 113 °F, using styrene ranging from 122 °F to 77 °F. Fouling factors of 0.0012 ft²h°F/BTU are required for both streams. The exchangers are 20 ft long, having a straight pipe annulus return, with the tubing specifications below:

Outer pipe	DN 150 mm sch 40S
Inner pipe	OD = 19.05 mm
	ID = 14.83 mm
No. of tubes	14
No. of fins	16
Fin height	6.35 mm
Fin thickness	0.889 mm

Rate the performance of this battery to evaluate the suitability for this service. Calculate (a) the inner and (b) outer heat transfer coefficients, (c) the overall heat transfer coefficient required by this service, (d) the clean and (e) design overall heat transfer coefficients yielded by the exchanger, the (f) tubes and (g) shell pressure drop, (h) the available fouling factor, and (i) over-surface. (j) Is this bank able to accomplish the specified service?

Answer:
(a) $h_i = 302.65$ BTU/(ft^{2}h°F)
(b) $h_o = 72.808$ BTU/(ft^{2}h°F)
(c) $U = 30.488$ BTU/(ft^{2}h°F)
(d) $U_c = 21.97$ BTU/(ft^{2}h°F)
(e) $U_d = 18.897$ BTU/(ft^{2}h°F)
(f) $\Delta P_i = 9.405$ psi
(g) $\Delta P_o = 0.74864$ psi
(h) -0.012717 ft^{2}h°F/BTU
(i) Over-surface $= -27.9\%$
(j) No

(9.6) 4,320 kg/h of phenol is cooled from 50 °C to 35 °C by untreated cooling water ranging from 23 °C to 45 °C. The cooling water must flow inside the tube to avoid corrosion of the external fins and annulus. A fouling factor of 0.00018 m^{2}K/W is required for each stream. It is available a few 6 m long carbon steel finned double pipe exchangers, having straight tube annulus return, with tubing and fins configuration given as:

Outer pipe	DN 65 mm sch 40S
Inner pipe	OD = 33.4 mm
	ID = 26.64 mm
No. of tubes	1
No. of fins	24
Fin height	12.7 mm
Fin thickness	0.889 mm

Assess if a bank of associated in series can perform the heat duty within the allowable pressure drop of 70 kPa for both fluids. Determine (a) the inner and (b) outer heat transfer coefficients, (c) the clean and (d) fouled overall heat transfer coefficients, (e) the number of exchangers in series required to meet the heat duty, the (f) tube, and (g) annulus pressure drop.

Answer:
(a) $h_i = 3193.3 \text{ W}/(\text{m}^2 \cdot \text{K})$
(b) $h_o = 147.42 \text{ W}/(\text{m}^2 \cdot \text{K})$
(c) $U_c = 82.692 \text{ W}/(\text{m}^2 \cdot \text{K})$
(d) $U_d = 72.126 \text{ W}/(\text{m}^2 \cdot \text{K})$
(e) $N_{hp} = 6$
(f) $\Delta P_i = 15.479 \text{ kPa}$
(g) $\Delta P_o = 82.752 \text{ kPa}$

(9.7) Considering that a bank of heat exchangers specified in Problem 9.6 associated in series cannot perform the required heat duty without exceeding the maximum pressure loss in the annulus, a viable design could be a series-parallel arrangement, where the phenol is divided in two parallel streams. Confirm if this alternative design is able to accomplish the service within the allowable pressure drop and calculate: (a) the fin efficiency, (b) the inner and (c) outer heat transfer coefficients, (d) the clean and (e) fouled overall heat transfer coefficients, (f) the number of exchangers in series required to meet the heat duty, the (g) tube, and (h) annulus pressure drop.

Answer:
(a) $\eta_f = 0.77488$
(b) $h_i = 3193.3 \text{ W}/(\text{m}^2 \cdot \text{K})$
(c) $h_o = 116.99 \text{ W}/(\text{m}^2 \cdot \text{K})$
(d) $U_c = 71.873 \text{ W}/(\text{m}^2 \cdot \text{K})$
(e) $U_d = 63.796 \text{ W}/(\text{m}^2 \cdot \text{K})$
(f) $N_{hp} = 10$
(g) $\Delta P_i = 25.96 \text{ kPa}$
(h) $\Delta P_o = 32.375 \text{ kPa}$

Appendix A Heat exchanger specifications

A.1 Piping data (DN, NB, NPS)

Tab. 1: Piping data – (diameter nominal (DN) | nominal bore (NB) | nominal pipe size (NPS) (adapted from Ref. [111].

NPS	DN	Schedule designations ANSI/ASME	OD		Wall thickness		ID		Flow area		Weight	
in	mm		in	mm	in	mm	in	mm	in^2	mm^2	lb/ft	kg/m
1/8	6	10/10S	0.405	10.3	0.049	1.24	0.307	7.82	0.074	48.0	0.19	0.28
1/8	6	STD/40/40S	0.405	10.3	0.068	1.73	0.269	6.84	0.057	36.7	0.24	0.36
1/8	6	XS/80/80S	0.405	10.3	0.095	2.41	0.215	5.48	0.036	23.6	0.31	0.47
1/4	8	10/10S	0.540	13.7	0.065	1.65	0.410	10.40	0.132	84.9	0.33	0.49
1/4	8	STD/40/40S	0.540	13.7	0.088	2.24	0.364	9.22	0.104	66.8	0.42	0.63
1/4	8	XS/80/80S	0.540	13.7	0.119	3.02	0.302	7.66	0.072	46.1	0.54	0.80
3/8	10	10/10S	0.675	17.1	0.065	1.65	0.545	13.80	0.233	149.6	0.42	0.63
3/8	10	STD/40/40S	0.675	17.1	0.091	2.31	0.493	12.48	0.191	122.3	0.57	0.84
3/8	10	XS/80/80S	0.675	17.1	0.126	3.20	0.423	10.70	0.141	89.9	0.74	1.10
1/2	15	5/5S	0.840	21.3	0.065	1.65	0.710	18.00	0.396	254.5	0.54	0.80
1/2	15	10/10S	0.840	21.3	0.083	2.11	0.674	17.08	0.357	229.1	0.67	1.00
1/2	15	STD/40/40S	0.840	21.3	0.109	2.77	0.622	15.76	0.304	195.1	0.85	1.27
1/2	15	XS/80/80S	0.840	21.3	0.147	3.73	0.546	13.84	0.234	150.4	1.09	1.62
1/2	15	160	0.840	21.3	0.188	4.78	0.464	11.74	0.169	108.2	1.31	1.95
1/2	15	XXS	0.840	21.3	0.294	7.47	0.252	6.36	0.050	31.8	1.71	2.55
3/4	20	5/5S	1.050	26.7	0.065	1.65	0.920	23.40	0.665	430.1	0.68	1.02
3/4	20	10/10S	1.050	26.7	0.083	2.11	0.884	22.48	0.614	396.9	0.86	1.28
3/4	20	STD/40/40S	1.050	26.7	0.113	2.87	0.824	20.96	0.533	345.0	1.131	1.68
3/4	20	XS/80/80S	1.050	26.7	0.154	3.91	0.742	18.88	0.432	280.0	1.47	2.19
3/4	20	160	1.050	26.7	0.219	5.56	0.612	15.58	0.294	190.6	1.94	2.89
3/4	20	XXS	1.050	26.7	0.308	7.82	0.434	11.06	0.148	96.1	2.44	3.63
1	25	5/5S	1.315	33.4	0.065	1.65	1.185	30.10	1.103	711.6	0.87	1.29
1	25	10/10S	1.315	33.4	0.109	2.77	1.097	27.86	0.945	609.6	1.40	2.09
1	25	STD/40/40S	1.315	33.4	0.133	3.38	1.049	26.64	0.864	557.4	1.68	2.50

https://doi.org/10.1515/9783110585872-010

Tab. 1 (continued)

NPS	DN	Schedule designations ANSI/ASME	OD		Wall thickness		ID		Flow area		Weight	
in	mm		in	mm	in	mm	in	mm	in^2	mm^2	lb/ft	kg/m
1	25	XS/80/80S	1.315	33.4	0.179	4.55	0.957	24.30	0.719	463.8	2.17	3.23
1	25	160	1.315	33.4	0.250	6.35	0.815	20.70	0.522	336.5	2.84	4.23
1	25	XXS	1.315	33.4	0.358	9.09	0.599	15.22	0.282	181.9	3.66	5.45
1-1/4	32	5/5S	1.660	42.2	0.065	1.65	1.530	38.90	1.839	1188.5	1.11	1.65
1-1/4	32	10/10S	1.660	42.2	0.109	2.77	1.442	36.66	1.633	1055.5	1.81	2.69
1-1/4	32	STD/40/40S	1.660	42.2	0.140	3.56	1.380	35.08	1.496	966.5	2.27	3.38
1-1/4	32	XS/80/80S	1.660	42.2	0.191	4.85	1.278	32.50	1.283	829.6	3.00	4.46
1-1/4	32	160	1.660	42.2	0.250	6.35	1.160	29.50	1.057	683.5	3.77	5.60
1-1/4	32	XXS	1.660	42.2	0.382	9.70	0.896	22.80	0.631	408.3	5.21	7.76
1-1/2	40	5/5S	1.900	48.3	0.065	1.65	1.770	45.00	2.461	1,590.4	1.27	1.90
1-1/2	40	10/10S	1.900	48.3	0.109	2.77	1.682	42.76	2.222	1,436.0	2.09	3.10
1-1/2	40	STD/40/40S	1.900	48.3	0.145	3.68	1.610	40.94	2.036	1,316.4	2.72	4.05
1-1/2	40	XS/80/80S	1.900	48.3	0.200	5.08	1.500	38.14	1.767	1,142.5	3.63	5.40
1-1/2	40	160	1.900	48.3	0.281	7.14	1.338	34.02	1.406	909.0	4.86	7.23
1-1/2	40	XXS	1.900	48.3	0.400	10.16	1.100	27.98	0.950	614.9	6.41	9.54
2	50	5/5S	2.375	60.3	0.065	1.65	2.245	57.00	3.958	2,551.8	1.60	2.39
2	50	10/10S	2.375	60.3	0.109	2.77	2.157	54.76	3.654	2,355.1	2.64	3.93
2	50	STD/40/40S	2.375	60.3	0.154	3.91	2.067	52.48	3.356	2,163.1	3.65	5.44
2	50	XS/80/80S	2.375	60.3	0.218	5.54	1.939	49.22	2.953	1,902.7	5.02	7.47
2	50	160	2.375	60.3	0.344	8.74	1.687	42.82	2.235	1,440.1	7.46	11.11
2	50	XXS	2.375	60.3	0.436	11.07	1.503	38.16	1.774	1,143.7	9.03	13.44
2-1/2	65	5/5S	2.875	73.0	0.083	2.11	2.709	68.78	5.764	3,715.5	2.48	3.68
2-1/2	65	10/10S	2.875	73.0	0.120	3.05	2.635	66.90	5.453	3,515.1	3.53	5.26
2-1/2	65	STD/40/40S	2.875	73.0	0.203	5.16	2.469	62.68	4.788	3,085.7	5.79	8.62
2-1/2	65	XS/80/80S	2.875	73.0	0.276	7.01	2.323	58.98	4.238	2,732.1	7.66	11.40
2-1/2	65	160	2.875	73.0	0.375	9.53	2.125	53.94	3.547	2,285.1	10.01	14.90
2-1/2	65	XXS	2.875	73.0	0.552	14.02	1.771	44.96	2.463	1,587.6	13.69	20.37
3	80	5/5S	3.500	88.9	0.083	2.11	3.334	84.68	8.730	5,631.9	3.03	4.51
3	80	10/10S	3.500	88.9	0.120	3.05	3.260	82.80	8.347	5,384.6	4.33	6.45

Tab. 1 (continued)

NPS	DN	Schedule designations ANSI/ASME	OD		Wall thickness		ID		Flow area		Weight	
in	mm		in	mm	in	mm	in	mm	in^2	mm^2	lb/ft	kg/m
3	80	STD/40/40S	3.500	88.9	0.216	5.49	3.068	77.92	7.393	4,768.6	7.58	11.27
3	80	XS/80/80S	3.500	88.9	0.300	7.62	2.900	73.66	6.605	4,261.4	10.25	15.25
3	80	160	3.500	88.9	0.438	11.13	2.624	66.64	5.408	3,487.9	14.32	21.31
3	80	XXS	3.500	88.9	0.600	15.24	2.300	58.42	4.155	2,680.5	18.58	27.65
3-1/2	90	5/5S	4.000	101.6	0.083	2.11	3.834	97.38	11.545	7,447.8	3.47	5.17
3-1/2	90	10/10S	4.000	101.6	0.120	3.05	3.760	95.50	11.104	7,163.0	4.97	7.40
3-1/2	90	STD/40/40S	4.000	101.6	0.226	5.74	3.548	90.12	9.887	6,378.7	9.11	13.56
3-1/2	90	XS/80/80S	4.000	101.6	0.318	8.08	3.364	85.44	8.888	5,733.4	12.50	18.60
3-1/2	90	XXS	4.000	101.6	0.636	16.15	2.728	69.30	5.845	3,771.9	22.85	34.01
4	100	5/5S	4.500	114.3	0.083	2.11	4.334	110.08	14.753	9,517.1	3.92	5.83
4	100	10/10S	4.500	114.3	0.120	3.05	4.260	108.20	14.253	9,194.8	5.61	8.35
4	100	STD/40/40S	4.500	114.3	0.237	6.02	4.026	102.26	12.730	8,213.0	10.79	16.06
4	100	XS/80/80S	4.500	114.3	0.337	8.56	3.826	97.18	11.497	7,417.3	14.98	22.29
4	100	120	4.500	114.3	0.438	11.13	3.624	92.04	10.315	6,653.4	19.00	28.28
4	100	160	4.500	114.3	0.531	13.49	3.438	87.32	9.283	5,988.5	22.51	33.50
4	100	XXS	4.500	114.3	0.674	17.12	3.152	80.06	7.803	5,034.1	27.54	40.99
5	125	5/5S	5.563	141.3	0.109	2.77	5.345	135.76	22.438	14,475.5	6.35	9.45
5	125	10/10S	5.563	141.3	0.134	3.40	5.295	134.50	22.020	14,208.0	7.77	11.56
5	125	STD/40/40S	5.563	141.3	0.258	6.55	5.047	128.20	20.006	12,908.2	14.62	21.76
5	125	XS/80/80S	5.563	141.3	0.375	9.53	4.813	122.24	18.194	11,735.9	20.78	30.93
5	125	120	5.563	141.3	0.500	12.70	4.563	115.90	16.353	10,550.1	27.04	40.24
5	125	160	5.563	141.3	0.625	15.88	4.313	109.54	14.610	9,424.0	32.96	49.05
5	125	XXS	5.563	141.3	0.750	19.05	4.063	103.20	12.965	8,364.7	38.55	57.37
6	150	5/5S	6.625	168.3	0.109	2.77	6.407	162.76	32.240	20,805.8	7.59	11.29
6	150	10/10S	6.625	168.3	0.134	3.40	6.357	161.50	31.739	20,485.0	9.29	13.82
6	150	STD/40/40S	6.625	168.3	0.280	7.11	6.065	154.08	28.890	18,645.9	18.97	28.23
6	150	XS/80/80S	6.625	168.3	0.432	10.97	5.761	146.36	26.067	16,824.2	28.57	42.52
6	150	120	6.625	168.3	0.562	14.27	5.501	139.76	23.767	15,341.1	36.39	54.16
6	150	160	6.625	168.3	0.719	18.26	5.187	131.78	21.131	13,639.2	45.35	67.49

Tab. 1 (continued)

NPS	DN	Schedule designations ANSI/ASME	OD		Wall thickness		ID		Flow area		Weight	
in	mm		in	mm	in	mm	in	mm	in^2	mm^2	lb/ft	kg/m
6	150	XXS	6.625	168.3	0.864	21.95	4.897	124.40	18.834	12,154.3	53.16	79.12
8	200	5S	8.625	219.1	0.109	2.77	8.407	213.56	55.510	35,820.3	9.91	14.75
8	200	10/10S	8.625	219.1	0.148	3.76	8.329	211.58	54.485	35,159.2	13.60	19.94
8	200	20	8.625	219.1	0.250	6.35	8.125	206.40	51.849	33,458.7	22.36	33.28
8	200	30	8.625	219.1	0.277	7.04	8.071	205.02	51.162	33,012.8	24.70	36.76
8	200	STD/40/40S	8.625	219.1	0.322	8.18	7.981	202.74	50.027	32,282.6	28.55	42.49
8	200	60	8.625	219.1	0.406	10.31	7.813	198.48	47.943	30,940.2	35.64	53.04
8	200	XS/80/80S	8.625	219.1	0.500	12.70	7.625	193.70	45.664	29,467.9	43.39	64.58
8	200	100	8.625	219.1	0.594	15.09	7.437	188.92	43.440	28,031.5	50.95	75.83
8	200	120	8.625	219.1	0.719	18.26	7.187	182.58	40.568	26,181.6	60.71	90.35
8	200	140	8.625	219.1	0.812	20.62	7.001	177.86	38.496	24,845.4	67.76	100.84
8	200	XXS	8.625	219.1	0.875	22.23	6.875	174.64	37.122	23,954.0	72.42	107.78
8	200	160	8.625	219.1	0.906	23.01	6.813	173.08	36.456	23,527.9	74.69	111.16
10	250	5S	10.75	273.1	0.134	3.40	10.482	266.30	86.294	55,697.1	15.19	22.61
10	250	10S	10.75	273.1	0.165	4.19	10.420	264.72	85.276	55,038.1	18.70	27.83
10	250	20	10.75	273.1	0.250	6.35	10.250	260.40	82.516	53,256.4	28.04	41.73
10	250	30	10.75	273.1	0.307	7.80	10.136	257.50	80.691	52,076.8	34.24	50.96
10	250	STD/40/40S	10.75	273.1	0.365	9.27	10.020	254.56	78.854	50,894.4	40.48	60.24
10	250	XS/60/80S	10.75	273.1	0.500	12.70	9.750	247.70	74.662	48,188.3	54.74	81.47
10	250	80	10.75	273.1	0.594	15.09	9.562	242.92	71.810	46,346.4	64.43	95.89
10	250	100	10.75	273.1	0.719	18.26	9.312	236.58	68.105	43,958.8	77.03	114.64
10	250	120	10.75	273.1	0.844	21.44	9.062	230.22	64.497	41,627.1	89.29	132.89
10	250	140/XX	10.75	273.1	1.000	25.40	8.750	222.30	60.132	38,812.2	104.13	154.97
10	250	160	10.75	273.1	1.125	28.58	8.500	215.94	56.745	36,623.2	115.64	172.10
12	300	5S	12.75	323.9	0.156	3.96	12.438	315.98	121.504	78,416.8	20.98	31.22
12	300	10S	12.75	323.9	0.180	4.57	12.390	314.76	120.568	77,812.4	24.20	36.02
12	300	20	12.75	323.9	0.250	6.35	12.250	311.20	117.859	76,062.2	33.38	49.68
12	300	30	12.75	323.9	0.330	8.38	12.090	307.14	114.800	74,090.5	43.77	65.14
12	300	STD/40S	12.75	323.9	0.375	9.53	12.000	304.84	113.097	72,985.0	49.56	73.76

Tab. 1 (continued)

NPS	DN	Schedule designations ANSI/ASME	OD		Wall thickness		ID		Flow area		Weight	
in	mm		in	mm	in	mm	in	mm	in^2	mm^2	lb/ft	kg/m
12	300	40	12.75	323.9	0.406	10.31	11.938	303.28	111.932	72,239.9	53.52	79.65
12	300	XS/80S	12.75	323.9	0.500	12.70	11.750	298.50	108.434	69,980.7	65.42	97.36
12	300	60	12.75	323.9	0.562	14.27	11.626	295.36	106.157	68,516.2	73.15	108.87
12	300	80	12.75	323.9	0.688	17.48	11.374	288.94	101.605	65,570.0	88.63	131.90
12	300	100	12.75	323.9	0.844	21.44	11.062	281.02	96.107	62,024.7	107.32	159.72
12	300	120/XX	12.75	323.9	1.000	25.40	10.750	273.10	90.763	58,577.8	125.49	186.76
12	300	140	12.75	323.9	1.125	28.58	10.500	266.74	86.590	55,881.3	139.67	207.86
12	300	160	12.75	323.9	1.312	33.32	10.126	257.26	80.531	51,979.8	160.27	238.52
14	350	10S	14	355.6	0.188	4.78	13.624	346.04	145.780	94,046.5	27.73	41.27
14	350	10	14	355.6	0.250	6.35	13.500	342.90	143.139	92,347.4	36.71	54.63
14	350	20	14	355.6	0.312	7.92	13.376	339.76	140.521	90,663.9	45.61	67.88
14	350	STD/30/40S	14	355.6	0.375	9.53	13.250	336.54	137.886	88,953.5	54.57	81.21
14	350	40	14	355.6	0.438	11.13	13.124	333.34	135.276	87,270.0	63.44	94.41
14	350	XS/80S	14	355.6	0.500	12.70	13.000	330.20	132.732	85,633.6	72.09	107.29
14	350	60	14	355.6	0.594	15.09	12.812	325.42	128.921	83,172.2	85.05	126.58
14	350	80	14	355.6	0.750	19.05	12.500	317.50	122.718	79,173.0	106.13	157.95
14	350	100	14	355.6	0.938	23.83	12.124	307.94	115.447	74,477.0	130.85	194.74
14	350	120	14	355.6	1.094	27.79	11.812	300.02	109.581	70,695.3	150.90	224.58
14	350	140	14	355.6	1.250	31.75	11.500	292.10	103.869	67,012.1	170.21	253.32
14	350	160	14	355.6	1.406	35.71	11.188	284.18	98.309	63,427.4	189.10	281.43
16	400	10S	16	406.4	0.188	4.78	15.624	396.84	191.723	123,686.1	31.75	47.25
16	400	10	16	406.4	0.250	6.35	15.500	393.70	188.692	121,736.5	42.05	62.58
16	400	20	16	406.4	0.312	7.92	15.376	390.56	185.685	119,802.4	52.27	77.79
16	400	STD/30/40S	16	406.4	0.375	9.53	15.250	387.34	182.654	117,835.1	62.58	93.13
16	400	XS/40/80S	16	406.4	0.500	12.70	15.000	381.00	176.715	114,009.2	82.77	123.18
16	400	60	16	406.4	0.656	16.66	14.688	373.08	169.440	109,318.5	107.50	159.99
16	400	80	16	406.4	0.844	21.44	14.312	363.52	160.876	103,787.8	136.61	203.31
16	400	100	16	406.4	1.031	26.20	13.938	354.00	152.578	98,423.0	164.82	245.29
16	400	120	16	406.4	1.219	30.96	13.562	344.48	144.457	93,200.4	192.43	286.38

Tab. 1 (continued)

NPS	DN	Schedule designations ANSI/ASME	OD		Wall thickness		ID		Flow area		Weight	
in	mm		in	mm	in	mm	in	mm	in^2	mm^2	lb/ft	kg/m
16	400	140	16	406.4	1.438	36.53	13.124	333.34	135.276	87,270.0	223.64	332.83
16	400	160	16	406.4	1.594	40.49	12.812	325.42	128.921	83,172.2	245.25	364.99
18	450	10S	18	457.2	0.188	4.78	17.624	447.64	243.949	157,379.3	35.76	53.22
18	450	10	18	457.2	0.250	6.35	17.500	444.50	240.528	155,179.2	47.39	70.53
18	450	20	18	457.2	0.312	7.92	17.376	441.36	237.132	152,994.5	58.94	87.72
18	450	STD/40S	18	457.2	0.375	9.53	17.250	438.14	233.705	150,770.3	70.59	105.06
18	450	30	18	457.2	0.438	11.13	17.124	434.94	230.303	148,576.0	82.15	122.26
18	450	XS/80S	18	457.2	0.500	12.70	17.000	431.80	226.980	146,438.5	93.45	139.08
18	450	40	18	457.2	0.562	14.27	16.876	428.66	223.681	144,316.4	104.67	155.78
18	450	60	18	457.2	0.750	19.05	16.500	419.10	213.825	137,951.1	138.17	205.63
18	450	80	18	457.2	0.938	23.83	16.124	409.54	204.190	131,729.3	170.92	254.37
18	450	100	18	457.2	1.156	29.36	15.688	398.48	193.297	124,710.5	207.96	309.50
18	450	120	18	457.2	1.375	34.93	15.250	387.34	182.654	117,835.1	244.14	363.34
18	450	140	18	457.2	1.562	39.67	14.876	377.86	173.805	112,137.7	274.22	408.11
18	450	160	18	457.2	1.781	45.24	14.438	366.72	163.721	105,623.1	308.50	459.13
20	500	10S	20	508.0	0.218	5.54	19.564	496.92	300.611	193,938.0	46.06	68.55
20	500	10	20	508.0	0.250	6.35	19.500	495.30	298.648	192,675.5	52.73	78.48
20	500	STD/20/40S	20	508.0	0.375	9.53	19.250	488.94	291.039	187,759.1	78.60	116.98
20	500	XS/30/80S	20	508.0	0.500	12.70	19.000	482.60	283.529	182,921.4	104.13	154.97
20	500	40	20	508.0	0.594	15.09	18.812	477.82	277.946	179,315.8	123.11	183.22
20	500	60	20	508.0	0.812	20.62	18.376	466.76	265.211	171,110.7	166.40	247.65
20	500	80	20	508.0	1.031	26.19	17.938	455.62	252.719	163,040.5	208.87	310.85
20	500	100	20	508.0	1.281	32.54	17.438	442.92	238.827	154,077.9	256.10	381.14
20	500	120	20	508.0	1.500	38.10	17.000	431.80	226.980	146,438.5	296.37	441.07
20	500	140	20	508.0	1.750	44.45	16.500	419.10	213.825	137,951.1	341.09	507.63
20	500	160	20	508.0	1.969	50.01	16.062	407.98	202.623	130,727.7	379.17	564.30
24	600	10/10S	24	609.6	0.250	6.35	23.500	596.90	433.736	279,829.2	63.41	94.37
24	600	STD/20/40S	24	609.6	0.375	9.53	23.250	590.54	424.557	273,897.8	94.62	140.82
24	600	XS/80S	24	609.6	0.500	12.70	23.000	584.20	415.476	268,048.3	125.49	186.76

Tab. 1 (continued)

NPS	DN	Schedule designations ANSI/ASME	OD		Wall thickness		ID		Flow area		Weight	
in	mm		in	mm	in	mm	in	mm	in^2	mm^2	lb/ft	kg/m
24	600	30	24	609.6	0.562	14.27	22.876	581.06	411.008	265,174.6	140.68	209.37
24	600	40	24	609.6	0.688	17.48	22.624	574.64	402.002	259,347.2	171.29	254.92
24	600	60	24	609.6	0.969	24.61	22.062	560.38	382.278	246,635.2	238.35	354.72
24	600	80	24	609.6	1.219	30.96	21.562	547.68	365.147	235,582.8	296.58	441.39
24	600	100	24	609.6	1.531	38.89	20.938	531.82	344.318	222,136.1	367.39	546.77
24	600	120	24	609.6	1.812	46.02	20.376	517.56	326.083	210,383.3	429.39	639.04
24	600	140	24	609.6	2.062	52.37	19.876	504.86	310.276	200,185.1	483.10	718.97
24	600	160	24	609.6	2.344	59.54	19.312	490.52	292.917	188,974.6	542.13	806.83
30	750	10	30	762.0	0.312	7.92	29.376	746.16	677.759	437,274.2	98.93	147.23
30	750	STD/40S	30	762.0	0.375	9.53	29.250	742.94	671.957	433,508.2	118.65	176.58
30	750	XS/20/80S	30	762.0	0.500	12.70	29.000	736.60	660.520	426,141.0	157.53	234.44
30	750	30	30	762.0	0.625	15.88	28.750	730.24	649.181	418,813.9	196.08	291.82
36	900	10	36	914.4	0.312	7.92	35.376	898.56	982.895	634,138.4	118.92	176.98
36	900	STD/40S	36	914.4	0.375	9.53	35.250	895.34	975.906	629,601.6	142.68	212.34
36	900	XS/80S	36	914.4	0.500	12.70	35.000	889.00	962.113	620,716.7	189.57	282.13

A.2 B.W.G.: Birmingham wire gauge

Tab. 2: B.W.G. – Birmingham wire gauge (adapted from Refs. [33, 112, 113]).

Birmingham wire gage B.W.G.

Gauge	Inches	mm
00000 (5/0)	0.500	12.7000
0000 (4/0)	0.454	11.5316
000 (3/0)	0.425	10.7950
00 (2/0)	0.380	9.6520
0	0.340	8.6360
1	0.300	7.6200

Tab. 2 (continued)

Birmingham wire gage B.W.G.

Gauge	Inches	mm
2	0.284	7.2136
3	0.259	6.5786
4	0.238	6.0452
5	0.220	5.5880
6	0.203	5.1562
7	0.180	4.5720
8	0.165	4.1910
9	0.148	3.7592
10	0.134	3.4036
11	0.120	3.0480
12	0.109	2.7686
13	0.095	2.4130
14	0.083	2.1082
15	0.072	1.8288
16	0.065	1.6510
17	0.058	1.4732
18	0.049	1.2446
19	0.042	1.0668
20	0.035	0.8890
21	0.032	0.8128
22	0.028	0.7112
23	0.025	0.6350
24	0.022	0.5588
25	0.020	0.5080
26	0.018	0.4572
27	0.016	0.4064
28	0.014	0.3556
29	0.013	0.3302
30	0.012	0.3048

Tab. 2 (continued)

Birmingham wire gage B.W.G.		
Gauge	Inches	mm
31	0.010	0.2540
32	0.009	0.2286
33	0.008	0.2032
34	0.007	0.1778
35	0.005	0.1270
36	0.004	0.1016

A.3 Heat transfer tube data (B.W.G)

Tab. 3: Heat transfer tube data (B.W.G – Birmingham wire gauge) (adapted from Refs. [33, 44]).

OD		BWG	Wall thickness		ID		Flow area	
in	mm		in	mm	in	mm	in^2	mm^2
1/2	12.70	12	0.109	2.77	0.282	7.16	0.062	40.3
1/2	12.70	13	0.095	2.41	0.310	7.87	0.075	48.7
1/2	12.70	14	0.083	2.11	0.334	8.48	0.088	56.5
1/2	12.70	15	0.072	1.83	0.356	9.04	0.100	64.2
1/2	12.70	16	0.065	1.65	0.370	9.40	0.108	69.4
1/2	12.70	17	0.058	1.47	0.384	9.75	0.116	74.7
1/2	12.70	18	0.049	1.24	0.402	10.21	0.127	81.9
1/2	12.70	19	0.042	1.07	0.416	10.57	0.136	87.7
1/2	12.70	20	0.035	0.89	0.430	10.92	0.145	93.7
3/4	19.05	10	0.134	3.40	0.482	12.24	0.182	117.7
3/4	19.05	11	0.120	3.05	0.510	12.95	0.204	131.8
3/4	19.05	12	0.109	2.77	0.532	13.51	0.222	143.4
3/4	19.05	13	0.095	2.41	0.560	14.22	0.246	158.9
3/4	19.05	14	0.083	2.11	0.584	14.83	0.268	172.8
3/4	19.05	15	0.072	1.83	0.606	15.39	0.288	186.1

Tab. 3 (continued)

OD		BWG	Wall thickness		ID		Flow area	
in	mm		in	mm	in	mm	in²	mm²
3/4	19.05	16	0.065	1.65	0.620	15.75	0.302	194.8
3/4	19.05	17	0.058	1.47	0.634	16.10	0.316	203.7
3/4	19.05	18	0.049	1.24	0.652	16.56	0.334	215.4
1	25.40	8	0.165	4.19	0.670	17.02	0.353	227.5
1	25.40	9	0.148	3.76	0.704	17.88	0.389	251.1
1	25.40	10	0.134	3.40	0.732	18.59	0.421	271.5
1	25.40	11	0.120	3.05	0.760	19.30	0.454	292.7
1	25.40	12	0.109	2.77	0.782	19.86	0.480	309.9
1	25.40	13	0.095	2.41	0.810	20.57	0.515	332.5
1	25.40	14	0.083	2.11	0.834	21.18	0.546	352.4
1	25.40	15	0.072	1.83	0.856	21.74	0.575	371.3
1	25.40	16	0.065	1.65	0.870	22.10	0.594	383.5
1	25.40	17	0.058	1.47	0.884	22.45	0.614	396.0
1	25.40	18	0.049	1.24	0.902	22.91	0.639	412.3
1 1/4	31.75	8	0.165	4.19	0.920	23.37	0.665	428.9
1 1/4	31.75	9	0.148	3.76	0.954	24.23	0.715	461.2
1 1/4	31.75	10	0.134	3.40	0.982	24.94	0.757	488.6
1 1/4	31.75	11	0.120	3.05	1.010	25.65	0.801	516.9
1 1/4	31.75	12	0.109	2.77	1.032	26.21	0.836	539.7
1 1/4	31.75	13	0.095	2.41	1.060	26.92	0.882	569.3
1 1/4	31.75	14	0.083	2.11	1.084	27.53	0.923	595.4
1 1/4	31.75	15	0.072	1.83	1.106	28.09	0.961	619.8
1 1/4	31.75	16	0.065	1.65	1.120	28.45	0.985	635.6
1 1/4	31.75	17	0.058	1.47	1.134	28.80	1.010	651.6
1 1/4	31.75	18	0.049	1.24	1.152	29.26	1.042	672.5
1 1/2	38.10	8	0.165	4.19	1.170	29.72	1.075	693.6
1 1/2	38.10	9	0.148	3.76	1.204	30.58	1.139	734.5
1 1/2	38.10	10	0.134	3.40	1.232	31.29	1.192	769.1

Tab. 3 (continued)

OD		BWG	Wall thickness		ID		Flow area	
in	mm		in	mm	in	mm	in^2	mm^2
1 1/2	38.10	11	0.120	3.05	1.260	32.00	1.247	804.4
1 1/2	38.10	12	0.109	2.77	1.282	32.56	1.291	832.8
1 1/2	38.10	13	0.095	2.41	1.310	33.27	1.348	869.6
1 1/2	38.10	14	0.083	2.11	1.334	33.88	1.398	901.7
1 1/2	38.10	15	0.072	1.83	1.356	34.44	1.444	931.7
1 1/2	38.10	16	0.065	1.65	1.370	34.80	1.474	951.0
1 1/2	38.10	17	0.058	1.47	1.384	35.15	1.504	970.6
1 1/2	38.10	18	0.049	1.24	1.402	35.61	1.544	996.0
1 3/4	44.45	7	0.180	4.57	1.390	35.31	1.517	979.0
1 3/4	44.45	8	0.165	4.19	1.420	36.07	1.584	1,021.7
1 3/4	44.45	9	0.148	3.76	1.454	36.93	1.660	1,071.2
1 3/4	44.45	10	0.134	3.40	1.482	37.64	1.725	1,112.9
1 3/4	44.45	11	0.120	3.05	1.510	38.35	1.791	1,155.3
1 3/4	44.45	12	0.109	2.77	1.532	38.91	1.843	1,189.3
2	50.80	4	0.238	6.05	1.524	38.71	1.824	1,176.9
2	50.80	5	0.220	5.59	1.560	39.62	1.911	1,233.1
2	50.80	6	0.203	5.16	1.594	40.49	1.996	1,287.5
2	50.80	7	0.180	4.57	1.640	41.66	2.112	1,362.8
2	50.80	8	0.165	4.19	1.670	42.42	2.190	1,413.2
2	50.80	9	0.148	3.76	1.704	43.28	2.280	1,471.3
2	50.80	10	0.134	3.40	1.732	43.99	2.356	1,520.0
2	50.80	11	0.120	3.05	1.760	44.70	2.433	1,569.6
2	50.80	12	0.109	2.77	1.782	45.26	2.494	1,609.1

A.4 Finned and multi-tube hairpin design data

Tab. 4: Bare multi-tube hairpin design data (adapted from Refs. [114, 115]).

	Bare multi-tube hairpin design data									
Spec. no.	Shell (Sch. 40)		Tube							
	NPS	DN	No. tubes	OD		ID		Th	
	in	mm		in	mm	in	mm	in	mm
1	2.0	50	4	0.625	15.88	0.505	12.83	0.060	1.52
2	2.0	50	2	0.750	19.05	0.584	14.83	0.083	2.11
3	2.5	65	7	0.625	15.88	0.505	12.83	0.060	1.52
4	2.5	65	4	0.750	19.05	0.584	14.83	0.083	2.11
5	3.0	80	7	0.625	15.88	0.505	12.83	0.060	1.52
6	3.0	80	7	0.750	19.05	0.584	14.83	0.083	2.11
7	3.0	80	3	1.000	25.40	0.782	19.86	0.109	2.77
8	3.5	90	12	0.625	15.88	0.505	12.83	0.060	1.52
9	3.5	90	7	0.750	19.05	0.584	14.83	0.083	2.11
10	3.5	90	7	0.875	22.23	0.709	18.01	0.083	2.11
11	3.5	90	4	1.000	25.40	0.782	19.86	0.109	2.77
12	4.0	100	19	0.625	15.88	0.505	12.83	0.060	1.52
13	4.0	100	7	0.750	19.05	0.584	14.83	0.083	2.11
14	4.0	100	12	0.750	19.05	0.584	14.83	0.083	2.11
15	4.0	100	7	0.875	22.23	0.709	18.01	0.083	2.11
16	4.0	100	7	1.000	25.40	0.782	19.86	0.109	2.77
17	5.0	125	31	0.625	15.88	0.505	12.83	0.060	1.52
18	5.0	125	12	0.750	19.05	0.584	14.83	0.083	2.11
19	5.0	125	19	0.750	19.05	0.584	14.83	0.083	2.11
20	5.0	125	9	1.000	25.40	0.782	19.86	0.109	2.77
21	6.0	150	42	0.625	15.88	0.505	12.83	0.060	1.52
22	6.0	150	24	0.750	19.05	0.584	14.83	0.083	2.11
23	6.0	150	31	0.750	19.05	0.584	14.83	0.083	2.11
24	6.0	150	14	1.000	25.40	0.782	19.86	0.109	2.77

Tab. 4 (continued)

Spec. no.	Shell (Sch. 40)		Bare multi-tube hairpin design data							
	NPS	DN	No. tubes	OD		ID		Th		
	in	mm		in	mm	in	mm	in	mm	
25	8.0	200	55	0.625	15.88	0.505	12.83	0.060	1.52	
26	8.0	200	73	0.625	15.88	0.505	12.83	0.060	1.52	
27	8.0	200	85	0.625	15.88	0.505	12.83	0.060	1.52	
28	8.0	200	37	0.750	19.05	0.584	14.83	0.083	2.11	
29	8.0	200	42	0.750	19.05	0.584	14.83	0.083	2.11	
30	8.0	200	44	0.750	19.05	0.584	14.83	0.083	2.11	
31	8.0	200	55	0.750	19.05	0.584	14.83	0.083	2.11	
32	8.0	200	37	0.875	22.23	0.709	18.01	0.083	2.11	
33	8.0	200	24	1.000	25.40	0.782	19.86	0.109	2.77	
34	8.0	200	31	1.000	25.40	0.782	19.86	0.109	2.77	
35	10.0	250	96	0.625	15.88	0.505	12.83	0.060	1.52	
36	10.0	250	121	0.625	15.88	0.505	12.83	0.060	1.52	
37	10.0	250	68	0.750	19.05	0.584	14.83	0.083	2.11	
38	10.0	250	85	0.750	19.05	0.584	14.83	0.083	2.11	
39	10.0	250	38	1.000	25.40	0.782	19.86	0.109	2.77	
40	10.0	250	42	1.000	25.40	0.782	19.86	0.109	2.77	
41	12.0	300	151	0.625	15.88	0.505	12.83	0.060	1.52	
42	12.0	300	174	0.625	15.88	0.505	12.83	0.060	1.52	
43	12.0	300	109	0.750	19.05	0.584	14.83	0.083	2.11	
44	12.0	300	121	0.750	19.05	0.584	14.83	0.083	2.11	
45	12.0	300	55	1.000	25.40	0.782	19.86	0.109	2.77	
46	12.0	300	64	1.000	25.40	0.782	19.86	0.109	2.77	
47	16.0	400	208	0.625	15.88	0.505	12.83	0.060	1.52	
48	16.0	400	258	0.625	15.88	0.505	12.83	0.060	1.52	
49	16.0	400	301	0.625	15.88	0.505	12.83	0.060	1.52	
50	16.0	400	151	0.750	19.05	0.584	14.83	0.083	2.11	

Tab. 4 (continued)

Spec. no.	Shell (Sch. 40)		Tube							
	NPS	DN	No. tubes	OD		ID		Th		
	in	mm		in	mm	in	mm	in	mm	
51	16.0	400	199	0.750	19.05	0.584	14.83	0.083	2.11	
52	16.0	400	85	1.000	25.40	0.782	19.86	0.109	2.77	
53	16.0	400	109	1.000	25.40	0.782	19.86	0.109	2.77	

Tab. 5: Finned double-pipe hairpin design data (adapted from Refs. [114, 115]).

Spec. no.	Shell (Sch. 40)		Tube						External fins		
	NPS	DN	OD		ID		Th		No. Fins	Fin height	
	in	mm	in	mm	in	mm	in	mm		in	mm
1	2.0	50	0.750	19.05	0.584	14.83	0.083	2.11	16	0.563	14.30
2	2.0	50	1.000	25.40	0.782	19.86	0.109	2.77	20	0.438	11.13
3	2.0	50	1.315	33.40	1.049	26.64	0.133	3.38	24	0.250	6.35
4	2.5	65	1.000	25.40	0.782	19.86	0.109	2.77	16	0.563	14.30
5	2.5	65	1.315	33.40	1.049	26.64	0.133	3.38	24	0.500	12.70
6	2.5	65	1.660	42.16	1.380	35.05	0.140	3.56	24	0.313	7.95
7	2.5	65	1.660	42.16	1.380	35.05	0.140	3.56	32	0.313	7.95
8	2.5	65	1.900	48.26	1.610	40.89	0.145	3.68	24	0.250	6.35
9	2.5	65	1.900	48.26	1.610	40.89	0.145	3.68	36	0.250	6.35
10	3.0	80	1.000	25.40	0.782	19.86	0.109	2.77	20	0.938	23.83
11	3.0	80	1.900	48.26	1.610	40.89	0.145	3.68	24	0.500	12.70
12	3.0	80	1.900	48.26	1.610	40.89	0.145	3.68	36	0.500	12.70
13	3.0	80	2.375	60.33	2.067	52.50	0.154	3.91	24	0.250	6.35
14	3.0	80	2.375	60.33	2.067	52.50	0.154	3.91	40	0.250	6.35
15	3.5	90	1.900	48.26	1.610	40.89	0.145	3.68	24	0.700	17.78
16	3.5	90	1.900	48.26	1.610	40.89	0.145	3.68	36	0.700	17.78
17	3.5	90	2.375	60.33	2.067	52.50	0.154	3.91	24	0.500	12.70

Tab. 5 (continued)

Spec. no.	Shell (Sch. 40)		Tube						External fins		
	NPS	DN	OD		ID		Th		No. Fins	Fin height	
	in	mm	in	mm	in	mm	in	mm		in	mm
18	3.5	90	2.375	60.33	2.067	52.50	0.154	3.91	40	0.500	12.70
19	4.0	100	1.900	48.26	1.610	40.89	0.145	3.68	24	1.000	25.40
20	4.0	100	1.900	48.26	1.610	40.89	0.145	3.68	36	1.000	25.40
21	4.0	100	2.375	60.33	2.067	52.50	0.154	3.91	24	0.750	19.05
22	4.0	100	2.375	60.33	2.067	52.50	0.154	3.91	40	0.750	19.05
23	4.0	100	2.875	73.03	2.469	62.71	0.203	5.16	24	0.500	12.70
24	4.0	100	2.875	73.03	2.469	62.71	0.203	5.16	40	0.500	12.70
25	5.0	125	2.875	73.03	2.469	62.71	0.203	5.16	24	1.000	25.40
26	5.0	125	2.875	73.03	2.469	62.71	0.203	5.16	48	1.000	25.40
27	5.0	125	3.500	88.90	3.068	77.93	0.216	5.49	24	0.625	15.88
28	5.0	125	3.500	88.90	3.068	77.93	0.216	5.49	40	0.625	15.88
29	5.0	125	3.500	88.90	3.068	77.93	0.216	5.49	56	0.625	15.88
30	5.0	125	4.000	101.6	3.548	90.12	0.226	5.74	56	0.438	11.13
31	6.0	150	4.000	101.6	3.548	90.12	0.226	5.74	24	0.938	23.83
32	6.0	150	4.000	101.6	3.548	90.12	0.226	5.74	36	0.938	23.83
33	6.0	150	4.000	101.6	3.548	90.12	0.226	5.74	64	0.938	23.83
34	6.0	150	4.500	114.3	4.026	102.26	0.237	6.02	24	0.688	17.48

Tab. 6: Finned multi-tube hairpin design data (adapted from Refs. [114, 115]).

Spec. no.	Shell (Sch. 40)		No. tubes	OD		ID		Th		No. fins	Fin height	
	NPS	DN		in	mm	in	mm	in	mm		in	mm
	in	mm										
1	2.0	50	3	0.750	19.05	0.584	14.83	0.083	2.11	16	0.313	7.95
2	4.0	100	7	0.750	19.05	0.584	14.83	0.083	2.11	16	0.250	6.35
3	4.0	100	7	0.875	22.23	0.709	18.01	0.083	2.11	16	0.210	5.33

Tab. 6 (continued)

Spec. no.	Shell (Sch. 40)			Tube								External fins		
	NPS	DN	No. tubes	OD		ID		Th				No. fins	Fin height	
	in	mm		in	mm	in	mm	in	mm				in	mm
4	5.0	125	7	1.000	25.40	0.782	19.86	0.109	2.77			16	0.313	7.95
5	5.0	125	7	1.000	25.40	0.782	19.86	0.109	2.77			20	0.313	7.95
6	6.0	150	14	0.750	19.05	0.584	14.83	0.083	2.11			16	0.250	6.35
7	6.0	150	7	1.000	25.40	0.782	19.86	0.109	2.77			20	0.438	11.13
8	6.0	150	10	1.000	25.40	0.782	19.86	0.109	2.77			16	0.250	6.35
9	6.0	150	10	1.000	25.40	0.782	19.86	0.109	2.77			20	0.250	6.35
10	8.0	200	19	0.750	19.05	0.584	14.83	0.083	2.11			16	0.375	9.53
11	8.0	200	31	0.750	19.05	0.584	14.83	0.083	2.11			16	0.210	5.33
12	8.0	200	19	0.875	22.23	0.709	18.01	0.083	2.11			16	0.313	7.95
13	8.0	200	19	1.000	25.40	0.782	19.86	0.109	2.77			16	0.210	5.33
14	8.0	200	19	1.000	25.40	0.782	19.86	0.109	2.77			20	0.210	5.33
15	10.0	250	42	0.750	19.05	0.584	14.83	0.083	2.11			16	0.250	6.35
16	10.0	250	31	1.000	25.40	0.782	19.86	0.109	2.77			20	0.250	6.35
17	12.0	300	44	1.000	25.40	0.782	19.86	0.109	2.77			20	0.250	6.35
18	16.0	400	55	1.000	25.40	0.782	19.86	0.109	2.77			20	0.250	6.35

Finned multi-tube hairpin design data

Appendix B Properties of materials

B.1 Absolute roughness of materials

Tab. 7: Absolute roughness of materials (adapted from Refs. [33, 116, 117]).

Material	Absolute roughness	
	mm	in
Aluminum, lead	0.001–0.002	3.9E-05–7.9E-05
Asphalted cast iron	0.012–1	4.7E-04–0.039
Brickwork, mature foul sewers	3–4	0.118–0.157
Carbon steel (badly corroded)	1–3	0.039–0.118
Carbon steel (cement-lined)	1.5–1.6	0.059–0.063
Carbon steel (moderately corroded)	0.15–1	5.9E-03–0.039
Carbon steel (new)	0.02–0.05	7.9E-04–2.0E-03
Carbon steel (slightly corroded)	0.05–0.15	2.0E-03–5.9E-03
Cast iron (new)	0.25–0.26	9.8E-03–0.010
Cast iron (old, sandblasted)	1–2	0.039–0.079
Concrete – fine (floated, brushed)	0.2–0.8	7.9E-03–0.031
Concrete – rough, form marks	0.8–3	0.031–0.118
Concrete – very smooth	0.025–0.2	9.8E-04–7.9E-03
Corroding cast iron	1.5–2.5	0.059–0.098
Drawn brass, copper, stainless steel (new)	0.0015–0.01	5.9E-05–3.9E-04
Drawn brass, drawn copper	0.0015	5.9E-05
Drawn tubing, glass, plastic	0.0015–0.01	5.9E-05–3.9E-04
Fiberglass	0.005	2.0E-04
Flexible rubber tubing – smooth	0.006–0.07	2.4E-04–2.8E-03
Flexible rubber tubing – wire reinforced	0.3–4	0.012–0.157
Galvanized iron	0.015–0.15	5.9E-04–5.9E-03
Galvanized steel	0.15	5.9E-03
New cast iron	0.25–0.8	9.8E-03–0.031
Ordinary concrete	0.3–3	0.012–0.118
Ordinary wood	5	0.197

https://doi.org/10.1515/9783110585872-011

Tab. 7 (continued)

Material	Absolute roughness	
	mm	in
PVC, plastic pipes	0.0015	5.9E-05
Riveted steel	0.91–9.1	0.036–0.358
Rusted steel	0.15–4	5.9E-03–0.157
Sheet metal ducts (with smooth joints)	0.02–0.1	7.9E-04–3.9E-03
Smooth cement	0.5–0.6	0.020–0.024
Smoothed cement	0.3	0.012
Stainless steel	0.015–0.03	5.9E-04–1.2E-03
Steel commercial pipe	0.045–0.09	1.8E-03–3.5E-03
Stretched steel	0.015	5.9E-04
Water mains with tuberculation	1.2–1.3	0.047–0.051
Weld steel	0.045	1.8E-03
Planed wood	0.18–0.9	7.1E-03–0.035
Wood stave	0.18–0.91	7.1E-03–0.036
Wood stave, used	0.25–1	9.8E-03–0.039
Worn cast iron	0.8–1.5	0.031–0.059
Wrought iron (new)	0.045	1.8E-03

B.2 Thermal conductivity of selected materials

Tab. 8: Thermal conductivity of selected materials (adapted from Refs. [118–122]).

Material	Thermal conductivity	
	W/(m K)	BTU/(ft h °F)
Aluminum, alloy	121–180	70–104
Aluminum, pure	220	127
Asbestos	0.16	0.09
Brass, red, 85%Cu–15%Zn	151	87
Brass, yellow, 65%Cu–35%Zn	119	69
Brick (alumina)	3.1	1.8

Tab. 8 (continued)

Material	Thermal conductivity	
	W/(m K)	BTU/(ft h °F)
Brick (building)	0.72	0.42
Clinkers (cement)	0.7	0.4
Concrete, heavy	1.3	0.8
Concrete, isolation	0.21	0.12
Concrete, light	0.42	0.24
Copper, alloy, 11,000	388	224
Copper, aluminum bronze, 95%Cu–5%Al	83	48
Copper, brass, 70%Cu–30%Zn	111	64
Copper, bronze, 75%Cu–25%Sn	26	15
Copper, constantan, 60%Cu–40%Ni	23	13
Copper, drawn wire	287	166
Copper, German silver, 62%Cu–15%Ni–22%Zn	25	14
Copper, pure	386	223
Copper, red brass, 85%Cu–9%Sn–6%Zn	61	35
Glass	0.94	0.54
Iron	55–80	32–46
Iron, cast	55	32
Iron, pure	71.8	41.5
Iron, wrought, 0.5%C	59	34
Lead, pure	35	20
Magnesium, Mg–Al, electrolytic, 8%Al–2%Zn	66	38
Magnesium, pure	171	99
Molybdenum	130	75
Nichrome, 80%Ni–20%Cr	12	7
Nickel, Ni–Cr, 80%Ni–20%Cr	12.6	7.3
Nickel, Ni–Cr, 90%Ni–10%Cr	17	10
Nickel, pure	99	57
Silver, pure	418–420	242–243

Tab. 8 (continued)

Material	Thermal conductivity	
	W/(m K)	**BTU/(ft h °F)**
Steel, carbon	36–54	21–31
Steel, carbon, 0.5%C	54	31
Steel, carbon, 1.0%C	43	25
Steel, carbon, 1.5%C	36	21
Steel, chrome	22–73	13–42
Steel, chrome–nickel, 18%Cr–8%Ni	16	9
Steel, Invar, 36%Ni	11	6
Steel, mild	45	26
Steel, nickel	10–73	6–42
Steel, SAE	59–64	34–37
Steel, stainless	8–16	5–9
Steel, stainless 304	8.1–16.3	4.7–9.4
Steel, stainless 310	14.2	8.2
Steel, stainless 316	15–16.3	8.7–9.4
Steel, stainless 430	8.1	4.7
Steel, tungsten	54–73	31–42
Tin, cast, hammered	62.5	36.1
Tin, pure	64	37
Titanium	15.6–21.0	9.0–12.1
Tungsten	174–180	101–104
Wood	0.17	0.10
Zinc, pure	112	65

Appendix C   Physical properties of chemical components

C.1 Index

C.1.1 List of gases (alphabetical)

C.1.2 List of vapors (alphabetical)

https://doi.org/10.1515/9783110585872-012

1-nitrobutane	267
1-nitropropane	283
1-nonene	267
1-octene	267
1-pentene	267
1-propanol	267
1-undecene	267
2,4,6-trinitrotoluene	283
2,4-dinitrotoluene	283
2-butanol	267
2-ethyl-m-xylene	267
2-heptanone	267
2-hexanone	267
2-methylindene	283
2-methylnaphthalene	283
2-pentanol	267
2-pentanone	267
3-heptanone	267
3-hexanone	267
Acetaldehyde	272
Acetic acid	272
Acetic anhydride	283
Acetone	272
Acetonitrile	272
Acetyl chloride	283
Acrylic acid	283
Acrylonitrile	267
Aniline	283
Anisole	283
Benzene	272
Benzoic acid	283
Butanal	272
Cumene	272
Cyclohexane	272
Cyclohexanol	283
Cyclohexanone	283
Cyclohexene	272
Cyclopentane	272
Diethanolamine	283
Diethyl ether	272
Diethyl sulfide	272
Diethylamine	272
Diisobutyl ketone	272
Ethanol	272
Ethyl acetate	272
Ethylacetylene	272
Ethylbenzene	272
Ethylcyclohexane	272
Ethylene oxide	272
Ethylenediamine	277

C.1.3 List of liquids (alphabetical)

1,1,2-trichloroethane	308
1,1-dichloroethane	308
1,2,4-trichlorobenzene	308
1,2-propylene oxide	308
1,4-butanediol	308
1,4-dioxane	308
1-butanol	293
1-heptanol	293
1-heptene	293
1-hexanol	293
1-hexene	293
1-nitrobutane	293
1-nitropropane	308
1-nonene	293
1-octene	293
1-pentene	293
1-propanol	293
1-undecene	293
2,4,6-trinitrotoluene	308
2,4-dinitrotoluene	308
2-butanol	293
2-ethyl-m-xylene	293
2-heptanone	293
2-hexanone	293
2-methylindene	308
2-methylnaphthalene	308
2-pentanol	293
2-pentanone	293
3-heptanone	293
3-hexanone	293
Acetaldehyde	298
Acetic acid	298
Acetic anhydride	308
Acetone	298
Acetonitrile	298
Acetyl chloride	308
Acrylic acid	308
Acrylonitrile	293
Aniline	308
Anisole	308
Benzene	298
Benzoic acid	308
Butanal	298
Cumene	298
Cyclohexane	298
Cyclohexanol	308
Cyclohexanone	308
Cyclohexene	298

C.2 Selected gases: part A

Component	
(1)	Acetylene
(2)	Air
(3)	Ammonia
(4)	Carbon monoxide
(5)	Ethane
(6)	Ethylene
(7)	Hydrogen sulfide
(8)	Methane
(9)	Neon
(10)	Nitric oxide
(11)	Nitrogen
(12)	Oxygen

C.2.1 Specific heat

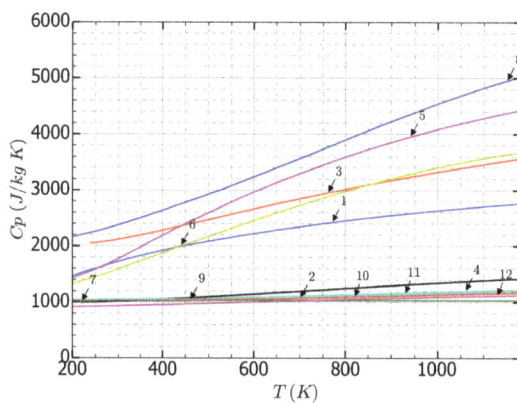

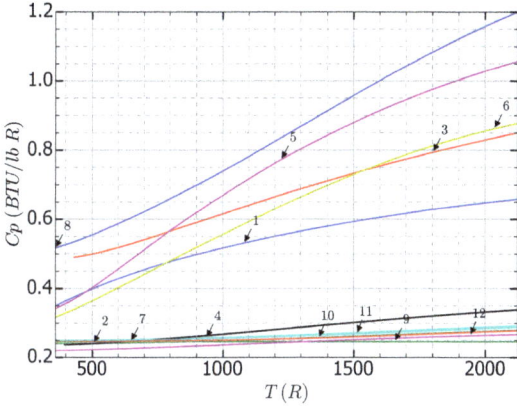

C.2.2 Thermal conductivity

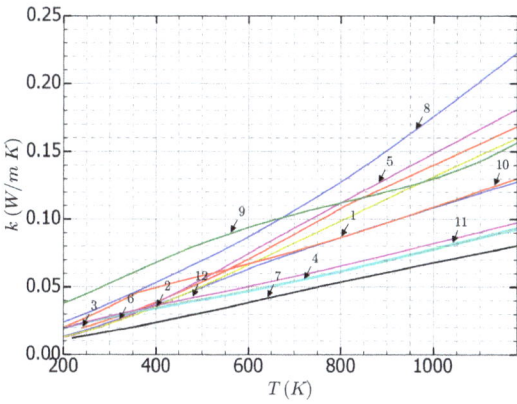

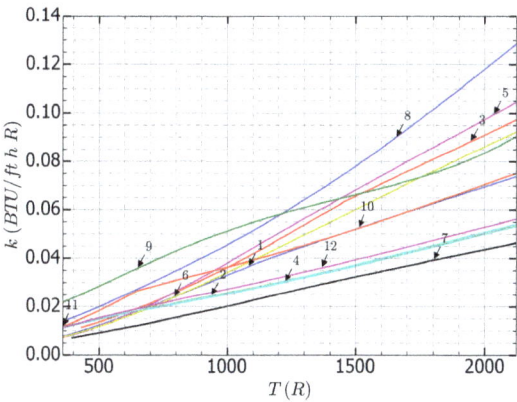

C.2.3 Prandtl number

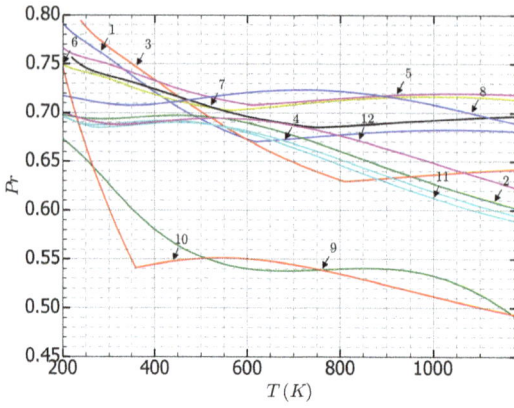

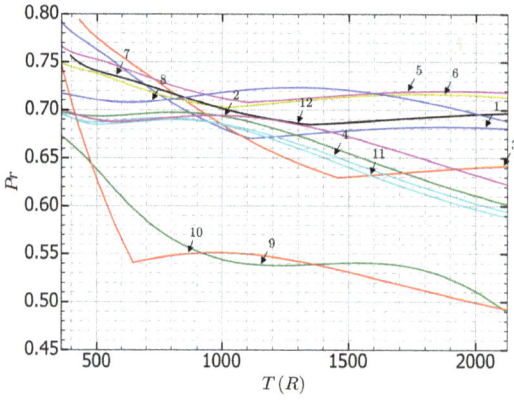

C.2.4 Specific mass

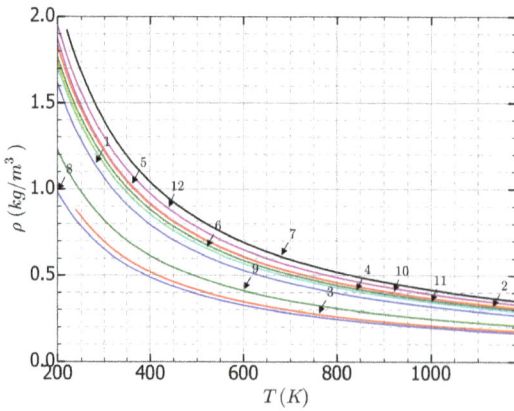

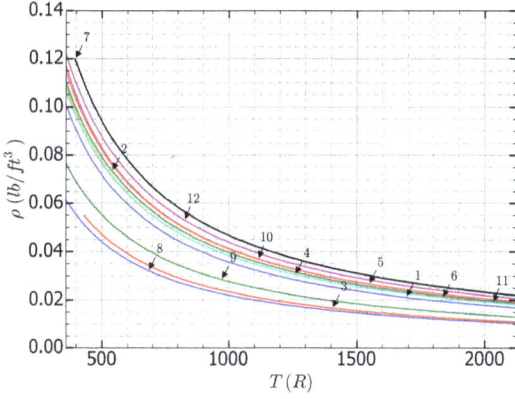

C.2.5 Dynamic viscosity

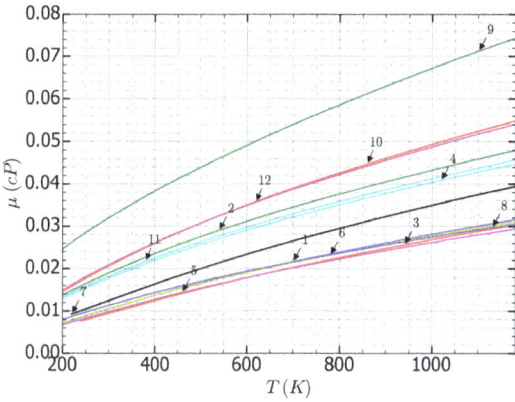

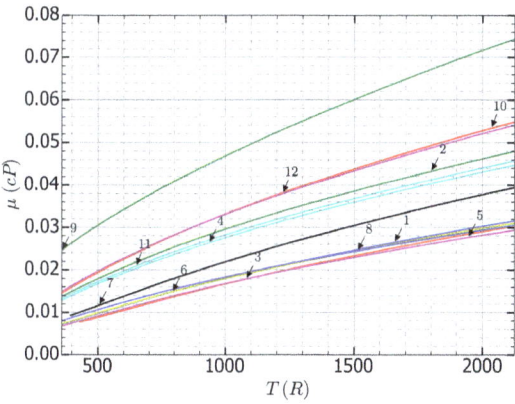

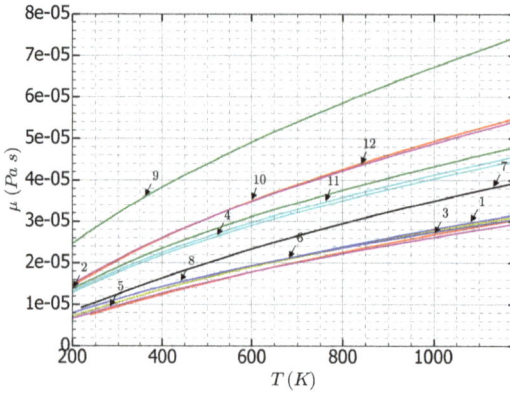

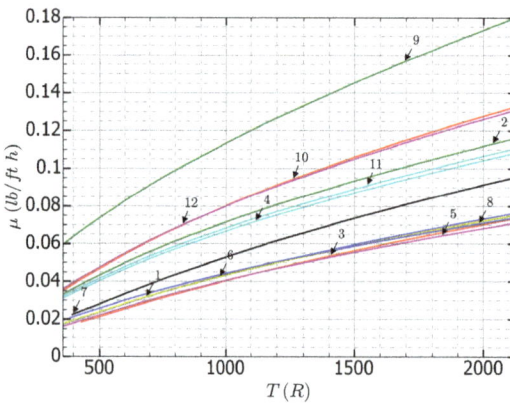

C.3 Selected gases: part B

Component	
(1)	Argon
(2)	Carbon dioxide
(3)	Dimethyl ether
(4)	Fluorine
(5)	Hydrogen chloride
(6)	Ketene
(7)	Nitrous oxide
(8)	Ozone

(continued)

Component	
(9)	Propadiene
(10)	Propane
(11)	Propylene

C.3.1 Specific heat

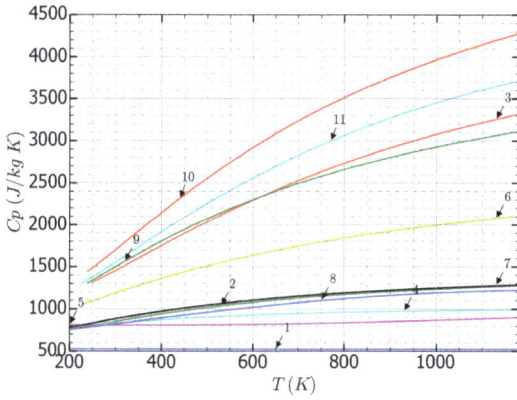

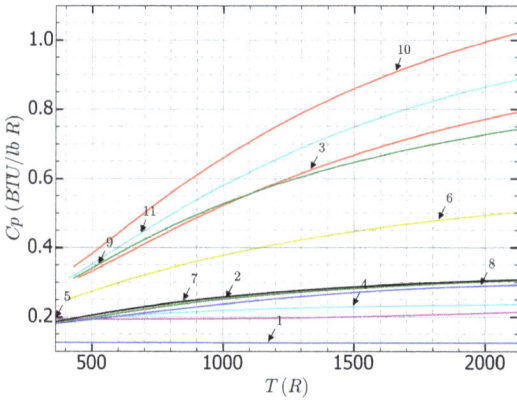

C.3.2 Thermal conductivity

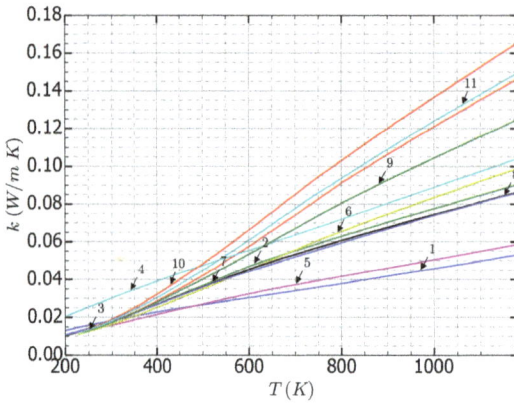

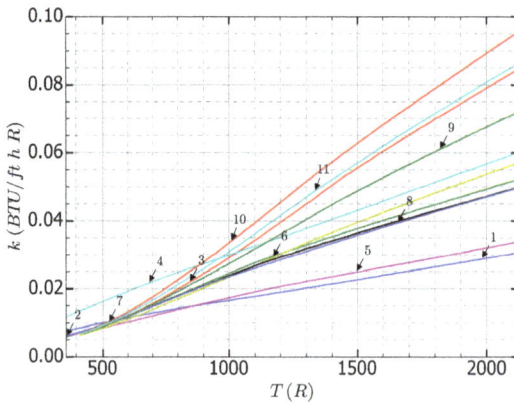

C.3.3 Prandtl number

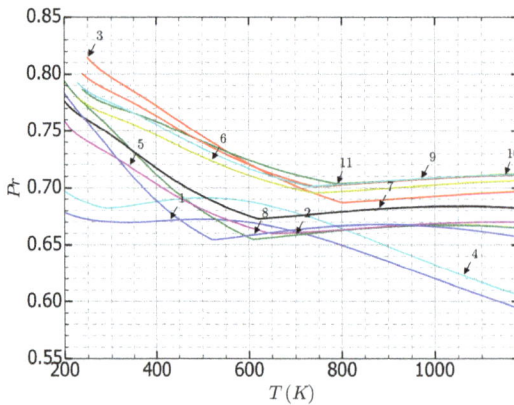

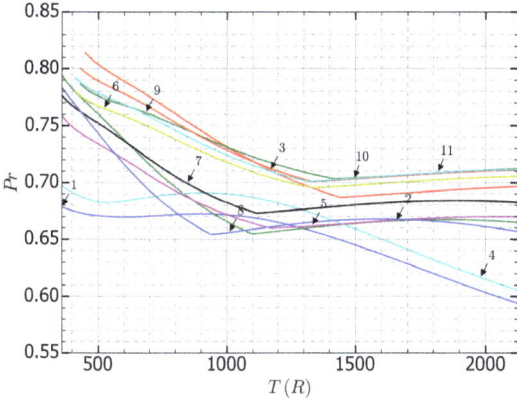

C.3.4 Specific mass

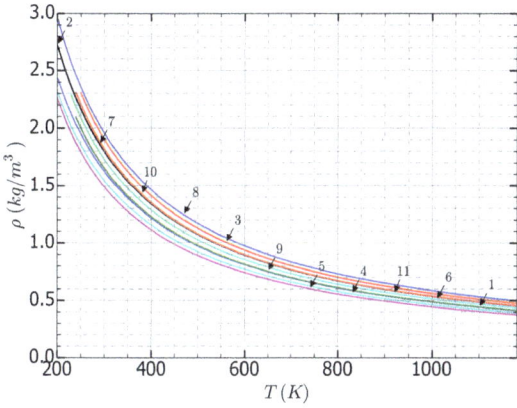

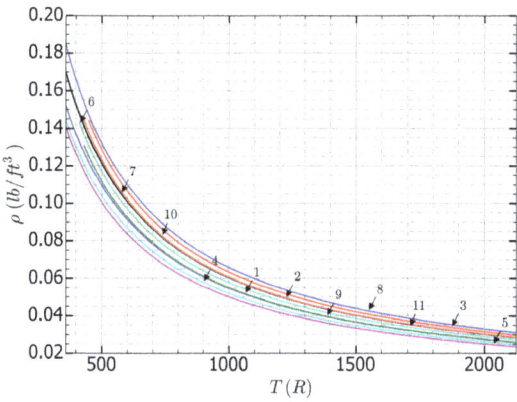

C.3.5 Dynamic viscosity

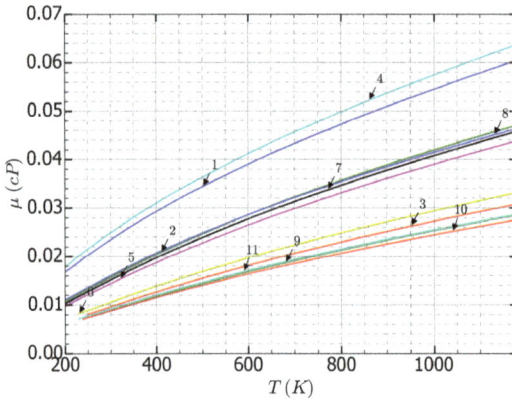

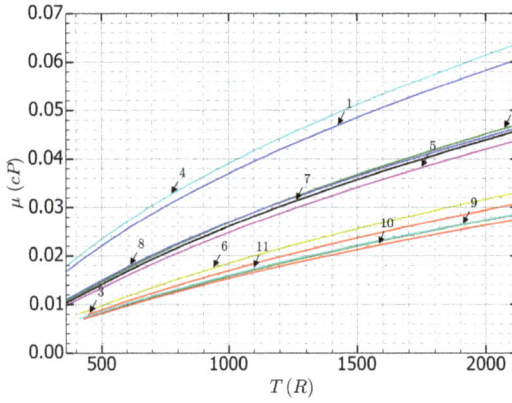

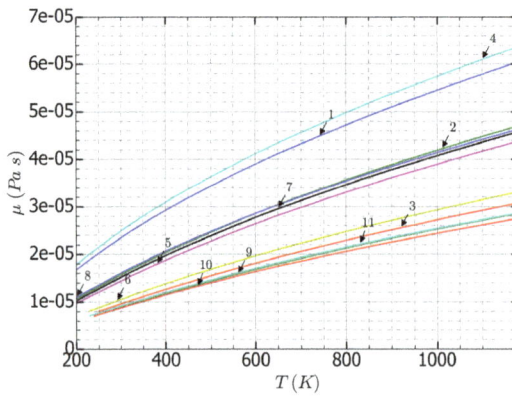

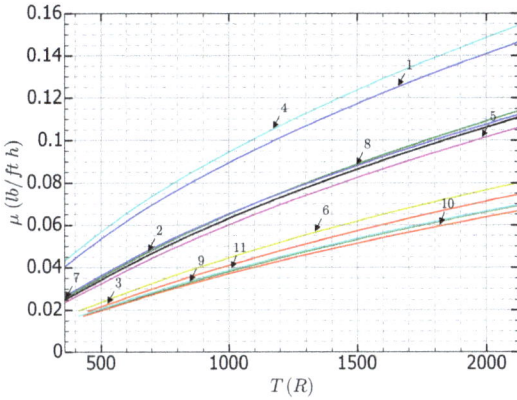

C.4 Selected vapors: part A

Component	
(1)	1-Butanol
(2)	1-Heptanol
(3)	1-Heptene
(4)	1-Hexanol
(5)	1-Hexene
(6)	1-Nitrobutane
(7)	1-Nonene
(8)	1-Octene
(9)	1-Pentene
(10)	1-Propanol
(11)	1-Undecene
(12)	2-Butanol
(13)	2-Ethyl-*m*-xylene
(14)	2-Heptanone
(15)	2-Hexanone
(16)	2-Pentanol
(17)	2-Pentanone

(continued)

Component	
(18)	3-Heptanone
(19)	3-Hexanone
(20)	Acrylonitrile

C.4.1 Heat capacity

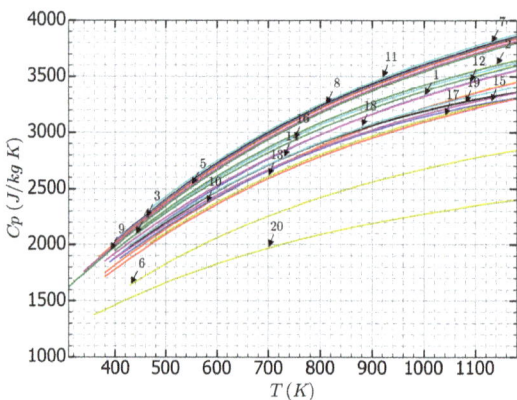

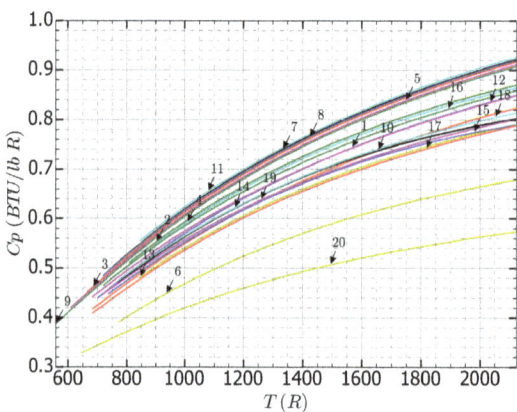

C.4.2 Thermal conductivity

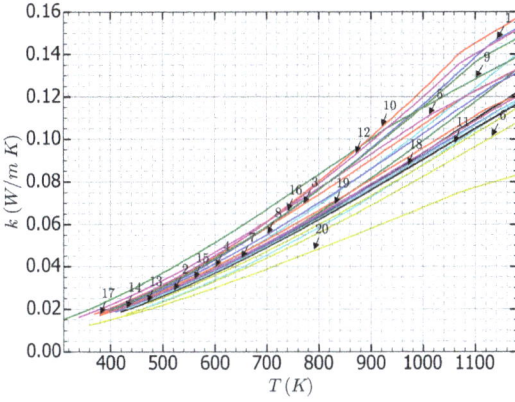

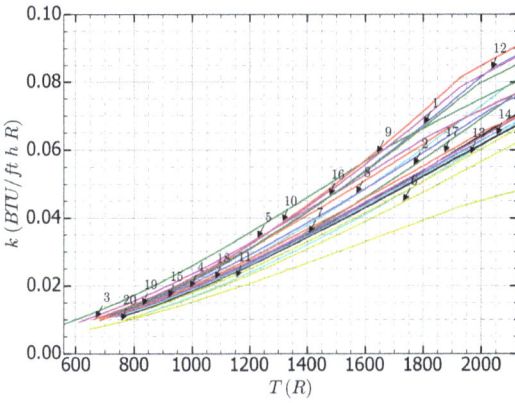

C.4.3 Prandtl number

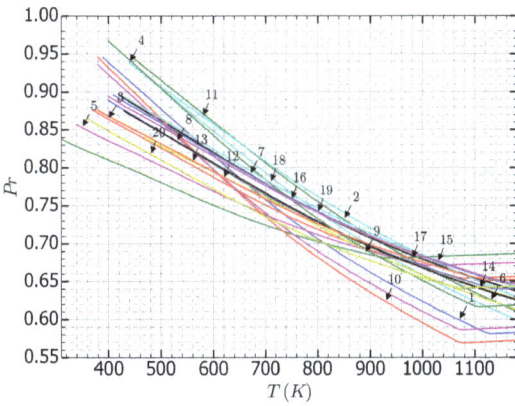

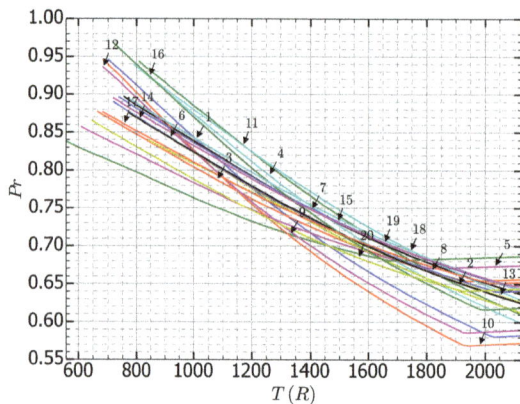

C.4.4 Specific mass

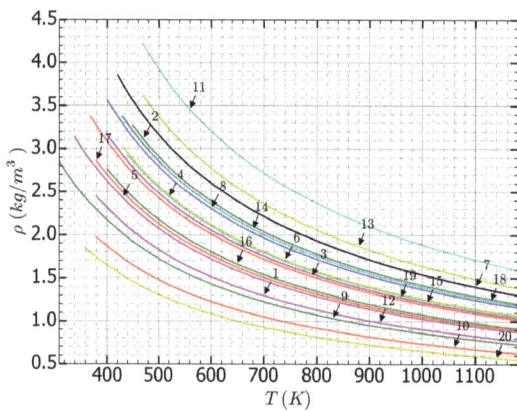

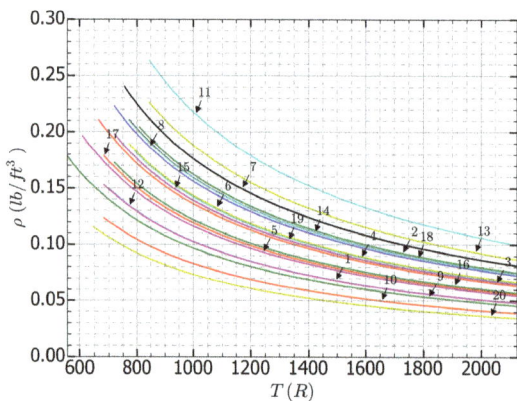

C.4.5 Dynamic viscosity

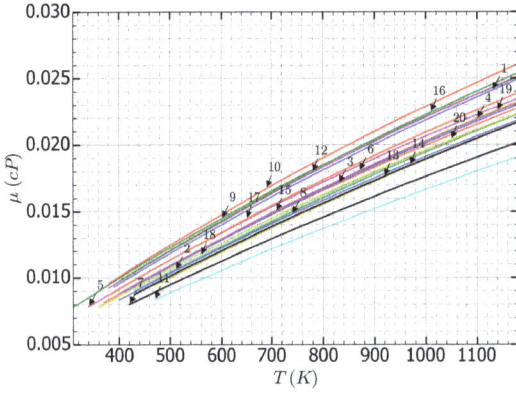

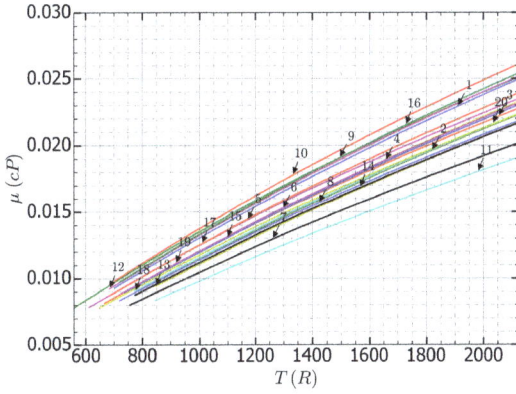

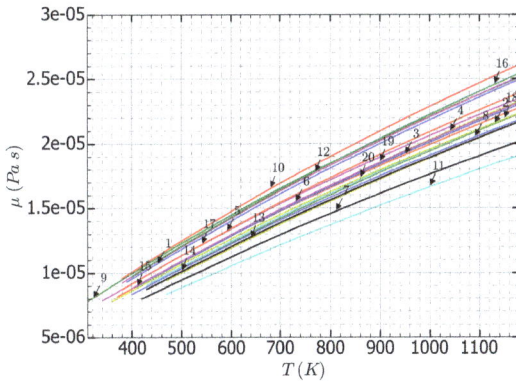

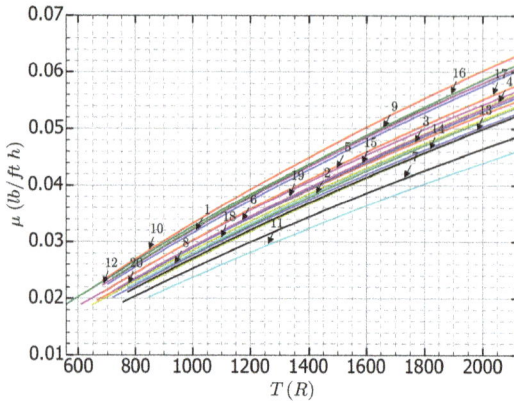

C.5 Selected vapors: part B

Component	
(1)	Acetaldehyde
(2)	Acetic acid
(3)	Acetone
(4)	Acetonitrile
(5)	Benzene
(6)	Butanal
(7)	Cumene
(8)	Cyclohexane
(9)	Cyclohexene
(10)	Cyclopentane
(11)	Diethyl ether
(12)	Diethyl sulfide
(13)	Diethylamine
(14)	Diisobutyl ketone
(15)	Ethanol
(16)	Ethyl acetate
(17)	Ethylacetylene

(continued)

Component	
(18)	Ethylbenzene
(19)	Ethylcyclohexane
(20)	Ethylene oxide

C.5.1 Specific heat

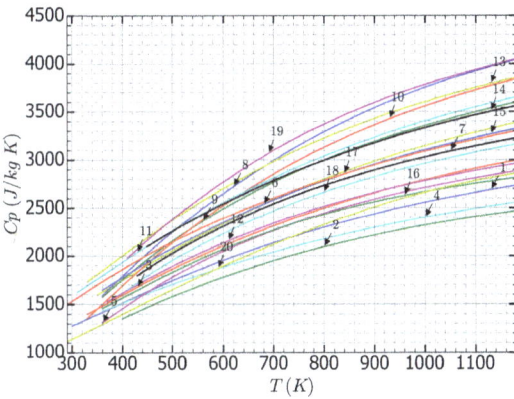

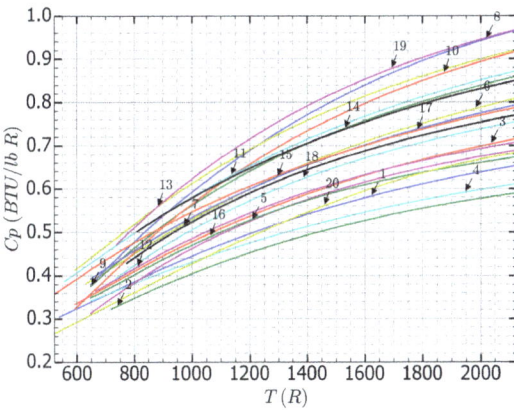

C.5.2 Thermal conductivity

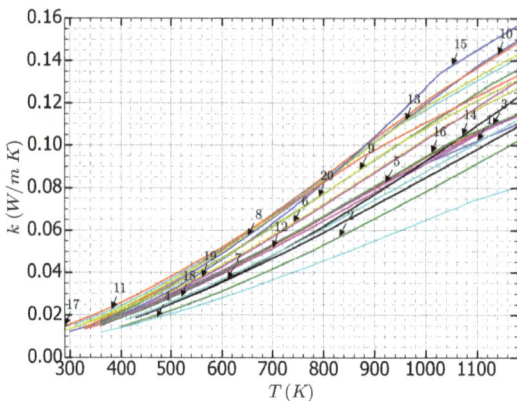

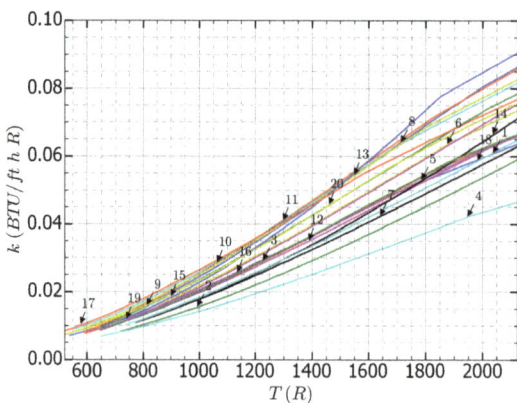

C.5.3 Prandtl number

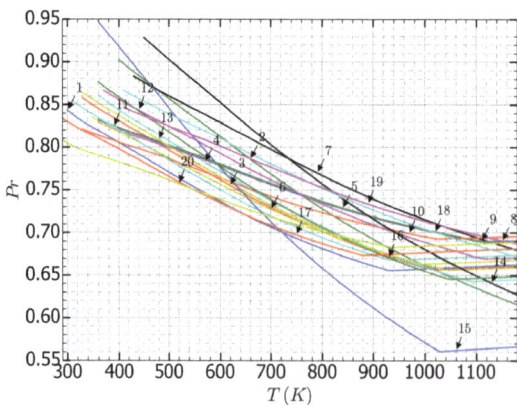

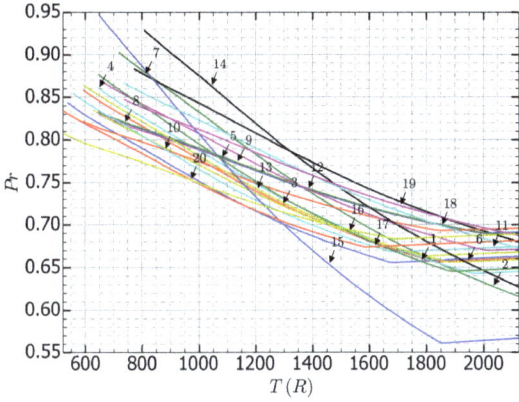

C.5.4 Specific mass

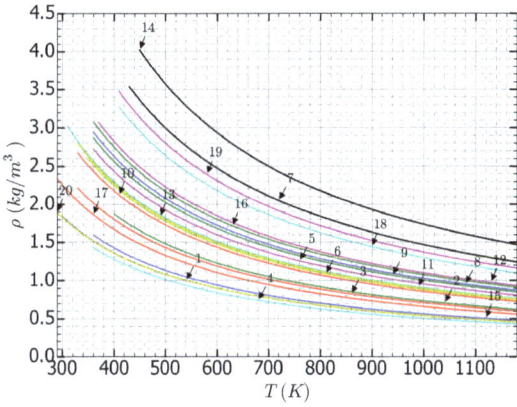

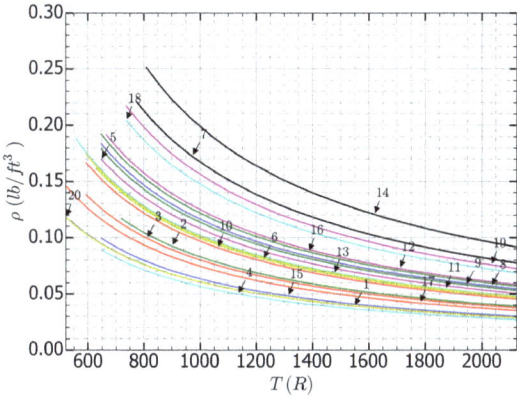

C.5.5 Dynamic viscosity

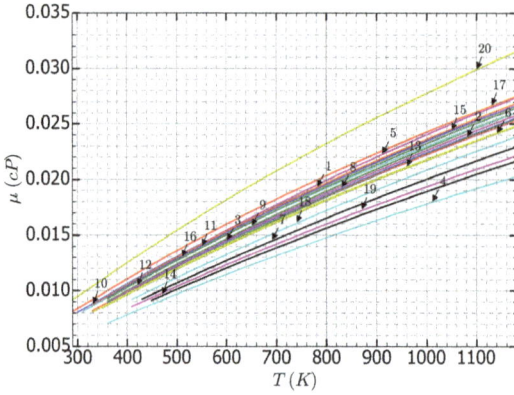

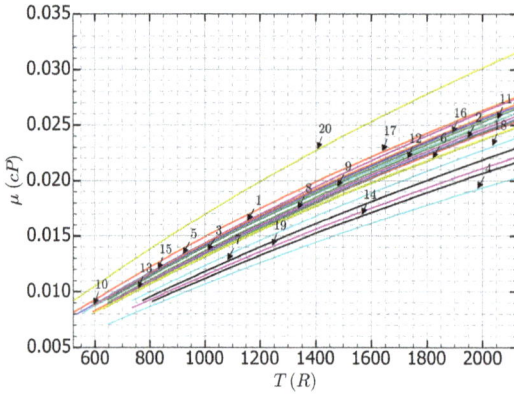

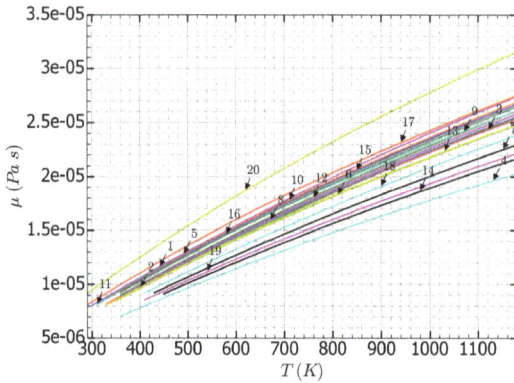

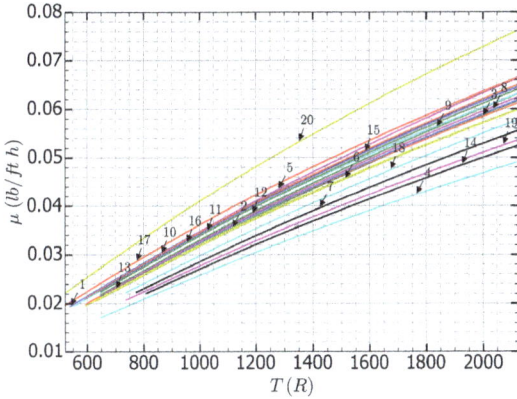

C.6 Selected vapors: part C

Component	
(1)	Ethylenediamine
(2)	Formaldehyde
(3)	Heptanal
(4)	Hexanal
(5)	Isobutyl acetate
(6)	Isobutylbenzene
(7)	Isopropanol
(8)	Methacrylonitrile
(9)	Methanol
(10)	Methyl acetate
(11)	Methyl ethyl ether
(12)	Methyl ethyl ketone
(13)	Methyl formate
(14)	Methyl isobutyl ether
(15)	Methyl isobutyl ketone
(16)	Methyl isopropyl ether
(17)	Methyl isopropyl ketone

(continued)

Component	
(18)	Methylacetylene
(19)	Methylamine
(20)	Methylcyclohexane

C.6.1 Specific heat

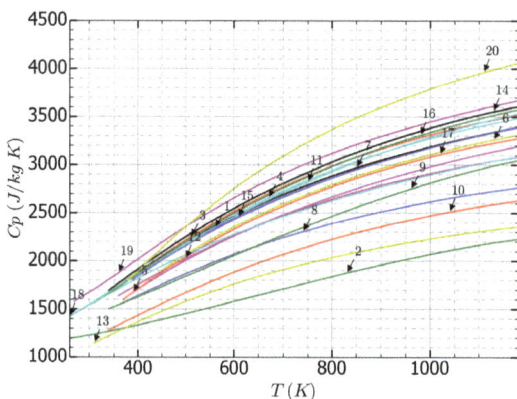

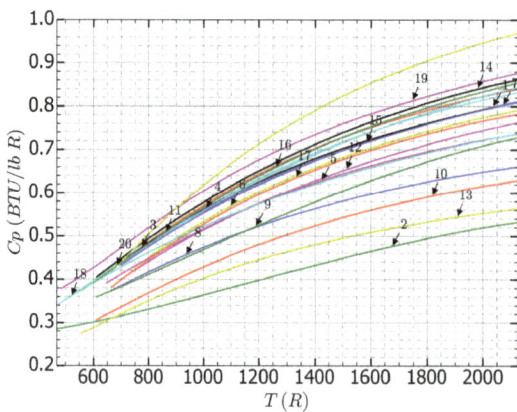

C.6.2 Thermal conductivity

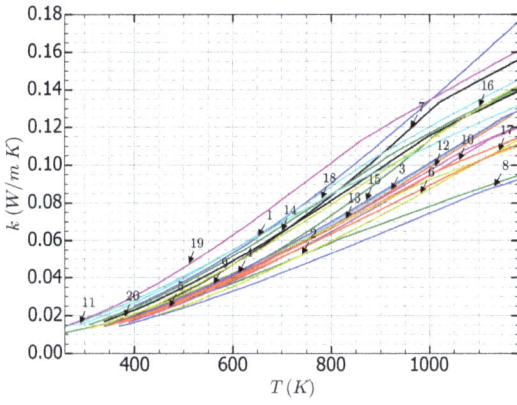

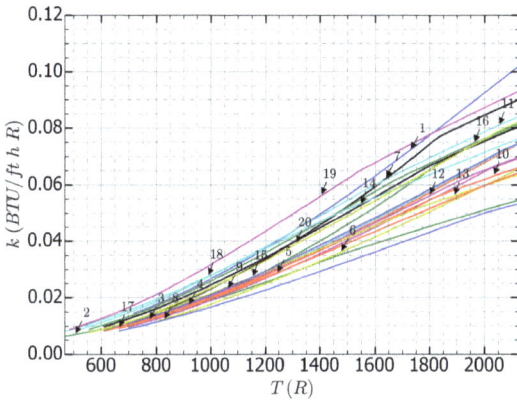

C.6.3 Prandtl number

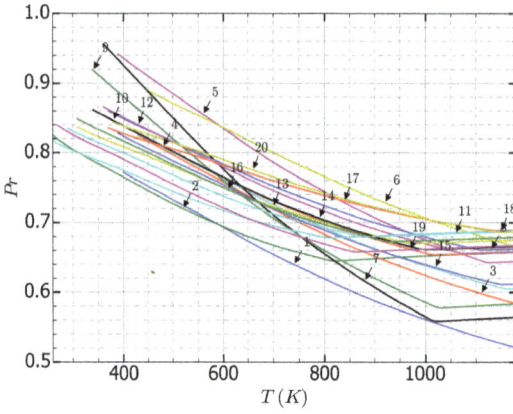

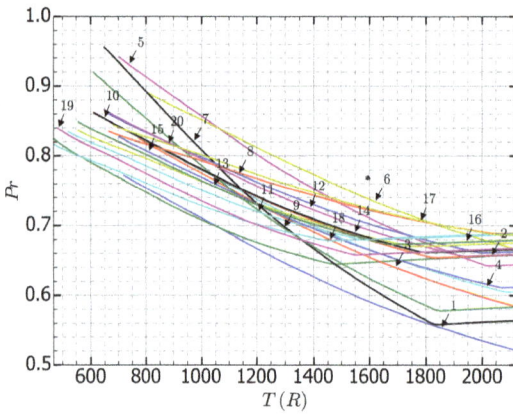

C.6.4 Specific mass

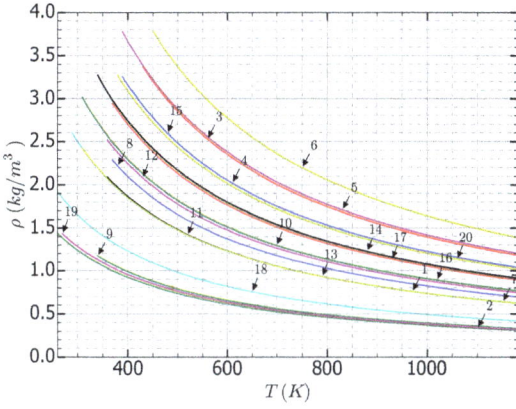

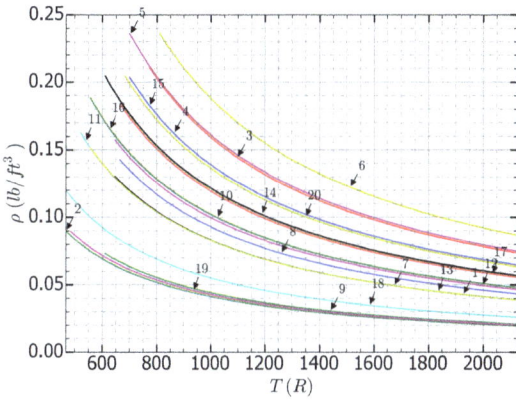

C.6.5 Dynamic viscosity

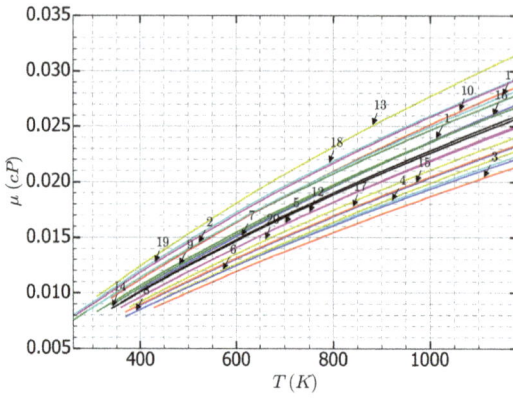

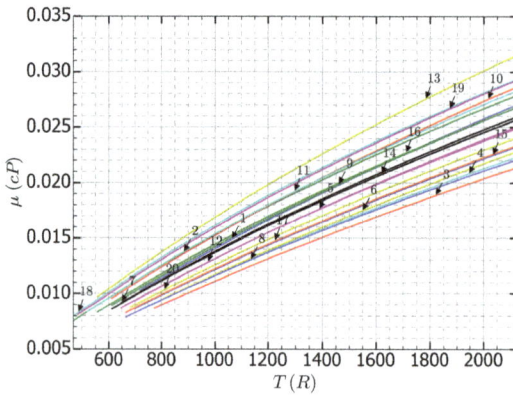

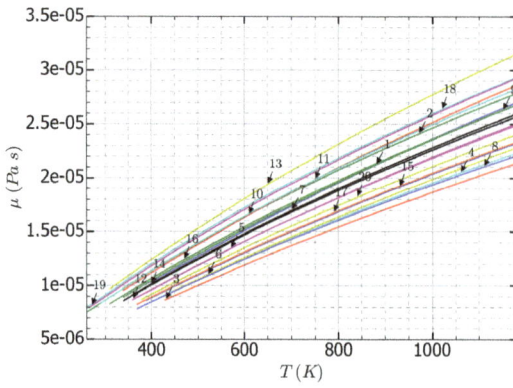

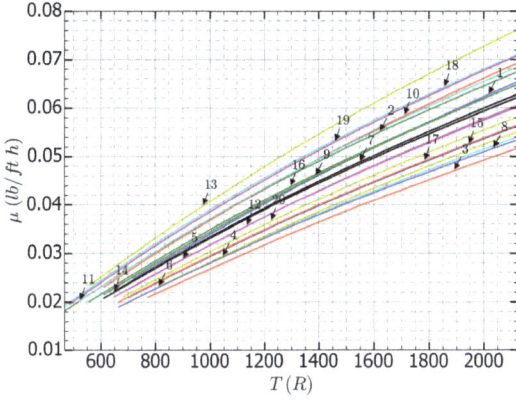

C.7 Selected vapors: part D

Component	
(1)	1,1,2-Trichloroethane
(2)	1,1-Dichloroethane
(3)	1,2,4-Trichlorobenzene
(4)	1,2-Propylene oxide
(5)	1,4-Butanediol
(6)	1,4-Dioxane
(7)	1-Nitropropane
(8)	2,4,6-Trinitrotoluene
(9)	2,4-Dinitrotoluene
(10)	2-Methylindene
(11)	2-Methylnaphthalene
(12)	Acetic anhydride
(13)	Acetyl chloride
(14)	Acrylic acid
(15)	Aniline
(16)	Anisole
(17)	Benzoic acid

(continued)

Component	
(18)	Cyclohexanol
(19)	Cyclohexanone
(20)	Diethanolamine

C.7.1 Specific heat

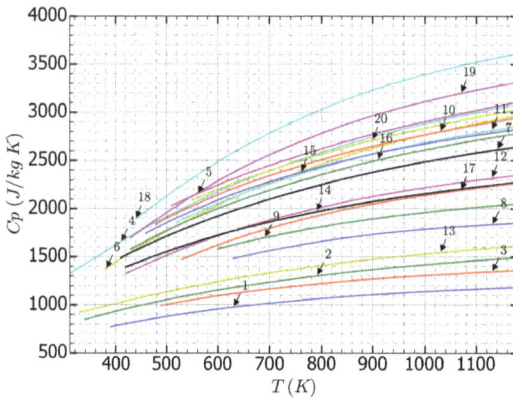

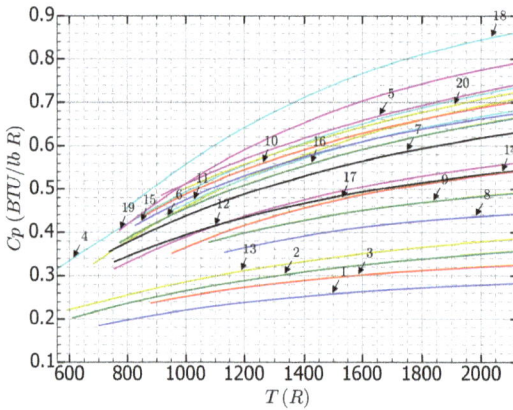

C.7.2 Thermal conductivity

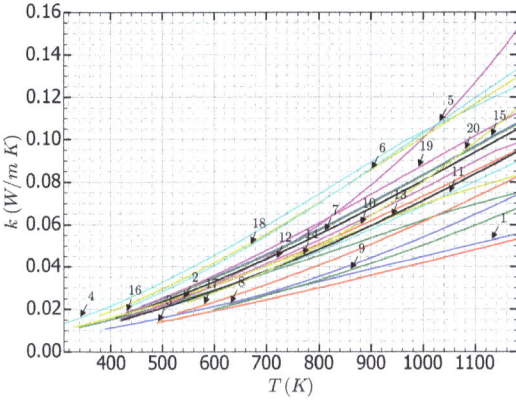

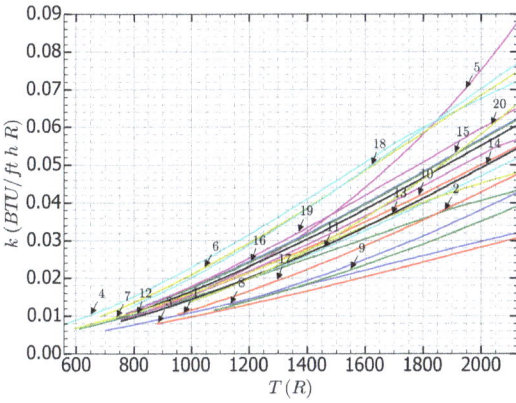

C.7.3 Prandtl number

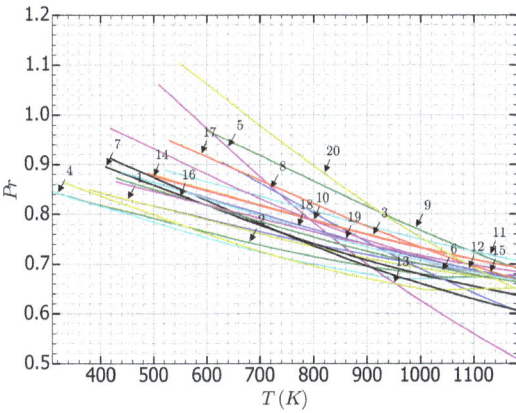

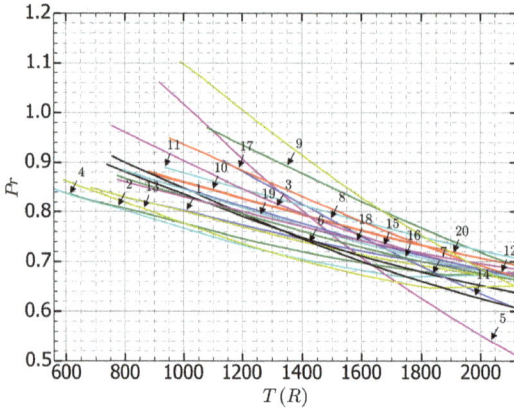

C.7.4 Specific mass

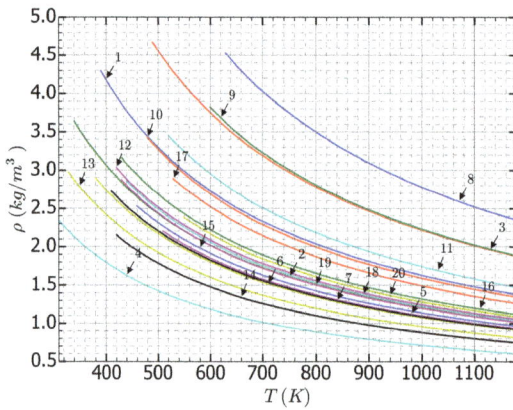

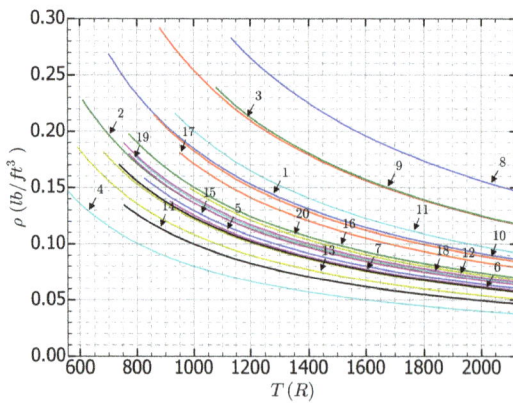

C.7.5 Dynamic viscosity

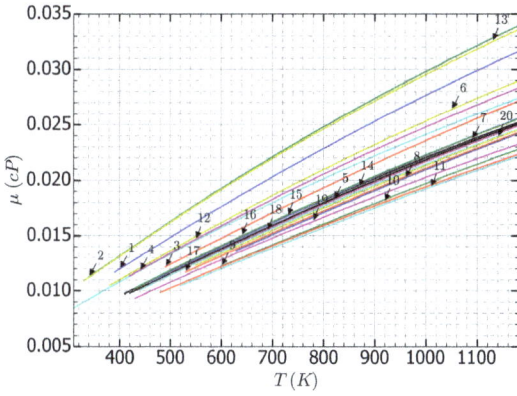

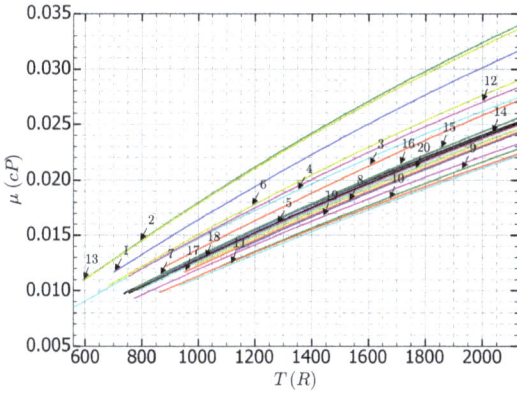

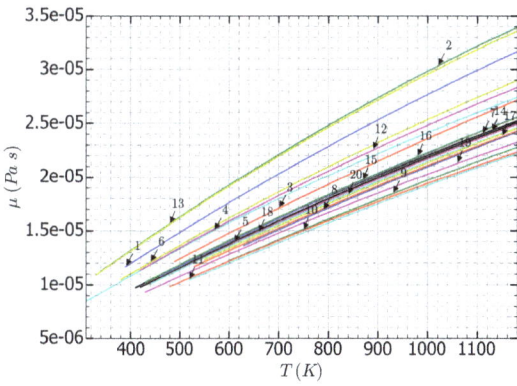

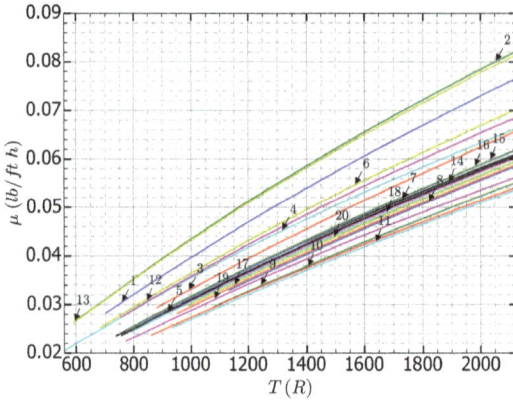

C.8 Selected vapors: part E

Component	
(1)	Monochlorobenzene
(2)	Monoethanolamine
(3)	N-Decane
(4)	N-Docosane
(5)	N-Dodecane
(6)	N-Hexane
(7)	N-Nonane
(8)	N-Octane
(9)	o-Xylene
(10)	p-Ethyltoluene
(11)	p-Nitrotoluene
(12)	p-Xylene
(13)	Phenol
(14)	Styrene
(15)	Thiophene
(16)	Toluene
(17)	Triethylamine

(continued)

Component	
(18)	Triethylene glycol
(19)	Trimethylamine
(20)	Water

C.8.1 Specific heat

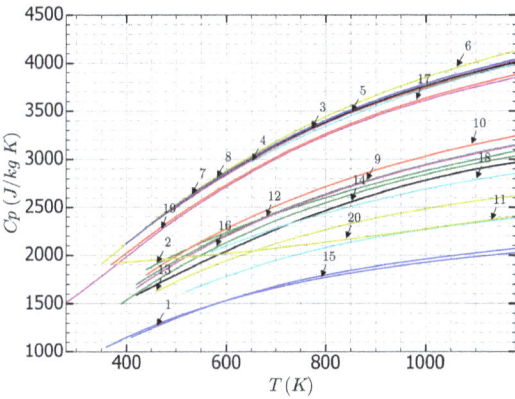

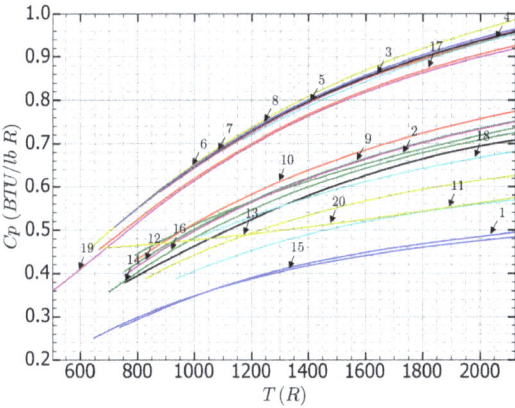

C.8.2 Thermal conductivity

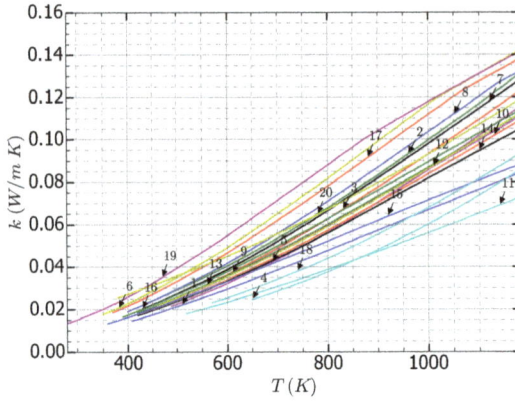

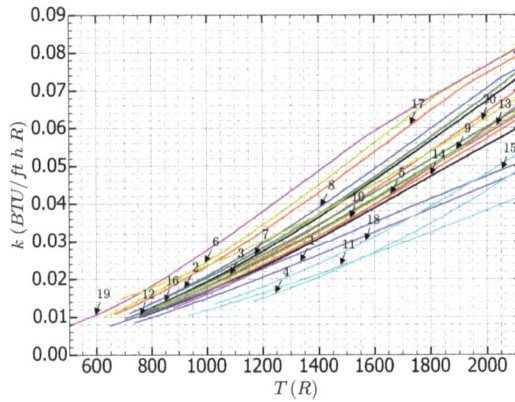

C.8.3 Prandtl number

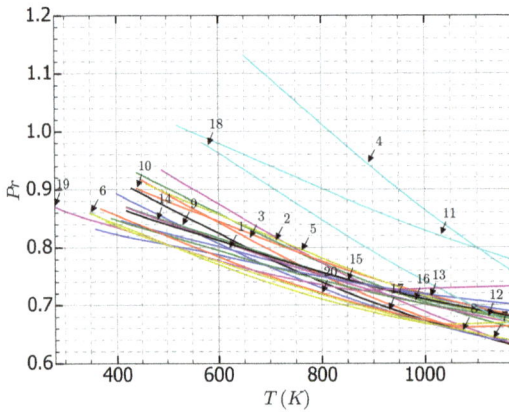

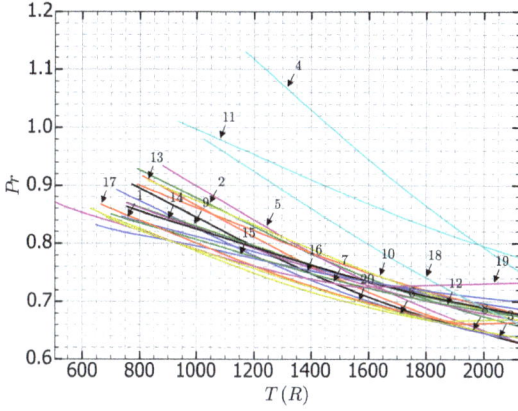

C.8.4 Specific mass

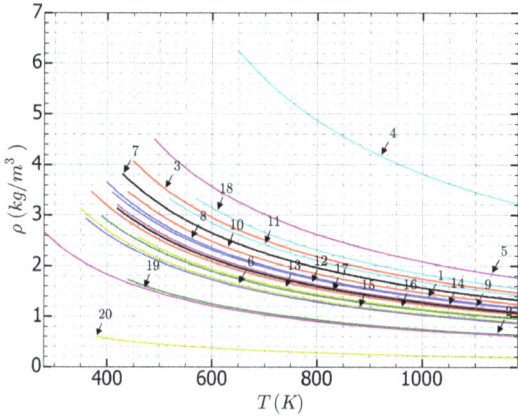

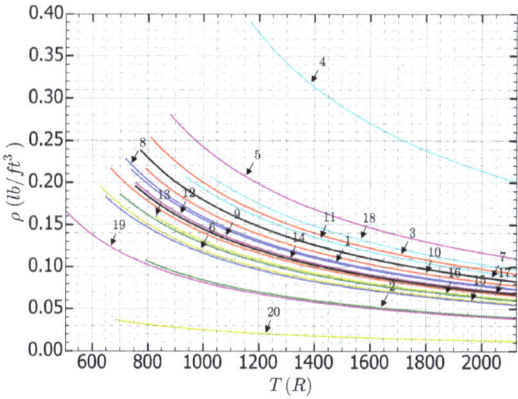

C.8.5 Dynamic viscosity

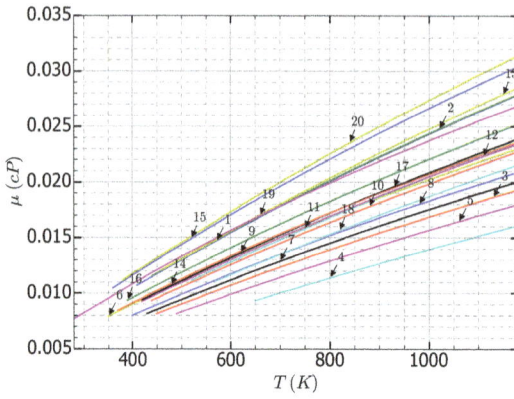

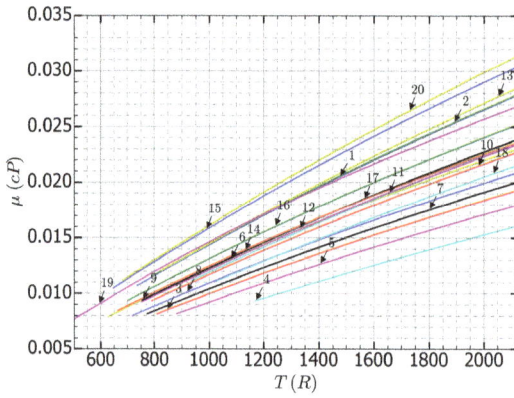

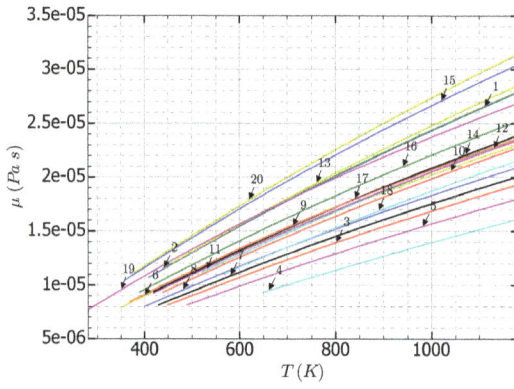

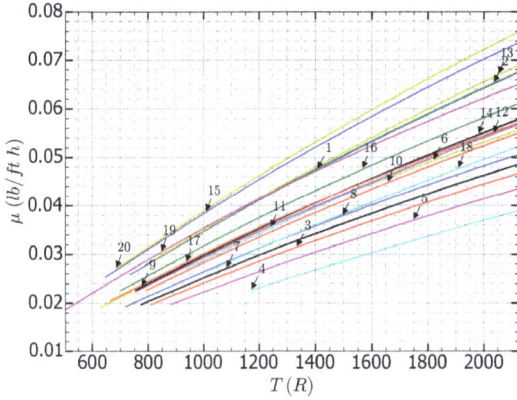

C.9 Selected liquids: part A

Component	
(1)	1-Butanol
(2)	1-Heptanol
(3)	1-Heptene
(4)	1-Hexanol
(5)	1-Hexene
(6)	1-Nitrobutane
(7)	1-Nonene
(8)	1-Octene
(9)	1-Pentene
(10)	1-Propanol
(11)	1-Undecene
(12)	2-Butanol
(13)	2-Ethyl-*m*-xylene
(14)	2-Heptanone
(15)	2-Hexanone
(16)	2-Pentanol
(17)	2-Pentanone

(continued)

Component	
(18)	3-Heptanone
(19)	3-Hexanone
(20)	Acrylonitrile

C.9.1 Specific heat

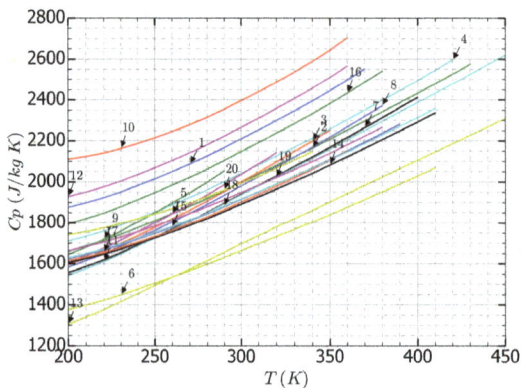

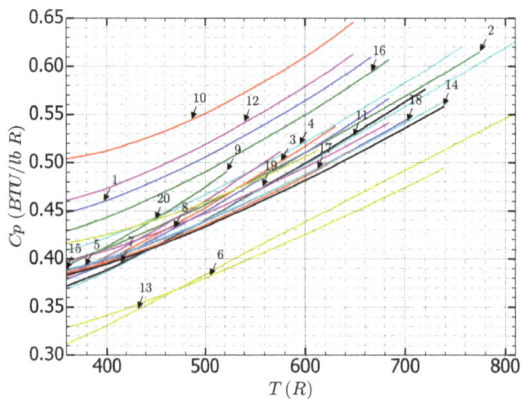

C.9.2 Thermal conductivity

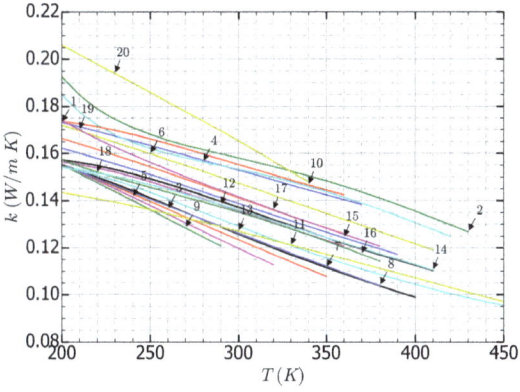

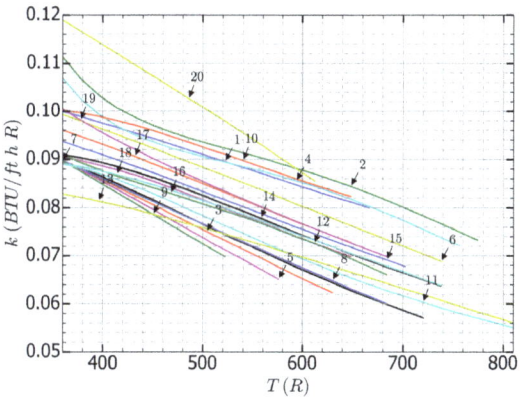

C.9.3 Prandtl number

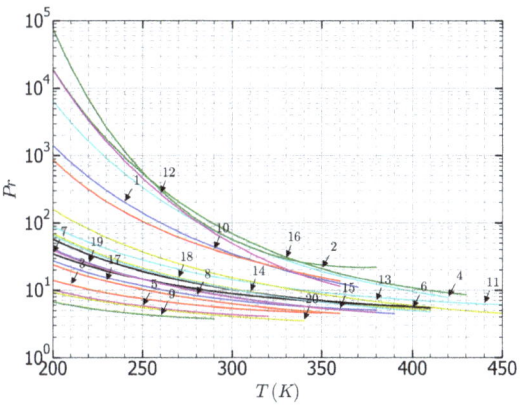

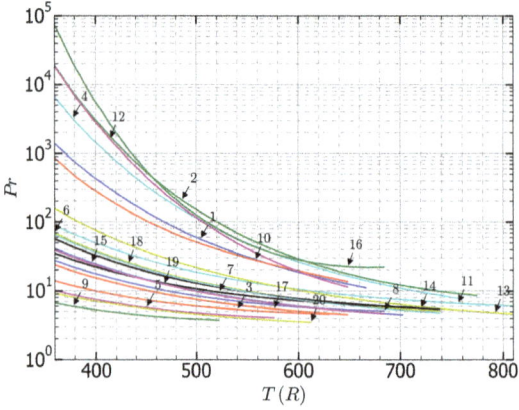

C.9.4 Specific mass

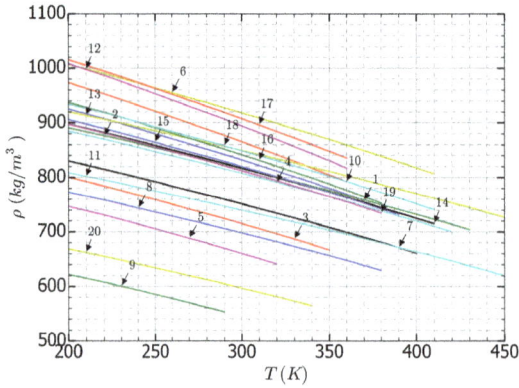

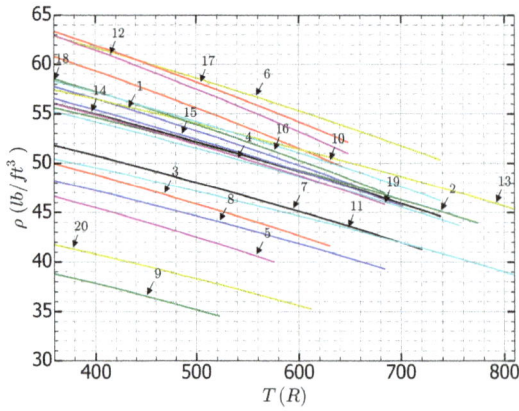

C.9.5 Dynamic viscosity

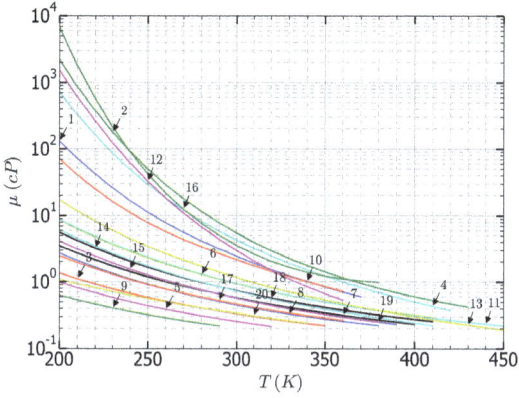

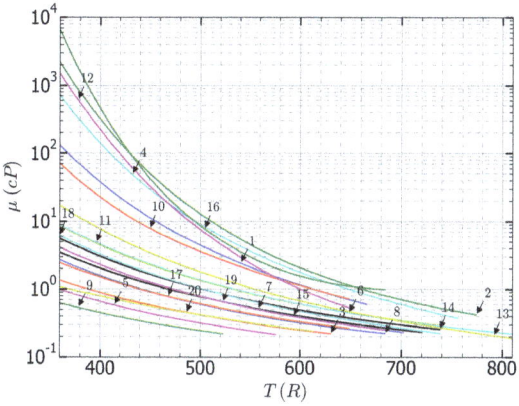

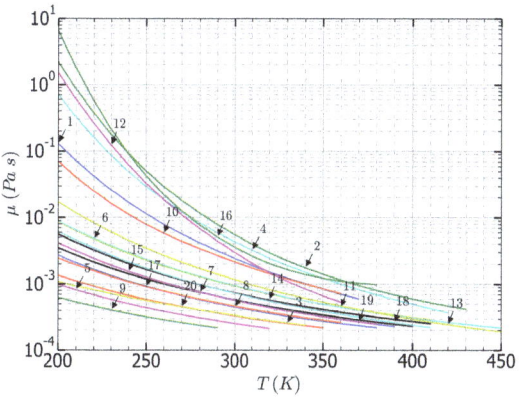

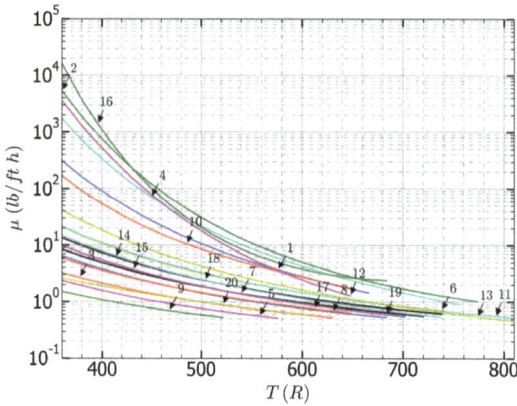

C.10 Selected liquids: part B

Component	
(1)	Acetaldehyde
(2)	Acetic acid
(3)	Acetone
(4)	Acetonitrile
(5)	Benzene
(6)	Butanal
(7)	Cumene
(8)	Cyclohexane
(9)	Cyclohexene
(10)	Cyclopentane
(11)	Diethyl ether
(12)	Diethyl sulfide
(13)	Diethylamine
(14)	Diisobutyl ketone
(15)	Ethanol
(16)	Ethyl acetate
(17)	Ethylacetylene

(continued)

Component	
(18)	Ethylbenzene
(19)	Ethylcyclohexane
(20)	Ethylene oxide

C.10.1 Specific heat

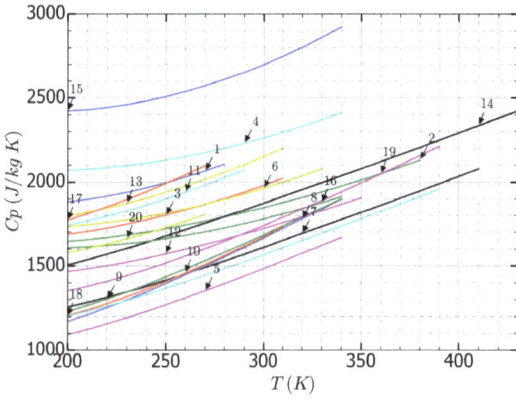

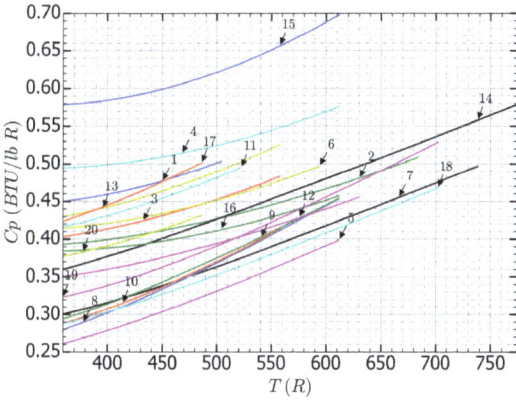

C.10.2 Thermal conductivity

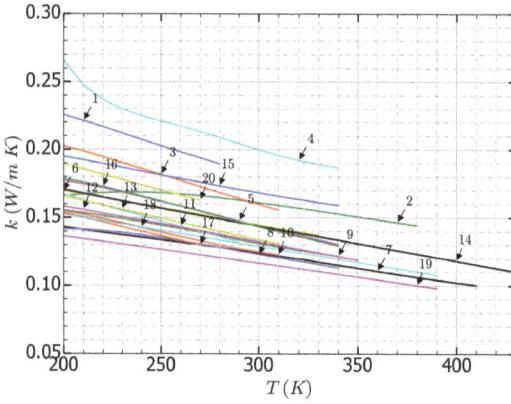

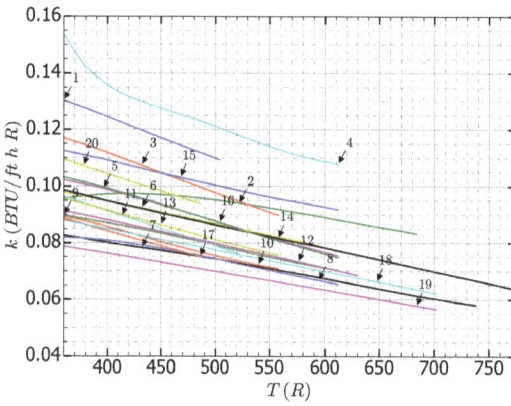

C.10.3 Prandtl number

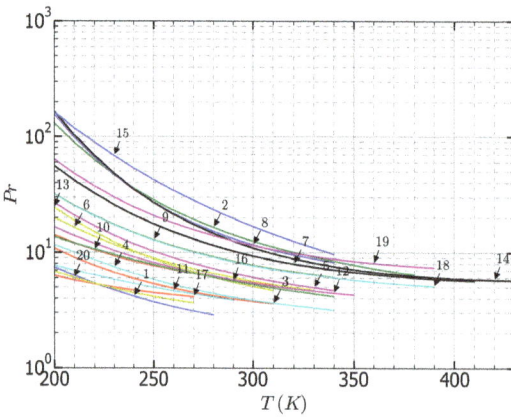

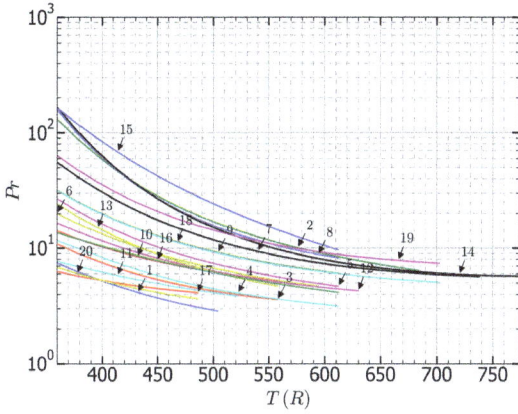

C.10.4 Specific mass

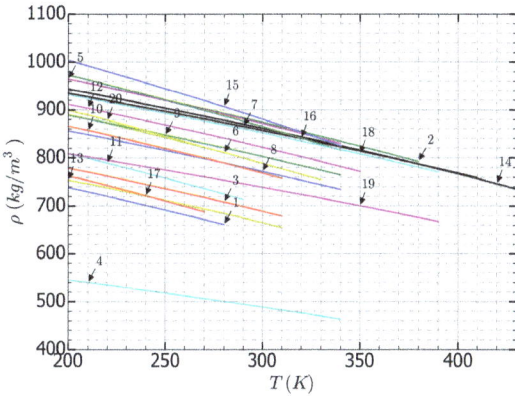

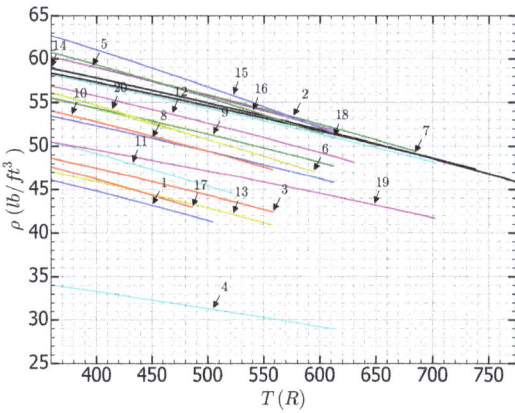

C.10.5 Dynamic viscosity

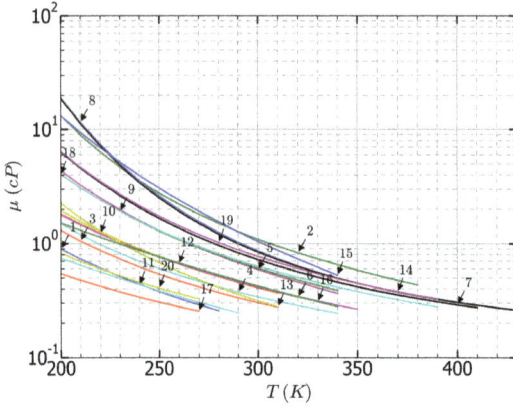

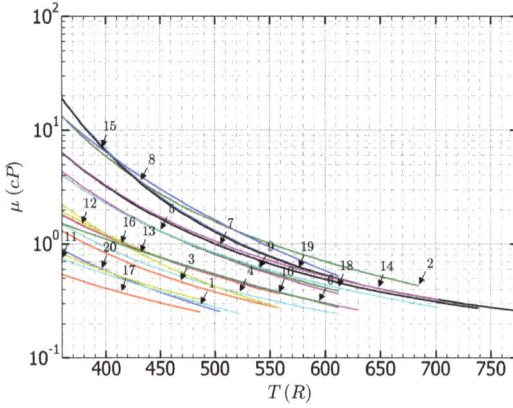

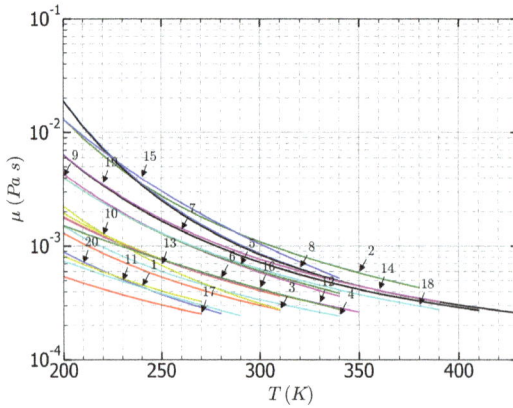

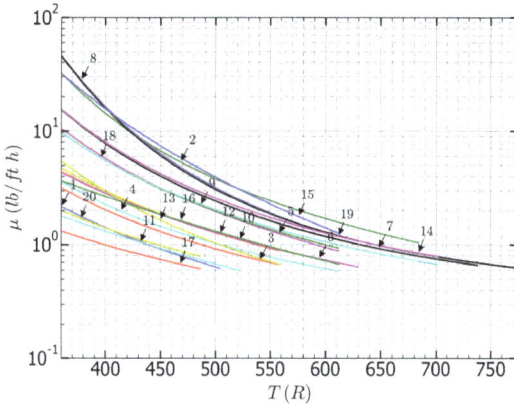

C.11 Selected liquids: part C

Component	
(1)	Ethylenediamine
(2)	Formaldehyde
(3)	Heptanal
(4)	Hexanal
(5)	Isobutyl acetate
(6)	Isobutylbenzene
(7)	Isopropanol
(8)	Methacrylonitrile
(9)	Methanol
(10)	Methyl acetate
(11)	Methyl ethyl ether
(12)	Methyl ethyl ketone
(13)	Methyl formate
(14)	Methyl isobutyl ether
(15)	Methyl isobutyl ketone
(16)	Methyl isopropyl ether
(17)	Methyl isopropyl ketone

(continued)

Component	
(18)	Methylacetylene
(19)	Methylamine
(20)	Methylcyclohexane

C.11.1 Specific heat

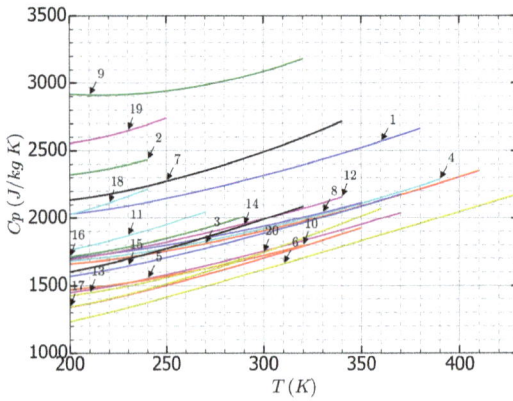

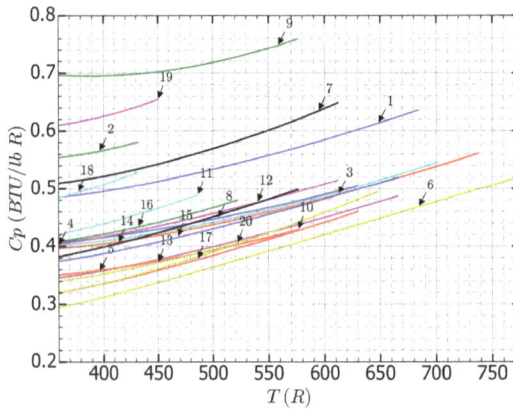

C.11.2 Thermal conductivity

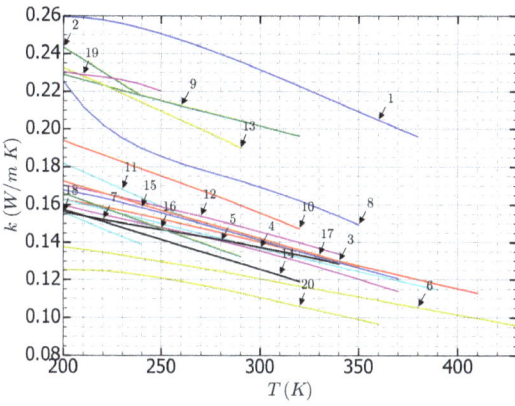

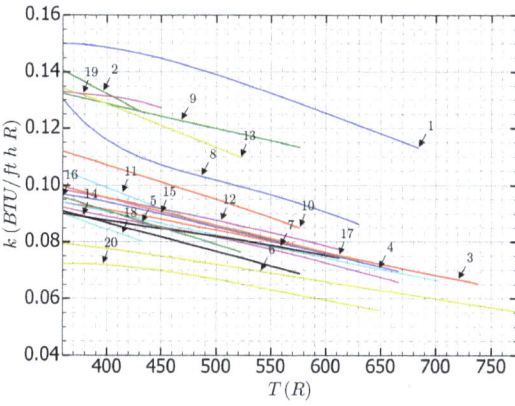

C.11.3 Prandtl number

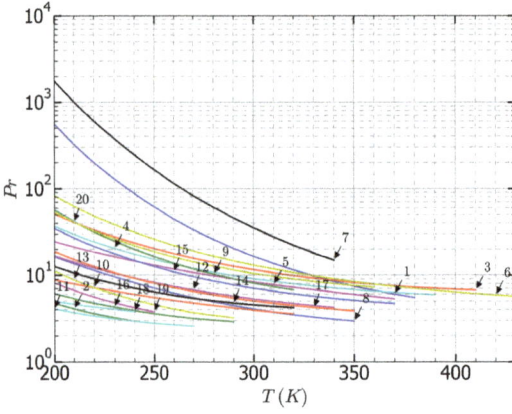

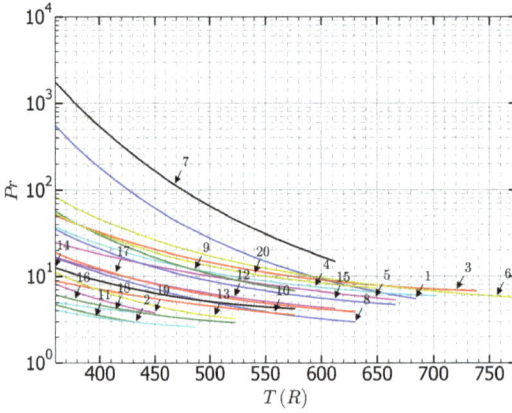

C.11.4 Specific mass

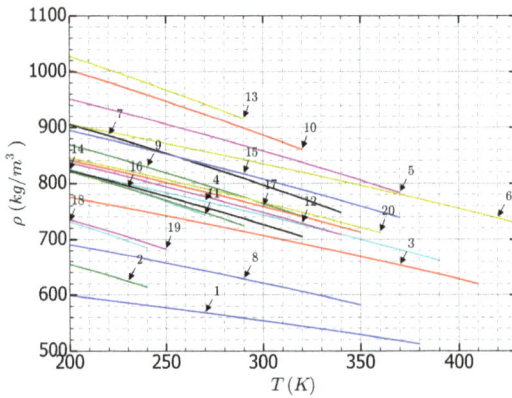

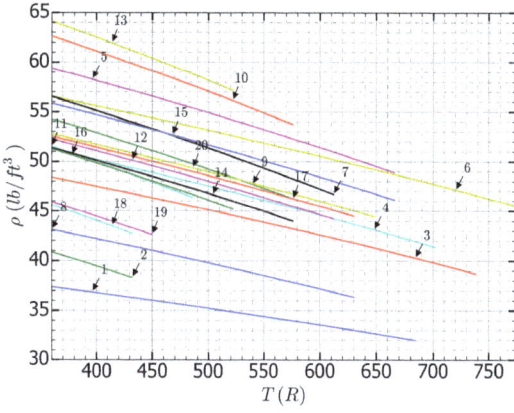

C.11.5 Dynamic viscosity

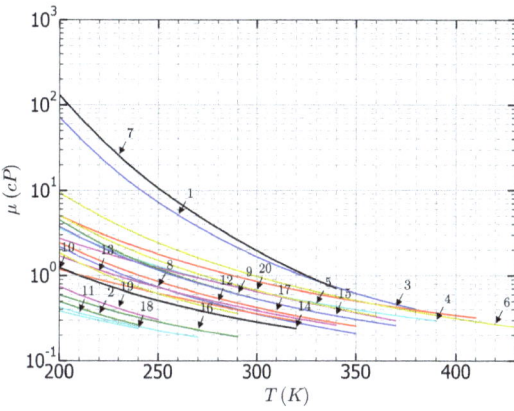

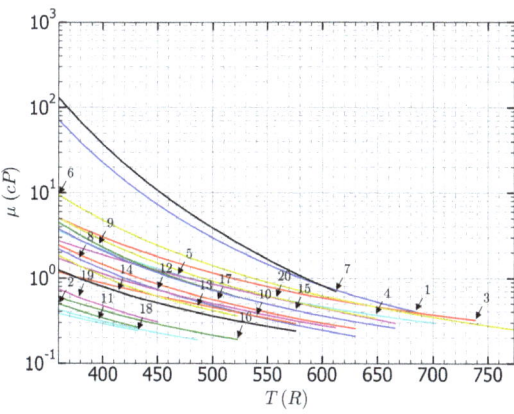

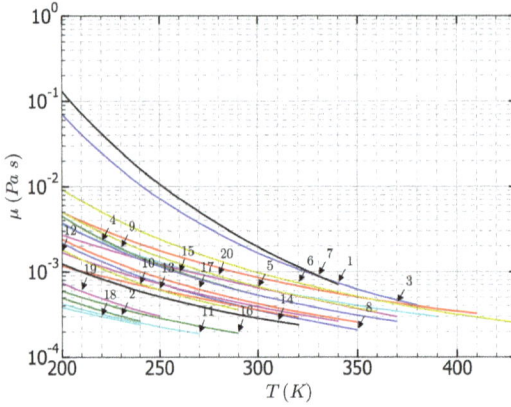

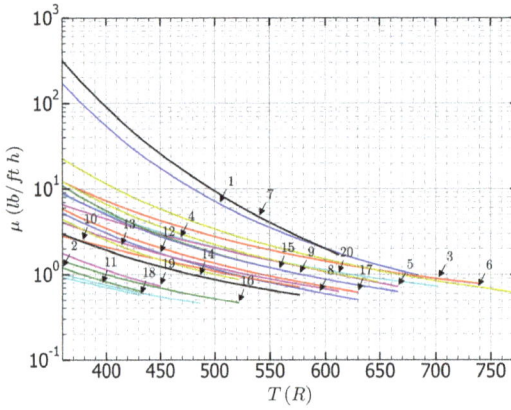

C.12 Selected liquids: part D

Component	
(1)	1,1,2-Trichloroethane
(2)	1,1-Dichloroethane
(3)	1,2,4-Trichlorobenzene
(4)	1,2-Propylene oxide
(5)	1,4-Butanediol
(6)	1,4-Dioxane
(7)	1-Nitropropane

(continued)

Component	
(8)	2,4,6-Trinitrotoluene
(9)	2,4-Dinitrotoluene
(10)	2-Methylindene
(11)	2-Methylnaphthalene
(12)	Acetic anhydride
(13)	Acetyl chloride
(14)	Acrylic acid
(15)	Aniline
(16)	Anisole
(17)	Benzoic acid
(18)	Cyclohexanol
(19)	Cyclohexanone
(20)	Diethanolamine

C.12.1 Specific heat

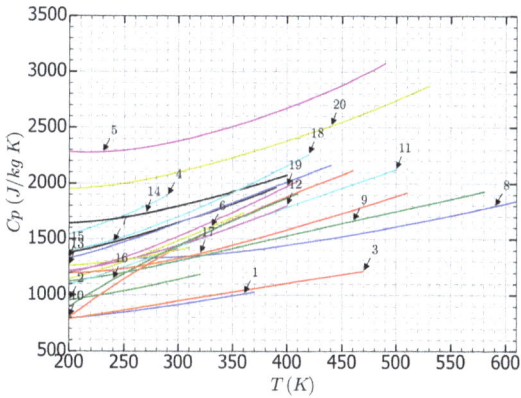

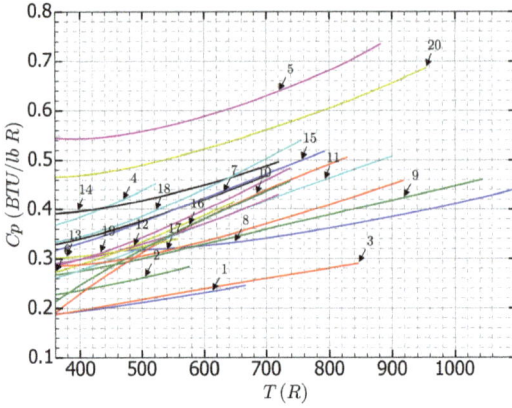

C.12.2 Thermal conductivity

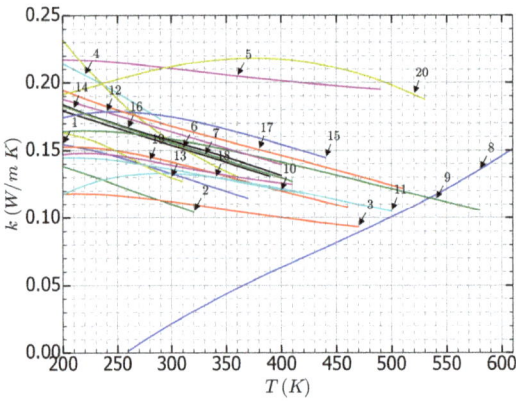

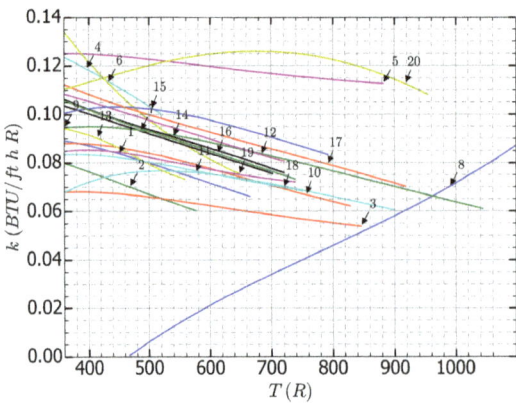

C.12.3 Prandtl number

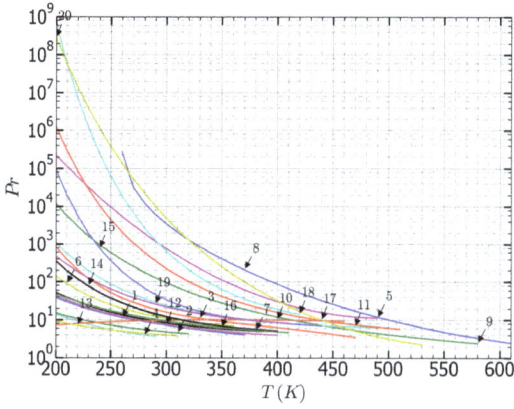

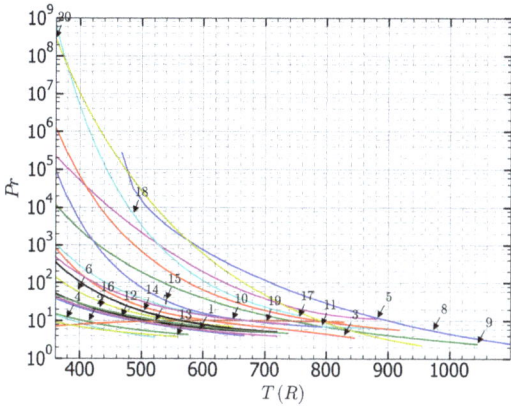

C.12.4 Specific mass

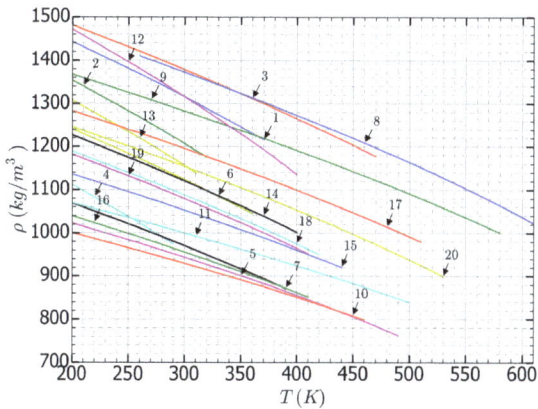

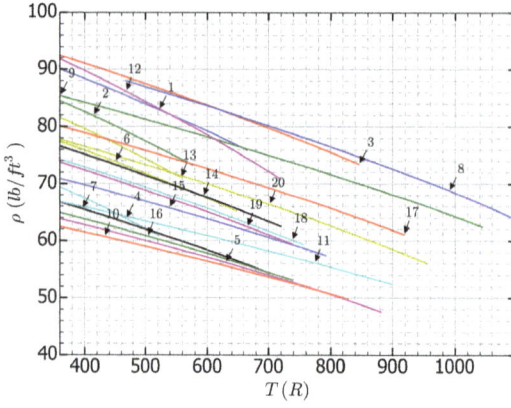

C.12.5 Dynamic viscosity

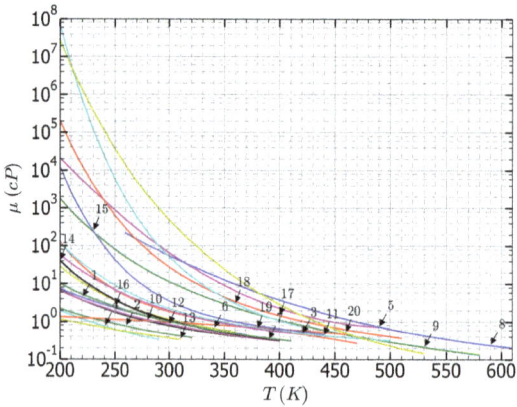

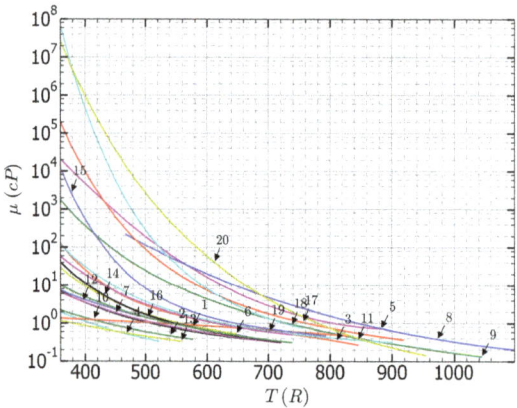

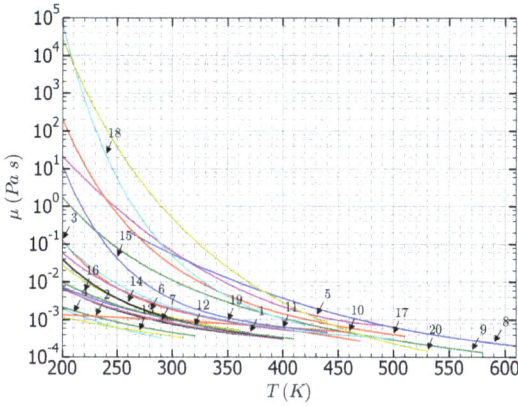

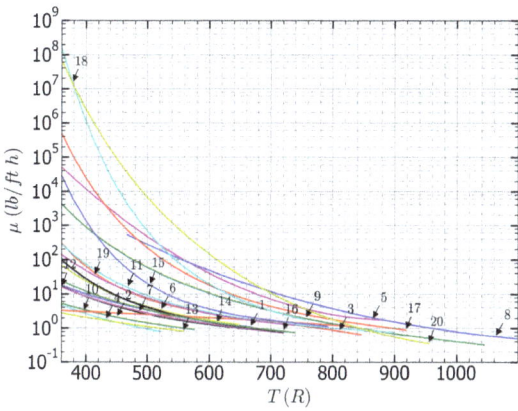

C.13 Selected liquids: part E

Component	
(1)	Monochlorobenzene
(2)	Monoethanolamine
(3)	N-Decane
(4)	N-Docosane
(5)	N-Dodecane
(6)	N-Hexane
(7)	N-Nonane

(continued)

Component	
(8)	N-Octane
(9)	o-Xylene
(10)	p-Ethyltoluene
(11)	p-Nitrotoluene
(12)	p-Xylene
(13)	Phenol
(14)	Styrene
(15)	Thiophene
(16)	Toluene
(17)	Triethylamine
(18)	Triethylene glycol
(19)	Trimethylamine
(20)	Water

C.13.1 Specific heat

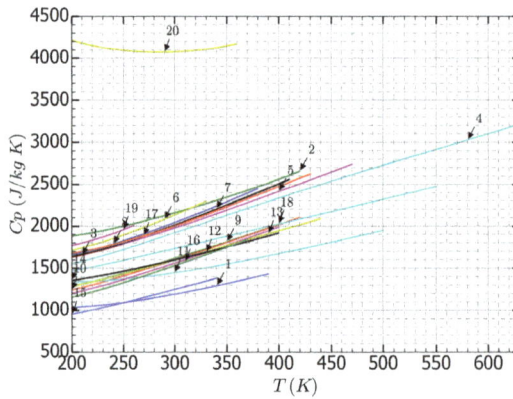

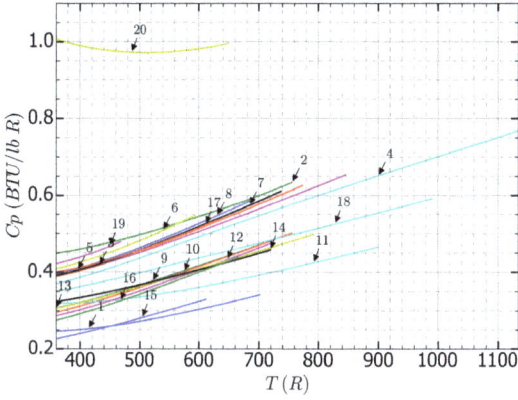

C.13.2 Thermal conductivity

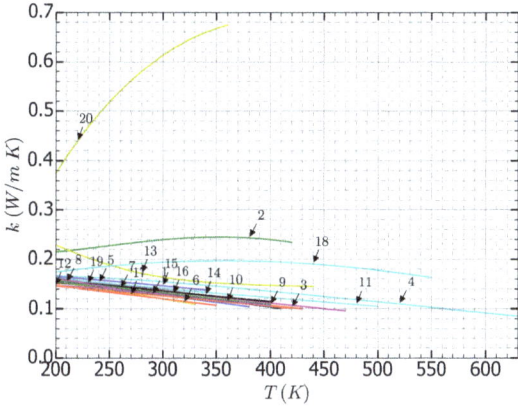

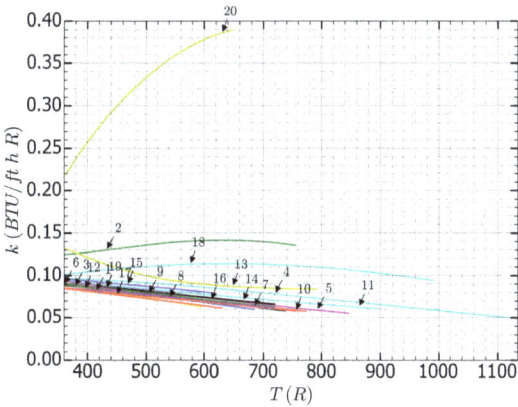

C.13.3 Prandtl number

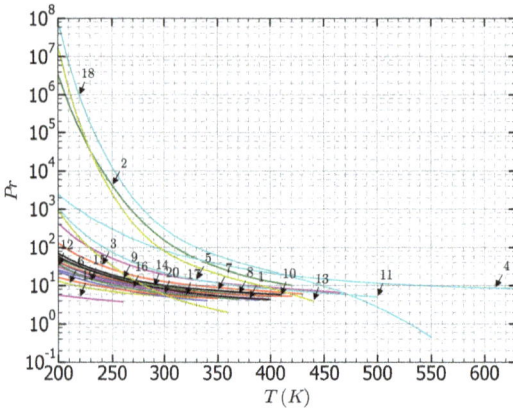

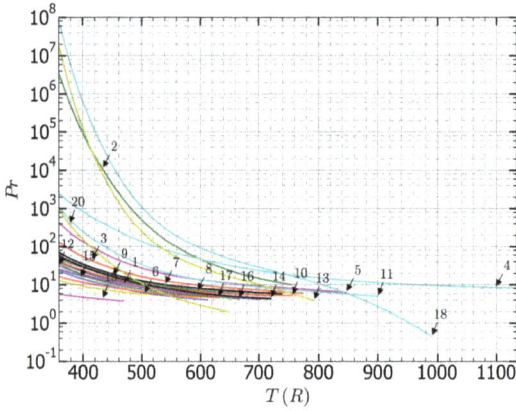

C.13.4 Specific mass

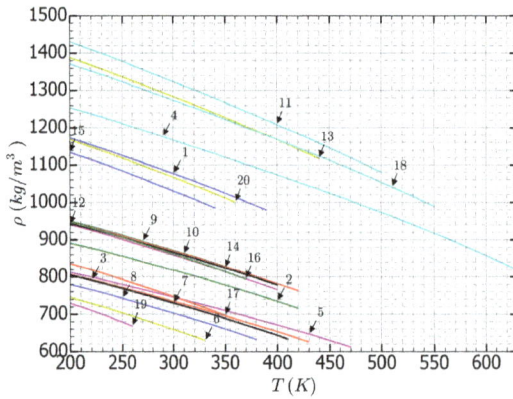

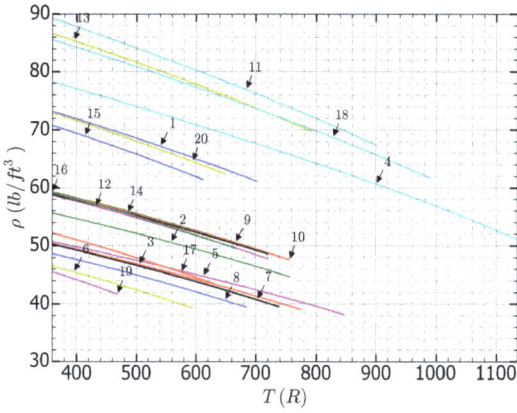

C.13.5 Dynamic viscosity

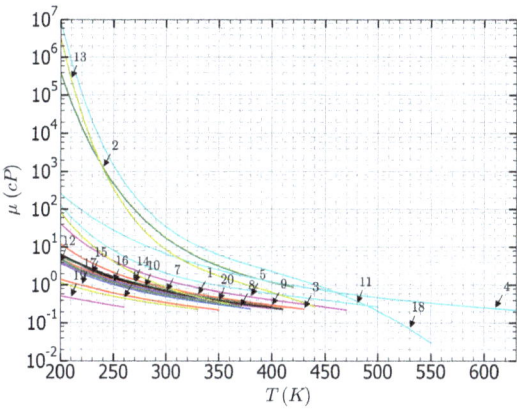

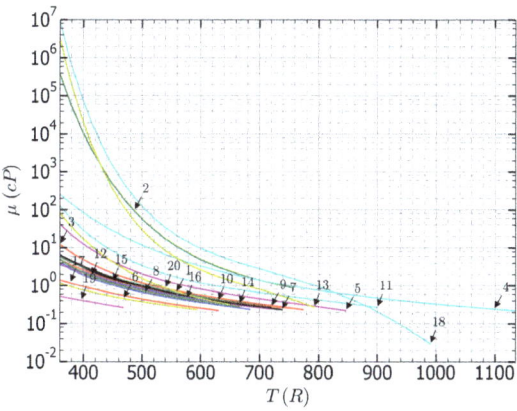

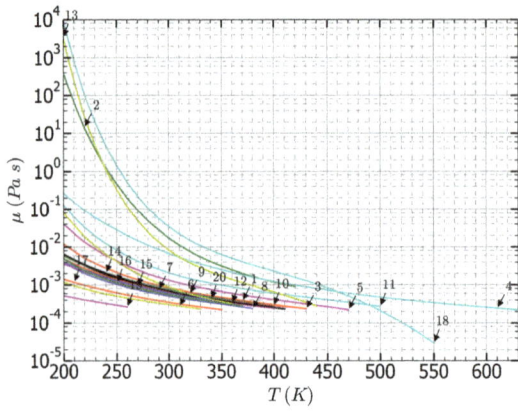

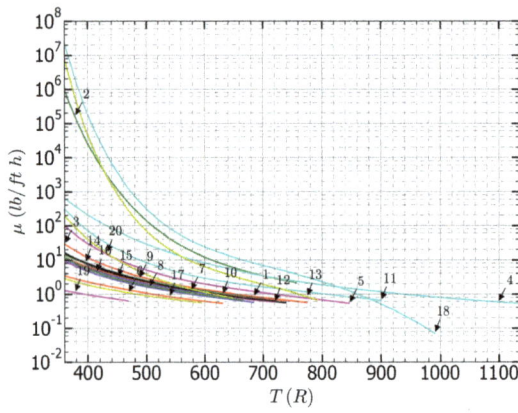

References

[1] "Stainless steel seamless pipe China." [Online]. Available at: http://www.nan-long.com/gpage/stainless_steel_pipe/Stainless_Steel_Seamless_Pipes.htm. [Accessed: 01 Mar 2015].

[2] "Carbon steel seamless pipe, best-selling seamless steel pipe from direct factory." [Online]. Available at: http://www.fsqysteelpipe.com/carbon-steel-seamless-pipe.html. [Accessed: 01 Mar 2015].

[3] "ERW-steel pipe types-China steel pipe manufacturer Shinestar Steel Pipe Corporation." [Online]. Available at: http://www.shinestarsteel.com/html/product/2014-6-9/917.html#p=1.

[4] "Standard steel pipe-standard tubulars-China steel pipe manufacturer Shinestar Steel Pipe Corporation." [Online]. Available at: http://www.shinestarsteel.com/html/product/2014-6-9/919.html#p=1. [Accessed: 01 Mar 2015].

[5] J. R. Couper, *Chemical Process Equipment: Selection and Design*. Gulf Professional Publishing, 2005.

[6] P. D. Hills, *Practical Heat Transfer*. Begell House Publishers, Incorporated, 2005.

[7] "Tube-in-tube heat exchanger – APV – product detail," 2013. [Online]. Available at: http://www.spx.com/en/apv/pd-040_mp-tubular-the/. [Accessed: 07 Aug 2013].

[8] "Heat exchangers – production – Mashzavod, LTD, Chernivtsy, Ukraine," 2013. [Online]. Available at: http://mashzavod.com/shell-and-tube-heat-exchangers.php?l=2. [Accessed: 07 Aug 2013].

[9] "Wine, tube-in-tube heat-exchangers," 2013. [Online]. Available at: http://www.biomashin.com/en/products/31/Wine/Tube-in-tube-Heat-Exchangers. [Accessed: 07 Aug 2013].

[10] "Products – KIESEL tube-in-tube heat exchangers – KIESEL tube-in-tube heat exchangers – cellar-tek," 2013. [Online]. Available at: http://www.cellartek.com/products/refrigeration_and_chillers/tube_in_tube.php. [Accessed: 07 Aug 2013].

[11] "Tube in tube chillers," 2013. [Online]. Available at: http://www.innovativecooling.com/tube_n_chillers/. [Accessed: 07 Aug 2013].

[12] "Hairpin Heat Exchanger | Heat Exchanger Design, Inc. | Worldwide Leaders in Heat Transfer Technology." [Online]. Available at: http://www.hed-inc.com/hairpin.html.

[13] "HeatExchanger » DSM – Optimization + Dynamic systems modeling & simulation." [Online]. Available at: http://www.sysmodeling.com/index.php?p=2_1. [Accessed: 22 Aug 2013].

[14] "Koch heat transfer company – designing, manufacturing and testing custom heat exchangers," 2013. [Online]. Available at: http://www.kochheattransfer.com/. [Accessed: 22 Aug 2013].

[15] "Hairpin heat exchanger heat exchanger design, Inc. Worldwide leaders in heat transfer technology." [Online]. Available at: http://www.hed-inc.com/hairpin.html. [Accessed: 21 Feb 2014].

[16] E. E. Ludwig, *Applied Process Design for Chemical and Petrochemical Plants (Vol 2): Separation*, vol. 2. Gulf Professional Publishing, 2010.

[17] C. Branan, *Rules of Thumb for Chemical Engineers: A Manual of Quick, Accurate Solutions to Everyday Process Engineering Problems*. Gulf Professional Publishing, 2005.

[18] R. K. Shah and D. P. Sekulić, *Fundamentals of Heat Exchanger Design*. John Wiley & Sons, 2003.

[19] G. Towler and R. K. Sinnott, *Chemical Engineering Design: Principles, Practice and Economics of Plant and Process Design*. Elsevier, 2012.

[20] E. Schlunder, *Heat Exchanger Design Handbook*. VDI-Verlag GmbH, 1986.

[21] "Koch Heat Transfer Company – Exemplary hairpin applications – Variety of solutions."

https://doi.org/10.1515/9783110585872-013

[22] "Why to consider using a hairpin heat exchanger." [Online]. Available at: http://www.pro cess-heating.com/articles/86928-why-to-consider-using-a-hairpin-heat-exchanger. [Accessed: 02 Mar 2014].

[23] T. F. Edgar, D. M. Himmelblau, and L. S. Lasdon, *Optimization of Chemical Processes*. McGraw-Hill, 2001.

[24] A. V. D. Bos, *Parameter Estimation for Scientists and Engineers*. John Wiley \& Sons, 2007.

[25] G. Corsano, J. M. Montagna, O. A. Iribarren, and P. A. Aguirre, *Mathematical Modeling Approaches for Optimization of Chemical Processes*. Nova Science Publishers, Inc., 2009.

[26] P. Englezos and N. Kalogerakis, *Applied Parameter Estimation for Chemical Engineers*. M. Dekker, 2001.

[27] S. S. Rao and S. S. Rao, *Engineering Optimization: Theory and Practice*. John Wiley \& Sons, 2009.

[28] R. W. Serth, *Process Heat Transfer: Principles and Applications*. Academic Press, 2007.

[29] "Hairpin Basco heat exchangers – Thiet bi trao doi nhiet – P.H.E Thermotechnics." [Online]. Available at: http://www.traodoinhiet.com/en/trao-doi-nhiet-kieu-hairpin-basco/. [Accessed: 07 Aug 2013].

[30] A. D. Kraus, A. Aziz, and J. R. Welty, *Extended Surface Heat Transfer*. John Wiley \& Sons, 2001.

[31] "Double pipe heat exchanger, double pipe heat exchanger manufacturers," 2013. [Online]. Available at: http://www.jcequipments.com/double-pipe-heat-exchanger.html. [Accessed: 07 Aug 2013].

[32] "Products – HRS heat exchangers," 2013. [Online]. Available at: http://www.hrs-heatexchangers.com/en/products/default.aspx. [Accessed: 07 Aug 2013].

[33] "Steel tube and pipe handbook." [Online]. Available at: http://www.steeltube.sk/. [Accessed: 20 Apr 2014].

[34] "Exchangers tube-in-tube heat Italian food materials and machinery," 2013. [Online]. Available at: http://italianfoodmaterialsandmachinery.com/exchangers-tube-in-tube-heat/. [Accessed: 13 Sep 2013].

[35] "Heat exchanger tube – Zeleziarne Podbrezova." [Online]. Available at: http://www.steel tube.sk/pipe/fitting.nsf/page/1_4_for_heat_exchangers. [Accessed: 22 Apr 2014].

[36] "Line pipe – Zeleziarne Podbrezova." [Online]. Available at: http://www.steeltube.sk/pipe/ fitting.nsf/page/1_7_line_pipes. [Accessed: 22 Apr 2014].

[37] R. A. Bowman, A. C. Mueller, and W. M. Nagle, "Mean temperature difference in design," *Transactions of ASME*, vol. 62, no. 4, pp. 283–294, 1940.

[38] W. M. Nagle, "Mean temperature differences in multipass heat exchangers," *Industrial & Engineering Chemistry*, vol. 25, no. 6, pp. 604–609, 1933.

[39] A. P. Colburn and E. du P. de, "Mean temperature difference and heat transfer coefficient in liquid heat exchangers," *Industrial & Engineering Chemistry*, vol. 25, no. 8, pp. 873–877, 1933.

[40] R. A. Bowman, "Mean temperature difference correction in multipass exchangers," *Industrial \& Engineering Chemistry*, vol. 28, no. 5, pp. 541–544, 1936.

[41] W. S. Janna, *Engineering Heat Transfer*. Boca Raton, Fla.: CRC Press, 2000.

[42] J. P. Holman, *Heat Transfer*. Boston, [Mass.]: McGraw Hill Higher Education, 2010.

[43] R. B. Bird, W. E. Stewart, and E. N. Lightfoot, *Transport Phenomena*. John Wiley & Sons, 2007.

[44] D. Q. Kern, *Process Heat Transfer*. McGraw-Hill, 1950.

[45] V.-G. V. und Chemieingenieurwesen and V. D. I. Gesellschaft, *VDI Heat Atlas*. Springer, 2010.

[46] S. J. M. Cartaxo and F. A. N. Fernandes, "Counterflow logarithmic mean temperature difference is actually the upper bound: A demonstration," *Applied Thermal Engineering*, vol. 31, no. 6–7, pp. 1172–1175, 2011.

[47] W. M. Rohsenow, J. P. Hartnett, and Y. I. Cho, *Handbook of Heat Transfer*. McGraw-Hill, 1998.

[48] A. Bejan and A. D. Kraus, *Heat Transfer Handbook*. J. Wiley, 2003.

[49] H. Schlichting, *Boundary Layer Theory*. McGraw-Hill, 1960.

[50] F. Kreith, R. M. Manglik, and M. S. Bohn, *Principles of Heat Transfer*. Cengage Learning, 2010.

[51] M. Michelsen and J. Villadsen, "The Graetz problem with axial heat conduction," *International Journal of Heat and Mass Transfer*, vol. 17, no. 11, pp. 1391–1402, 1974.

[52] F. W. Dittus and L. M. K. Boelter, "University of California publications on engineering," *University of California Publications in Engineering*, vol. 2, p. 371, 1930.

[53] W. H. McAdams, *Heat Transmission*, vol. 3. McGraw-Hill New York, 1954.

[54] R. Winterton, "Where did the Dittus and Boelter equation come from?," *International Journal of Heat and Mass Transfer*, vol. 41, no. 4, pp. 809–810, 1998.

[55] R. S. Brodkey and H. C. Hershey, *Transport Phenomena: A Unified Approach*. Brodkey Publishing, 1988.

[56] C. A. Sleicher and M. W. Rouse, "A convenient correlation for heat transfer to constant and variable property fluids in turbulent pipe flow," *International Journal of Heat and Mass Transfer*, vol. 18, no. 5, pp. 677–683, 1975.

[57] Y. A. Çengel, *Heat Transfer: A Practical Approach*. McGraw-Hill, 2003.

[58] R. H. Perry and D. W. Green, *Perry's Chemical Engineers' Handbook (7th)*, 7th ed. McGraw-Hill, 1997.

[59] J. J. Lorentz, D. T. Yung, C. B. Panchal, and G. E. Layton, "An assessment of heat-transfer correlations for turbulent water flow through a pipe at Prandtl numbers of 6.0 and 11.6," *NASA STI/Recon Technical Report N*, vol. 82, p. 33677, 1982.

[60] J. A. Malina and E. Sparrow, "Variable-property, constant-property, and entrance-region heat transfer results for turbulent flow of water and oil in a circular tube," *Chemical Engineering Science*, vol. 19, no. 12, pp. 953–962, 1964.

[61] R. W. Allen and E. R. G. Eckert, "Friction and heat-transfer measurements to turbulent pipe flow of water (Pr = 7 and 8) at uniform wall heat flux," *Journal of Heat Transfer*, vol. 86, p. 301, 1964.

[62] E. N. Sieder and G. E. Tate, "Heat transfer and pressure drop of liquids in tubes," *Industrial & Engineering Chemistry*, vol. 28, no. 12, pp. 1429–1435, 1936.

[63] R. H. Perry and D. W. Green, *Perry's Chemical Engineers' Handbook (8th)*, 8th ed. New York: McGraw-Hill, 2008.

[64] B. S. Petukhov and V. N. Popov, "Theoretical calculation of heat exchange and frictional resistance in turbulent flow in tubes of an incompressible fluid with variable physical properties (Heat exchange and frictional resistance in turbulent flow of liquids with variable physical properties through tubes)," *High Temperature*, vol. 1, pp. 69–83, 1963.

[65] R. A. Seban and T. T. Shimazaki, "Temperature distributions for air flowing turbulently in a smooth heated pipe," *General Discussion on Heat Transfer*, pp. 122–26, 1951.

[66] E. M. Sparrow and R. Siegel, "Unsteady turbulent heat transfer in tubes," *Journal of Heat Transfer (US)*, vol. 82, 1960.

[67] C. A. Sleicher and M. Tribus, *Transaction of ASME*, vol. 79, no. 4, pp. 789–797, 1957.

[68] G. Filonenko, "Hydraulic resistance in pipes," *Teploenergetika*, vol. 1, no. 4, pp. 40–44, 1954.

[69] R. Notter and C. Sleicher, "A solution to the turbulent Graetz problem – III Fully developed and entry region heat transfer rates," *Chemical Engineering Science*, vol. 27, no. 11, pp. 2073–2093, 1972.

[70] V. Gnielinski, "New equations for heat and mass transfer in turbulent pipe and channel flow," *International Chemical Engineering*, vol. 16, no. 2, pp. 359–368, 1976.

[71] H. Hausen, "Darstellung des Warmeuberganges in Rohren durch verallgemeinerte Potenzbeziehungen," *Z. VDI Beihefte Verfahrenstech*, no. 4, pp. 91–98, 1943.

[72] B. S. Petukhov, "Heat transfer and friction in turbulent pipe flow with variable physical properties," *Advances in Heat Transfer*, vol. 6, pp. 503–564, 1970.

[73] H. Schlichting, J. Kestin, H. Schlichting, and H. Schlichting, *Boundary-layer Theory*, vol. 539. McGraw-Hill New York, 1968.

[74] H. Hausen, "New equations for heat transfer with free or forced convection," *Allg. Waermetech*, vol. 9, pp. 75–79, 1959.

[75] A. Eagle and R. M. Ferguson, "On the coefficient of heat transfer from the internal surface of tube walls," *Proceedings of the Royal Society of London. Series A, Containing Papers of a Mathematical and Physical Character*, vol. 127, no. 806, pp. 540–566, 1930.

[76] H. Li and J. Liu, "Revolutionizing heat transport enhancement with liquid metals: Proposal of a new industry of water-free heat exchangers," *Frontiers in Energy*, vol. 5, no. 1, pp. 20–42, 2011.

[77] E. Skupinski, J. Tortel, and L. Vautrey, "Determination des coefficients de convection d'un alliage sodium-potassium dans un tube circulaire," *International Journal of Heat and Mass Transfer*, vol. 8, no. 6, pp. 937–951, 1965.

[78] B. S. Petukhov and L. I. Roizen, "Generalized relationships for heat transfer in a turbulent flow of gas in tubes of annular section(Heat transfer coefficients and adiabatic temperatures obtained for turbulent gas flow in tubes with annular section)," *High Temperature*, vol. 2, pp. 65–68, 1964.

[79] V. Gnielinski, "Berechnung des Druckverlustes in glatten konzentrischen Ringspalten bei ausgebildeter laminarer und turbulenter isothermer Strömung," *Chemie Ingenieur Technik*, vol. 79, no. 1–2, pp. 91–95, 2007.

[80] Crane-Co, "Flow of Fluids Through Valves, Fittings, and Pipe (TP-410 (US edition))," Crane, 1982.

[81] W. B. Hooper, *Chemical Engineering*, vol. 24, pp. 96–100, 1981.

[82] R. Darby, "Correlate pressure drops through fittings," *Chemical Engineering*, vol. 108, no. 4, pp. 127–130, 2001.

[83] R. Darby, *Chemical Engineering Fluid Mechanics*. Marcel Dekker, 2001.

[84] S. Haaland, "Simple and explicit formulas for the friction factor in turbulent pipe flow," *Journal of Fluids Engineering*, vol. 105, no. 1, pp. 89–90, 1983.

[85] C. F. Colebrook, "Turbulent Flow in Pipes, with particular reference to the Transition Region between the Smooth and Rough Pipe Laws.," *Journal of the ICE*, vol. 11, no. 4, pp. 133–156, 1939.

[86] S. W. Churchill, "Comprehensive correlating equations for heat, mass and momentum transfer in fully developed flow in smooth tubes," *Industrial \& Engineering Chemistry Fundamentals*, vol. 16, no. 1, pp. 109–116, 1977.

[87] S. W. Churchill and R. Usagi, "A standardized procedure for the production of correlations in the form of a common empirical equation," *Industrial \& Engineering Chemistry Fundamentals*, vol. 13, no. 1, pp. 39–44, 1974.

[88] M. Bhatti and R. Shah, "Turbulent and transition flow convective heat transfer in ducts," *Handbook of Single-Phase Convective Heat Transfer*, pp. 4–1, 1987.

[89] "Pressure drop in pipe fittings and valves." [Online]. Available at: http://www.katmarsoft ware.com/articles/pipe-fitting-pressure-drop.htm. [Accessed: 27 Feb 2014].

[90] A. K. Coker, *Fortran Programs for Chemical Process Design, Analysis, and Simulation*. Gulf Professional Publishing, 1995.

[91] H. Institute, "Pipe Friction Manual," *New York, NY (3rd ed., 1961)*, 1954.

[92] L. L. Simpson, "Sizing piping for process plants," *Chemical Engineering*, vol. 75, no. 13, pp. 192–214, 1968.

[93] R. Mukherjee, "Effectively design shell-and-tube heat exchangers," *Chemical Engineering Progress*, vol. 94, no. 2, pp. 21–37, 1998.

[94] R. Mukherjee, *Practical Thermal Design of Shell-And-Tube Heat Exchangers*. Begell House, 2004.

[95] J. P. Gupta, *Working with Heat Exchangers: Questions and Answers.* Hemisphere Publishing Company, 1990.

[96] "Heat exchanger fluid allocation: Shellside or tubeside?," *Smart Process Design.* [Online]. Available at: http://smartprocessdesign.com/heat-exchanger-fluid-allocation-shellside-tubeside/. [Accessed: 10 Apr 2015].

[97] C. A. Bennett, R. S. Kistler, T. G. Lestina, and D. C. King, "Improving heat exchanger designs," *Chemical Engineering Progress,* vol. 103, no. 4, pp. 40–45, 2007.

[98] "Aspen Plus® – AspenTech." [Online]. Available at: https://www.aspentech.com/products/aspen-plus.aspx. [Accessed: 23 Apr 2014].

[99] "Aspen HYSYS® – AspenTech." [Online]. Available at: https://www.aspentech.com/products/aspen-hysys.aspx. [Accessed: 23 Apr 2014].

[100] "SimSci PRO/II – Comprehensive process simulation," *Invensys Software.* [Online]. Available at: http://software.invensys.com/products/simsci/design/pro-ii/. [Accessed: 23Apr 2014].

[101] "Chemstations' integrated suite chemical process engineering software." [Online]. Available at: http://www.chemstations.com/Products/What_is_CHEMCAD/. [Accessed: 23 Apr 2014].

[102] "COCO – the CAPE-OPEN to CAPE-OPEN simulator." [Online]. Available at: http://www.cocosimulator.org/. [Accessed: 23 Apr 2014].

[103] "EMSO – ALSOC PROJECT – EMSO." [Online]. Available at: http://www.enq.ufrgs.br/trac/alsoc/wiki/EMSO. [Accessed: 23 Apr 2014].

[104] "DWSIM Wiki." [Online]. Available at: http://dwsim.inforside.com.br/wiki/index.php?title=Main_Page. [Accessed: 23 Apr 2014].

[105] "ASCEND." [Online]. Available at: http://ascend4.org/. [Accessed: 23 Apr 2014].

[106] "List of chemical process simulators," *Wikipedia, the free encyclopedia,* 2014. [Online]. Available at: http://en.wikipedia.org/w/index.php?title=List_of_chemical_process_simulators&oldid=604716973. [Accessed: 23 Apr 2014].

[107] "List of computer simulation software," *Wikipedia, the free encyclopedia,* 2014. [Online]. Available at: http://en.wikipedia.org/w/index.php?title=List_of_computer_simulation_software&oldid=600120405. [Accessed: 23 Apr 2014].

[108] "DWSIM – Open Source Process Simulator Free Science & Engineering software downloads at SourceForge.net." [Online]. Available at: http://sourceforge.net/projects/dwsim/. [Accessed: 23 Apr 2014].

[109] *TEMA 9th ed. 2007 standards.* TEMA, 2007.

[110] "Koch Heat Transfer Company – Designing, manufacturing and testing custom heat exchangers," 2013. [Online]. Available at: http://www.kochheattransfer.com/. [Accessed: 22 Aug 2013].

[111] "Pipe Dimension (Diameter Nominal (DN) | Nominal Bore (NB) | Nominal Pipe Size (NPS)." [Online]. Available at: http://www.nan-long.com/Metal%20Steel%20Technology%20Summary/Diameter_Nominal-Nominal_Bore-Nominal_Pipe_Size.htm. [Accessed: 15 Apr 2014].

[112] "B.W.G. – Birmingham Wire Gauge Sunny Steel Enterprise Ltd.," 2013. [Online]. Available at: http://www.sunnysteel.com/blog/index.php/bwg-birmingham-wire-gauge/. [Accessed: 09 May 2013].

[113] "Birmingham wire gauge – Neutrium." [Online]. Available at: http://neutrium.net/piping/birmingham-wire-gauge/. [Accessed: 19 Apr 2014].

[114] "R.W. Holand, Inc." [Online]. Available at: http://www.rwholland.com/. [Accessed: 02 May 2015].

[115] "Shell & tube heat exchanger heat exchanger design, Inc. Worldwide leaders in heat transfer technology." [Online]. Available at: http://www.hed-inc.com/shell-tube.html. [Accessed: 21 Feb 2014].

[116] "Absolute roughness of pipe material – Neutrium." [Online]. Available at: http://neutrium. net/fluid_flow/absolute-roughness/. [Accessed: 28 May 2014].

[117] "Absolute pipe roughness EnggCyclopedia." [Online]. Available at: http://www.enggcyclope dia.com/2011/09/absolute-roughness/. [Accessed: 28 May 2014].

[118] "Thermal conductivity of common materials – Neutrium." [Online]. Available at: https://neu trium.net/heat_transfer/thermal-conductivity-of-common-materials/. [Accessed: 26 Nov 2014].

[119] "Thermal conductivity of some common materials and gases." [Online]. Available at: http:// www.engineeringtoolbox.com/thermal-conductivity-d_429.html. [Accessed: 26 Nov 2014].

[120] "List of thermal conductivities – Wikipedia, the free encyclopedia." [Online]. Available at: http://en.wikipedia.org/wiki/List_of_thermal_conductivities. [Accessed: 26 Nov 2014].

[121] "Thermal properties of metals, conductivity, thermal expansion, specific heat – engineers edge." [Online]. Available at: http://www.engineersedge.com/properties_of_metals.htm. [Accessed: 26 Nov 2014].

[122] "Conductive materials or metal conductivity – TIBTECH innovations -." [Online]. Available at: http://www.tibtech.com/conductivity.php. [Accessed: 26 Nov 2014].

[123] K. J. Bell and A. C. Mueller, *Wolverine Heat Transfer Engineering Data book II*. Huntsville, AL, USA: Wolverine Tube, Inc, 2001.

[124] P. H. Oosthuizen and D. Naylor, *An Introduction to Convective Heat Transfer Analysis*. WCB/ McGraw Hill, 1999.

List of gases

https://doi.org/10.1515/9783110585872-014

List of vapors

1,1,2-trichloroethane 283
1,1-dichloroethane 283
1,2,4-trichlorobenzene 283
1,2-propylene oxide 283
1,4-butanediol 283
1,4-dioxane 283
1-butanol 267
1-heptanol 267
1-heptene 267
1-hexanol 267
1-hexene 267
1-nitrobutane 267
1-nitropropane 283
1-nonene 267
1-octene 267
1-pentene 267
1-propanol 267
1-undecene 267

2,4,6-trinitrotoluene 283
2,4-dinitrotoluene 283
2-butanol 267
2-ethyl-m-xylene 267
2-heptanone 267
2-hexanone 267
2-methylindene 283
2-methylnaphthalene 283
2-pentanol 267
2-pentanone 267

3-heptanone 267
3-hexanone 267

Acetaldehyde 272
Acetic acid 272
Acetic anhydride 283
Acetone 272
Acetonitrile 272
Acetyl chloride 283
Acrylic acid 283
Acrylonitrile 267
Aniline 283
Anisole 283

Benzene 272
Benzoic acid 283
Butanal 272

Cumene 272
Cyclohexane 272
Cyclohexanol 283
Cyclohexanone 283
Cyclohexene 272
Cyclopentane 272

Diethanolamine 283
Diethyl ether 272
Diethyl sulfide 272
Diethylamine 272
Diisobutyl ketone 272

Ethanol 272
Ethyl acetate 272
Ethylacetylene 272
Ethylbenzene 272
Ethylcyclohexane 272
Ethylene oxide 272
Ethylenediamine 277

Formaldehyde 277

Heptanal 277
Hexanal 277

Isobutyl acetate 277
Isobutylbenzene 277
Isopropanol 277

Methacrylonitrile 277
Methanol 277
Methyl acetate 277
Methyl ethyl ether 277
Methyl ethyl ketone 277
Methyl formate 277
Methyl isobutyl ether 277
Methyl isobutyl ketone 277
Methyl isopropyl ether 277

https://doi.org/10.1515/9783110585872-015

List of liquids

1,1,2-trichloroethane 308
1,1-dichloroethane 308
1,2,4-trichlorobenzene 308
1,2-propylene oxide 308
1,4-butanediol 308
1,4-dioxane 308
1-butanol 293
1-heptanol 293
1-heptene 293
1-hexanol 293
1-hexene 293
1-nitrobutane 293
1-nitropropane 308
1-nonene 293
1-octene 293
1-pentene 293
1-propanol 293
1-undecene 293

2,4,6-trinitrotoluene 308
2,4-dinitrotoluene 308
2-butanol 293
2-ethyl-m-xylene 293
2-heptanone 293
2-hexanone 293
2-methylindene 308
2-methylnaphthalene 308
2-pentanol 293
2-pentanone 293

3-heptanone 293
3-hexanone 293

Acetaldehyde 298
Acetic acid 298
Acetic anhydride 308
Acetone 298
Acetonitrile 298
Acetyl chloride 308
Acrylic acid 308
Acrylonitrile 293
Aniline 308
Anisole 308

Benzene 298
Benzoic acid 308
Butanal 298

Cumene 298
Cyclohexane 298
Cyclohexanol 308
Cyclohexanone 308
Cyclohexene 298
Cyclopentane 298

Diethanolamine 308
Diethyl ether 298
Diethyl sulfide 298
Diethylamine 298
Diisobutyl ketone 298

Ethanol 298
Ethyl acetate 298
Ethylacetylene 298
Ethylbenzene 298
Ethylcyclohexane 298
Ethylene oxide 298
Ethylenediamine 303

Formaldehyde 303

Heptanal 303
Hexanal 303

Isobutyl acetate 303
Isobutylbenzene 303
Isopropanol 303

Methacrylonitrile 303
Methanol 303
Methyl acetate 303
Methyl ethyl ether 303
Methyl ethyl ketone 303
Methyl formate 303
Methyl isobutyl ether 303
Methyl isobutyl ketone 303
Methyl isopropyl ether 303

https://doi.org/10.1515/9783110585872-016

Keyword index

https://doi.org/10.1515/9783110585872-017

www.ingramcontent.com/pod-product-compliance
Lightning Source LLC
Chambersburg PA
CBHW080903220326
41598CB00034B/5460